Facing Gaia

For Ulysse and Maya

Facing Gaia

Eight Lectures on the
New Climatic Regime

Bruno Latour

Translated by Catherine Porter

polity

First published in French as *Face à Gaïa. Huit conférences sur le nouveau régime climatique* © Éditions La Découverte, Paris, 2015.

This edition copyright © Bruno Latour 2017

The right of Bruno Latour to be identified as Author of this Work has been asserted in accordance with the UK Copyright, Designs and Patents Act 1988.

23 words from p. 186 of James Lovelock, *Gaia: The Practical Science of Planetary Medicine* (2001) used as an epigraph. © Gaia Books Ltd. By permission of Oxford University Press, USA.

First published in English in 2017 by Polity Press

Reprinted 2017 (twice)

Polity Press
65 Bridge Street
Cambridge CB2 1UR, UK

Polity Press
101 Station Landing
Suite 300,
Medford, MA 02155
USA

All rights reserved. Except for the quotation of short passages for the purpose of criticism and review, no part of this publication may be reproduced, stored in a retrieval system or transmitted, in any form or by any means, electronic, mechanical, photocopying, recording or otherwise, without the prior permission of the publisher.

ISBN-13: 978-0-7456-8433-8
ISBN-13: 978-0-7456-8434-5 (pb)

A catalogue record for this book is available from the British Library.

Library of Congress Cataloging-in-Publication Data

Names: Latour, Bruno, author.
Title: Facing Gaia : eight lectures on the new climatic regime / Bruno Latour.
Description: Cambridge, UK ; Malden, MA : Polity, 2017. | Includes bibliographical references and index.
Identifiers: LCCN 2016057696 (print) | LCCN 2017022305 (ebook) | ISBN 9780745684369 (Mobi) | ISBN 9780745684376 (Epub) | ISBN 9780745684338 (hardback) | ISBN 9780745684345 (paperback)
Subjects: LCSH: Gaia hypothesis. | Philosophy of nature. | Nature in literature. | Climatic changes–Philosophy. | Philosophical anthropology. | Nature–Effect of human beings on. | BISAC: SCIENCE / Philosophy & Social Aspects.
Classification: LCC QH331 (ebook) | LCC QH331 .L3313 2017 (print) | DDC 570.1–dc23
LC record available at https://lccn.loc.gov/2016057696

Typeset in 10.5 on 12 pt Sabon by Toppan Best-set Premedia Limited
Printed and bound in the UK by CPI Group (UK) Ltd, Croydon, CR0 4YY

The publisher has used its best endeavours to ensure that the URLs for external websites referred to in this book are correct and active at the time of going to press. However, the publisher has no responsibility for the websites and can make no guarantee that a site will remain live or that the content is or will remain appropriate.

Every effort has been made to trace all copyright holders, but if any have been inadvertently overlooked the publisher will be pleased to include any necessary credits in any subsequent reprint or edition.

For further information on Polity, visit our website: politybooks.com

Contents

Introduction — 1

First Lecture: On the instability of the (notion of) nature — 7
A mutation of the relation to the world • Four ways to be driven crazy by ecology • The instability of the nature/culture relation • The invocation of human nature • The recourse to the "natural world" • On a great service rendered by the pseudo-controversy over the climate • "Go tell your masters that the scientists are on the warpath!" • In which we seek to pass from "nature" to the world • How to face up

Second Lecture: How not to (de-)animate nature — 41
Disturbing "truths" • Describing in order to warn • In which we concentrate on agency • On the difficulty of distinguishing between humans and nonhumans • "And yet it moves!" • A new version of natural law • On an unfortunate tendency to confuse cause and creation • Toward a nature that would no longer be a religion?

Third Lecture: Gaia, a (finally secular) figure for nature — 75
Galileo, Lovelock: two symmetrical discoveries • Gaia, an exceedingly treacherous mythical name for a scientific theory • A parallel with Pasteur's microbes • Lovelock too makes micro-actors proliferate • How to avoid the idea of a system? • Organisms make their own environment, they do not adapt to it • On a slight complication of Darwinism • Space, an offspring of history

Fourth Lecture: The Anthropocene and the destruction of (the image of) the Globe 111

The Anthropocene: an innovation • *Mente et Malleo* • A debatable term for an uncertain epoch • An ideal opportunity to disaggregate the figures of Man and Nature • Sloterdijk, or the theological origin of the image of the Sphere • Confusion between Science and the Globe • Tyrrell against Lovelock • Feedback loops do not draw a Globe • Finally, a different principle of composition • *Melancholia*, or the end of the Globe

Fifth Lecture: How to convene the various peoples (of nature)? 146

Two Leviathans, two cosmologies • How to avoid war between the gods? • A perilous diplomatic project • The impossible convocation of a "people of nature" • How to give negotiation a chance? • On the conflict between science and religion • Uncertainty about the meaning of the word "end" • Comparing collectives in combat • Doing without any natural religion

Sixth Lecture: How (not) to put an end to the end of times? 184

The fateful date of 1610 • Stephen Toulmin and the scientific counter-revolution • In search of the religious origin of "disinhibition" • The strange project of achieving Paradise on Earth • Eric Voegelin and the avatars of Gnosticism • On an apocalyptic origin of climate skepticism • From the religious to the terrestrial by way of the secular • A "people of Gaia"? • How to respond when accused of producing "apocalyptic discourse"

Seventh Lecture: The States (of Nature) between war and peace 220

The "Great Enclosure" of Caspar David Friedrich • The end of the State of Nature • On the proper dosage of Carl Schmitt • "We seek to understand the normative order of the Earth" • On the difference between war and police work • How to turn around and face Gaia? • Human versus Earthbound • Learning to identify the struggling territories

Eighth Lecture: How to govern struggling (natural) territories? 255
  In the Theater of Negotiations, Les Amandiers, May 2015
  • Learning to meet without a higher arbiter • Extension of the Conference of the Parties to Nonhumans • Multiplication of the parties involved • Mapping the critical zones • Rediscovering the meaning of the State • *Laudato Sí* • Finally, facing Gaia • "Land ho!"

References 293

Index 315

"In mythical language, the *earth* became known as mother of law ... This is what the poet means when he speaks of the infinitely just earth: *justissima tellus*."

Carl Schmitt
The Nomos *of the Earth*, 42

"It is no longer politics *sans phrase* that is destiny, but rather climate politics."

Peter Sloterdijk
Spheres, vol. 2, *Globes*, 333

"I would sooner expect to see a goat to succeed as a gardener than expect humans to become responsible stewards of the Earth."

James Lovelock
Gaia: The Practical Science of Planetary Medicine, 186

"Nature is but a name for excess."

William James
A Pluralistic Universe, 148

Introduction

It all began with the idea of a dance movement that captured my attention, some ten years ago. I couldn't shake it off. A dancer is rushing backwards to get away from something she must have found frightening; as she runs, she keeps glancing back more and more anxiously, as if her flight is accumulating obstacles behind her that increasingly impede her movements, until she is forced to turn around. And there she stands, suspended, frozen, her arms hanging loosely, looking at something coming towards her, something even more terrifying than what she was first seeking to escape – until she is forced to recoil. Fleeing from one horror, she has met another, partly created by her flight.

Figure 0.1 Still from the dance "The Angel of Geostory," by Stéphanie Ganachaud, filmed by Jonathan Michel, February 12, 2013.

I became convinced that this dance expressed the spirit of the times, that it summed up in a single situation, one very disturbing to me, the one the Moderns had first fled – the archaic horror of the past – and what they had to face today – the emergence of an enigmatic figure, the source of a horror that was now in front of them rather than behind. I had first noted the emergence of this monster, half cyclone, half Leviathan, under an odd name: "Cosmocolossus."[1] The figure merged very quickly in my mind with another highly controversial figure that I had been thinking about as I read James Lovelock: the figure of Gaia. Now, I could no longer escape: I needed to understand what was coming at me in the harrowing form of a force that was at once mythical, scientific, political, and probably religious as well.

Since I knew nothing about dance, it took me several years to find, in Stéphanie Ganachaud, the ideal interpreter of this brief movement.[2] Meanwhile, not knowing what to do with the obsessional figure of the Cosmocolossus, I persuaded some close friends to create a play about it, which has since become the *Gaia Global Circus*.[3] It was at this point, in one of those coincidences that shouldn't surprise anyone who has been gripped by an obsession, that the Gifford Lecture committee asked me to come to Edinburgh in 2013 to give a series of six talks under the intriguing heading of "natural religion." How could I resist an offer that William James, Alfred North Whitehead, John Dewey, Henri Bergson, Hannah Arendt, and many others had accepted?[4] Wasn't this the ideal opportunity to develop through argument what dance and theater had first compelled me to explore? At least this medium wasn't too foreign to me, especially since I had just finished writing an inquiry into the modes of existence that turned

---

[1] See Bruno Latour, *Kosmokoloss* (2013d), a radio play broadcast in Germany (in German). The text of the play and most of my own articles cited in this book are accessible in their final or provisional versions at www.bruno-latour.fr.

[2] The movement was performed on February 12, 2013, and filmed by Jonathan Michel; see www.vimeo.com/60064456.

[3] A collective project carried out starting in the spring of 2010 with Chloë Latour and Frédérique Aït-Touati, directors, and Claire Astruc, Jade Collinet, Matthieu Protin, and Luigi Cerri, actors. Pierre Daubigny wrote the text, *Gaia Global Circus*, which led to performances in Toulouse in the context of the Novela, a festival celebrating new knowledge and culture, in October 2013, and in Reims at the Comédie in December of the same year, before the cast went on tour in France and abroad.

[4] The six talks are available on video at the site of the Gifford Lectures at the University of Edinburgh and in text form on my website (2013c). On the history of these lectures, and on the field of "natural religion," a rather enigmatic term, see Larry Witham, *The Measure of God* (2005).

Introduction 3

out to be under the more and more pervasive shadow of Gaia.⁵ These lectures, reworked, expanded, and completely rewritten, are the basis for the present book.

If I retain the genre, style, and tone of the lectures in publishing them, it is because the anthropology of the Moderns that I have been pursuing for forty years turns out to resonate increasingly with what can be called the *New Climate Regime*.⁶ I use this term to summarize the present situation, in which the physical framework that the Moderns had taken for granted, the ground on which their history had always been played out, has become unstable. As if the décor had gotten up on stage to share the drama with the actors. From this moment on, everything changes in the way stories are told, so much so that the political order now includes everything that previously belonged to nature – a figure that, in an ongoing backlash effect, becomes an ever more undecipherable enigma.

For years, my colleagues and I tried to come to grips with this intrusion of nature and the sciences into politics; we developed a number of methods for following and even mapping ecological controversies. But all this specialized work never succeeded in shaking the certainties of those who continued to imagine a social world without objects set off against a natural world without humans – and without scientists seeking to know that world. While we were trying to unravel some of the knots of epistemology and sociology, the whole edifice that had distributed the functions of these fields was falling to the ground – or, rather, was falling, literally, back down to Earth. We were still discussing possible links between humans and nonhumans, while in the meantime scientists were inventing a multitude of ways to talk about the same thing, but on a completely different scale: the "Anthropocene," the "great acceleration," "planetary limits," "geohistory," "tipping points," "critical zones," all these astonishing terms that we shall encounter as we go along, terms that scientists had to invent in their attempt to understand this Earth that seems to react to our actions.

My original discipline, science studies, finds itself reinforced today by the widely accepted understanding that the old constitution, the

---

⁵Bruno Latour, *An Inquiry into Modes of Existence: An Anthropology of the Moderns* ([2012] 2013b).
⁶The expression is derived from the term "climatic regime" introduced by Stefan Aykut and Amy Dahan, in *Gouverner le climat? Vingt ans de négociations internationales* (2014), to designate a very particular and, in their view, not very effective way to try to "govern the climate" as if CO_2 were another case of pollution. Their work, unfortunately not translated, plays an important role in the present book.

one that distributed powers between science and politics, has become obsolete. As if we had really passed from an Old Regime to a new one marked by the emergence in multiple forms of the question of *climates* and, even more strangely, of their link to *government*. I am using these terms (which historians of geography have generally abandoned except with reference to Montesquieu's "climate theory," itself long since deemed obsolete) in their broadest sense. All a sudden, everyone senses that another *Spirit of the Laws of Nature*[7] is in the process of emerging and that we had better start writing it down if we want to survive the forces unleashed by the New Regime. The present volume seeks to contribute to this collective work of exploration.

Gaia is presented here as the occasion for a return to Earth that allows for a differentiated version of the respective qualities that can be required of sciences, politics, and religions, as these are finally reduced to more modest and more earthbound definitions of their former vocations. The lectures come in pairs. The first two deal with the notion of *agency* (in the sense of "power to act"), an indispensable concept for allowing exchanges between heretofore distinct fields and disciplines; the next two introduce the principal characters – first *Gaia*, then the *Anthropocene*; the fifth and sixth lectures define the peoples who are struggling to occupy the Earth and the epoch in which they find themselves; and the last two explore the geopolitical question of the territories involved in the struggle.

The potential audience for a book is even more difficult to pin down than the audience for a lecture, but, since we have actually entered a period of history that is at once geological and human, I would like to address readers with diverse skills. It is impossible to understand what is happening to us without turning to the sciences – the sciences have been the first to sound the alarm. And yet, to understand them, it is impossible to settle for the image offered by the old epistemology; the sciences are now and will remain from now on so intermingled with the entire culture that we need to turn to the humanities to understand how they really function. Hence a hybrid style for a hybrid subject addressed to a necessarily hybrid audience.

Such a book is hybrid in its composition, too, as you might imagine. Once the six Gifford Lectures had been drafted for delivery in Edinburgh in February 2013, they were translated into French by Franck Lemonde, along with another talk given in 2013.[8] But then I

---

[7]*Trans.*: This imagined title refers to a work on political theory by Charles de Secondat, Baron de Montesquieu, *The Spirit of the Laws* ([1748] 1989).
[8]The second lecture includes parts of my "Agency at the Time of the Anthropocene" (2014a).

put the text through what translators hate most when they have the misfortune of needing to translate into an author's mother tongue: I thoroughly modified the French version and added two new chapters, reshaping it to such an extent that it is an entirely different text, now translated once more for publication in English. The English version differs from the French only in some footnotes, several of the works cited, and a few cosmetic changes.

If writers can flatter themselves that their readers are the same from the beginning to the end of a book, and that these readers will be learning as they proceed from chapter to chapter, the same cannot be said for speakers, who must address a partly different audience every time. That is why each of the eight lectures can be read on its own and they can be perused in any order. The more specialized points have been shifted to the notes.

*

I owe thanks to too many people to name them all here; I attempt to acknowledge my debt, instead, in the bibliographical references.

Still, it would be unfair not to cite first and foremost the members of the Gifford Lecture committee, who allowed me to address the theme of "natural religion," without forgetting the audience in the Santa Cecilia Room during those six marvelous days in February 2013 in sun-drenched Edinburgh.

It is thanks to Isabelle Stengers that I first became interested in what she has called the intrusion of Gaia, and it was as usual by going to Simon Schaffer for help that I tried to sort out Gaia's impossible character, sharing my anxieties with Clive Hamilton, Dipesh Chakrabarty, Déborah Danowski, Eduardo Viveiros de Castro, Donna Haraway, Bronislaw Szerzynski, and many other colleagues.

But I would like to offer special thanks to Jérôme Gaillardet and Jan Zalasiewicz, who confirmed for me that there has been, since the Anthropocene, a common ground for the natural sciences and the humanities that we all share.

I unquestionably owe much more than they imagine to the students who created and produced *Make it Work* at the Théâtre des Amandiers in Nanterre in May 2015; I am equally indebted to the creators of the *Anthropocene Monument* exhibit at the Abattoirs museum in Toulouse in October 2014, as well as to the students in the course titled "Political Philosophy of Nature."

Finally, I want to thank Philippe Pignarre, whose editorial work has supported me for a very long time. I don't think he has ever published a book that makes such direct reference to the name of his collection⁹ – because, contrary to what people too often think, Gaia is actually not global at all. Gaia is unquestionably the great *empêcheur de penser en rond*, the grand inhibitor of circular thinking, a great impetus to thinking outside the box...[10]

---

⁹Trans.: *Les empêcheurs de penser en rond* is the name of a publishing house founded by Philippe Pignarre in 1989, taken over as a collection devoted to the humanities and social sciences by Seuil in 2000 and then by La Découverte in 2008. The term plays on the familiar French expression *empêcheur de tourner en rond*, literally someone who interferes with a smoothly running operation, metaphorically someone who "throws sand in the gears," a "spoilsport," a "killjoy," a "party pooper."
[10]The very important doctoral thesis by Sébastien Dutreuil, "Gaïa: hypothèse, programme de recherche pour le système terre, ou philosophie de la nature?," defended in 2016 at Université de Paris I, was completed too late for me to use it in his book. Once published, it will significantly renew the history of Lovelock and Gaia and their place in earth science.

FIRST LECTURE

On the instability of the (notion of) nature

> A mutation of the relation to the world • Four ways to be driven crazy by ecology • The instability of the nature/culture relation • The invocation of human nature • The recourse to the "natural world" • On a great service rendered by the pseudo-controversy over the climate • "Go tell your masters that the scientists are on the warpath!" • In which we seek to pass from "nature" to the world • How to face up

It doesn't stop; every morning it begins all over again. One day, it's rising water levels; the next, it's soil erosion; by evening, it's the glaciers melting faster and faster; on the 8 p.m. news, between two reports on war crimes, we learn that thousands of species are about to disappear before they have even been properly identified. Every month, the measurements of carbon dioxide in the atmosphere are even worse than the unemployment statistics. Every year, we are told that it is the hottest since the first weather recording stations were set up; sea levels keep on rising; the coastline is increasingly threatened by spring storms; as for the ocean, every new study finds it more acidic than before. This is what the press calls living in the era of an "ecological crisis."

Alas, talking about a "crisis" would be just another way of reassuring ourselves, saying that "this too will pass," the crisis "will soon be behind us." If only it were just a crisis! If only it had been just a crisis! The experts tell us we should be talking instead about a "mutation": we were used to one world; we are now tipping, mutating, into another. As for the adjective "ecological," we use that word for reassurance as well, all too often, as a way of distancing ourselves from the troubles with which we're threatened: "Ah, if you're talking about ecological questions, fine! They don't really concern us, of course." We behave just like people in the twentieth century when they talked about "the environment," using that term to designate the beings of nature considered from afar, through the shelter of bay windows. But today, according to the experts, all of us are affected, on the inside, in the intimacy of our precious little existences, by these news bulletins that warn us directly about what we ought to eat and drink, about our land use, our modes of transportation, our clothing choices. As we hear one piece of bad news after another, you might expect us to feel that we had shifted from a mere ecological crisis into what should instead be called *a profound mutation in our relation to the world*.

And yet this is surely not the case. For we receive all this news with astonishing calm, even with an admirable form of stoicism. If a radical mutation were really at issue, we would all have already modified the bases of our existence from top to bottom. We would have begun to change our food, our habitats, our means of transportation, our cultural technologies, in short, our mode of production. Every time we heard the sirens we would have rushed out of our shelters to invent new technologies equal to the threat. The inhabitants of the wealthy countries would have been as inventive as they were earlier in times of war, and, as they did in the twentieth century, they would have solved the problem in four or five years, by a massive transformation of their ways of life. Thanks to their vigorous actions, the quantity of CO_2 captured at the Mauna Loa observatory in Hawaii would already be starting to stabilize;[1] well-watered soil would be swarming with earthworms, and the sea, rich in plankton, would again be full of fish; even the Arctic ice might have slowed its decline (unless it has been on an irreversible slope, shifting for millennia toward a new state).[2]

In any case, *we would already have acted*. Beginning some thirty years ago, the crisis would already be over. We would be looking back at the era of "the great ecological war" with the pride of people who

---

[1] This observatory has been providing measurements of atmospheric CO_2 longer than any other. On the history of these measurements, see Charles David Keeling, "Rewards and Penalties of Recording the Earth" (1998). I shall come back to this example a number of times.
[2] See David Archer, *The Long Thaw* (2010b).

had nearly succumbed, but who had figured out how to turn the situation around to their advantage by reacting rapidly and mobilizing the totality of their powers of invention. We might even be taking our grandchildren to visit museums devoted to this struggle, hoping that they would be as stunned by our progress as they are today when they see how the Second World War gave rise to the Manhattan Project, the refinement of penicillin, and the dramatic progress of radar and air travel.

But here we are: what could have been just a passing crisis has turned into a profound alteration of our relation to the world. It seems as though we have become the people *who could have acted* thirty or forty years ago – and who did nothing, or far too little.³ A strange situation: we crossed a series of thresholds, we went through total war, and we hardly noticed a thing! So that now we're bending under the weight of a gigantic event that has crept up on us behind our backs without our really realizing it, without our putting up a fight. Just imagine: hidden behind the profusion of world wars, colonial wars, and nuclear threats, there was, in the twentieth century, that "classic century of war," another war, also worldwide, also total, also colonial, that we lived through without experiencing it. Whereas we are now preparing ourselves quite nonchalantly to take an interest in the fate of "future generations" (as they used to say), just imagine what it would be like if everything had already been done by the previous generations! Just imagine that something has happened that is not ahead of us, as a threat to come, but rather *behind* us, behind those who have already been born. How can we not feel rather ashamed that we have made a situation irreversible because we moved along like sleepwalkers when the alarms sounded?

And yet we haven't lacked for warnings. The sirens have been blaring all along. Awareness of ecological disasters has been long-standing, active, supported by arguments, documentation, proofs, from the very beginning of what is called the "industrial era" or the "machine age." We can't say that we didn't know.⁴ It's just that there are many ways of knowing and not knowing at the same time. Usually, when it's a question of paying attention to oneself, to one's own survival, to the well-being of those we care about, we tend rather

---

³This is the object of the frightening little exercise in science fiction produced by historian of science Naomi Oreskes and her colleague Erik M. Conway, *The Collapse of American Civilization: A View from the Future* (2014).
⁴This is the theme addressed by Jean-Baptiste Fressoz in his important book *L'apocalypse joyeuse: une histoire du risque technologique* (2012), and again in Christophe Bonneuil and Jean-Baptiste Fressoz, eds, *The Shock of the Anthropocene: The Earth, History, and Us* (2016).

to err in the direction of security: when our children have the sniffles, we check with the pediatrician; at the slightest threat to our plantings, we call for insecticide; if there is any doubt about the safety of our property, we take out insurance and install surveillance cameras; to prevent a potential invasion, we assemble armies at our borders. The overly celebrated precautionary principle is applied abundantly as soon as it is a matter of protecting our surroundings and our belongings, even if we are not too sure about the diagnosis and even if the experts are still quibbling about the scope of the dangers.[5] Now, for this worldwide crisis, no one invokes the precautionary principle in order to plunge bravely into action. This time, our very old, cautious, tentative humanity, which usually advances only by groping, tapping each obstacle with its white cane like a blind person, making careful adjustments at every sign of risk, pulling back as soon as it feels resistance, rushing ahead as soon as the horizon opens up before hesitating once again as soon as a new obstacle appears, this humanity has remained impassive. None of its old peasant, bourgeois, artisanal, working-class, political virtues seem to come into play here. The alarms have sounded; they've been disconnected one after another. People have opened their eyes, they have seen, they have known, and they have forged straight ahead with their eyes shut tight![6] If we are astonished, reading Christopher Clark's *The Sleepwalkers*, to see Europe in 1914 hurtling toward the Great War with its eyes wide open,[7] how can we not be astonished to learn retrospectively with what precise knowledge of the causes and effects Europeans (and all those that have followed the same path since) have rushed headlong into this other Great War about which we are learning, stunned, that it has already taken place – and that we have probably lost it?

*

"An alteration of the relation to the world": this is the scholarly term for madness. We understand nothing about ecological mutations if we

---

[5] The precautionary principle is often misinterpreted: it is a question not of abstaining from action when one is uncertain but, on the contrary, of acting even when one does not have complete certainty: "Better to be safe than sorry." It is a principle of action and research and not, as its enemies would have it, a principle of obscurantism.
[6] This is why, in *L'apocalypse joyeuse*, Fressoz uses the term "disinhibition": "The word disinhibition condenses the two phases of moving into action: that of reflexivity and that of going beyond; that of taking danger into account and that of normalizing danger. Modernity was a process of reflexive disinhibition…" (p. 16). In the sixth lecture, I shall look more closely at this term in search of its religious origin.
[7] Christopher M. Clark, *The Sleepwalkers: How Europe Went to War in 1914* (2013).

don't measure the extent to which they throw everyone into a panic. Even if they have several different ways of driving us crazy!

One segment of the public – some intellectuals, some journalists, helped occasionally by certain experts – has decided to plunge little by little into a parallel world in which there is no longer either any agitated nature or any real threat. If they remain calm, it is because they are sure that scientific data have been manipulated by dark forces or, in any case, have been so exaggerated that we must courageously resist the opinions of those whom they call "catastrophists"; we must learn, as they say, "to keep our heads" and go on living as before, without worrying too much. This madness sometimes takes on fanatical form, as it does with the so-called climate skeptics – and even sometimes "climate deniers" – who adhere in varying degrees to a conspiracy theory and who, like many elected American officials, see in the issue of ecology a devious way of imposing socialism on the United States![8] This view is much more widespread in the world at large, however, in the form of a low-level madness that can be characterized as *quietist*, with reference to a religious tradition in which the faithful trusted in God to take care of their salvation. Climate quietists, like the others, live in a parallel universe, but, because they have disconnected all the alarms, no strident announcement forces them up from the soft pillow of doubt: "We'll wait and see. The climate has always varied. Humanity has always come through. We have other things to worry about. The important thing is to wait, and above all not to panic." A strange diagnosis: these people are crazy by dint of staying calm! Some of them don't even hesitate to stand up in a political meeting and invoke the covenant in Genesis where God promises Noah that He will send no more floods: "Never again will I curse the ground because of man, even though every inclination of his heart is evil from childhood, and never again will I destroy all living creatures, as I have done" (Gen. 8: 21).[9] With such solid assurance, it would be wrong indeed to worry!

Others, fortunately fewer in number, have heard the warning sirens but have reacted with such panic that they have plunged into a differ-

---

[8] There is now an abundant literature on the origins of climate skepticism, starting with the classic book by Naomi Oreskes and Erik M. Conway, *Merchants of Doubt: How a Handful of Scientists Obscured the Truth on Issues from Tobacco Smoke to Global Warming* (2010). This phenomenon occupies an important place in my own study, and I shall come back to it often in these lectures.

[9] Cited by Congressman John Shimkus of Illinois on March 25, 2009, during a meeting of the United States Energy Subcommittee on Environment and Economy; see Shawn Lawrence Otto, *Fool Me Twice: Fighting the Assault on Science in America* (2011), p. 295.

ent frenzy: "Since the threats are so serious and the transformations we have caused in the planet are so radical," they argue, "let's come to grips with the entire terrestrial system, which we can conceive as a vast machine that has stopped working properly only because we have not controlled it *completely enough*." And there they are, seized by a new urge for total domination over a nature always perceived as recalcitrant and wild. In the great delirium that they call, modestly, *geo-engineering*, they mean to embrace the Earth as a whole.[10] To recover from the nightmares of the past, they propose to increase still further the dosage of megalomania needed for survival in this world, which in their eyes has become a clinic for patients with frayed nerves. Modernization has led us into an impasse? Let's be even more resolutely modern! If the members of the first group of climate skeptics have to be shaken up to keep them from sleeping, those in this second group need to be strait-jacketed to keep them from doing too many foolish things.[11]

How can we begin to list all the nuances of depression that strike a third group of people, much more numerous, who carefully observe the rapid transformations of the Earth and who have decided that these can neither be ignored nor, alas, be remedied by any radical measures? Sadness, the blues, melancholia, neurasthenia? Yes, they've lost their nerve, their throats are tightening; they can hardly bring themselves to read a newspaper; they're stirred from their lethargy only by their rage at seeing others even crazier than they are. But once this fit of anger has subsided, they end up prostrate under huge doses of antidepressants.

The craziest of all are those who appear to believe that they can do something despite the odds, that it isn't too late, that the rules of collective action are surely going to work here again, that one has to be able to act rationally, with eyes wide open, even in the face of threats as serious as these, while respecting the framework of existing institutions.[12] But the people in this group are probably bipolar, full of energy in the manic phase, before the letdown that gives them a terrible urge to jump out of the window – or to toss their adversaries out instead.

---

[10]In Clive Hamilton's book *Earthmasters: The Dawn of Climate Engineering* (2013), the presentation of the solutions proposed is enough to make one's hair stand on end.
[11]In *The Planet Remade: How Geoengineering Could Change the World* (2015), Oliver Morton tries to draw a fine line between *hubris* and sanity.
[12]This is what Stefan Aykut and Amy Dahan, in *Gouverner le climat?* (2014), call the "denial of reality" on the part of international organizations; they analyze the negotiation procedure that has worked to limit certain instances of pollution as it is applied to a much thornier problem.

Are there still a few people left who are able to escape these symptoms? Yes, but don't think for a moment that that means they're of sound mind! They are most likely artists, hermits, gardeners, explorers, activists, or naturalists, looking in near total isolation for other ways of resisting anguish: *esperados*, to use Romain Gary's humorous label[13] (unless they are like me, and manage to shed their anguish only because they have found clever ways to induce it in others!).

No doubt about it, ecology drives people crazy; this has to be our point of departure – not with the goal of finding a cure, just so we can learn to survive without getting carried away by denial, or hubris, or depression, or hope for a reasonable solution, or retreat into the desert. There is no cure for the condition of belonging to the world. But, by taking care, we can cure ourselves of believing that we do not belong to it, that the essential question lies elsewhere, that what happens to the world does not concern us. The time is past for hoping to "get through it." We are indeed, as they say, "in a tunnel," except that we won't see light at the end. In these matters, hope is a bad counselor, since we are not in a crisis. We can no longer say "this, too, will pass." We're going to have to get used to it. *It's definitive.*

The imperative confronting us, therefore, is to discover *a course of treatment* – but without the illusion that a cure will come quickly. In this sense, it would not be impossible to make progress, but it would be progress in reverse: this would mean rethinking the idea of progress, *retrogressing*, discovering a different way of experiencing the passage of time. Instead of speaking of hope, we would have to explore a rather subtle way of "dis-hoping"; this doesn't mean "despairing" but, rather, not trusting in hope alone as a way of engaging with passing time.[14] The hope of no longer counting on hope? Admittedly, that doesn't sound very encouraging.

---

[13] Romain Gary, interview by Pierre Dumayet, in *Lecture pour tous*, December 19, 1956. For me, the model is George Monbiot, a journalist with *The Guardian* whose blog (www.monbiot.com) is as depressing as it is invigorating, but also Gilles Clément, a "planetary gardener," a renowned landscape architect who has held a chair in artistic creation at the Collège de France.

[14] The relation to hope is the object of Clive Hamilton's book *Requiem for a Species: Why We Resist the Truth about Climate Change* (2010). I shall come back to it in the fifth and sixth lectures when we approach the question of the "end time." The link between paradoxical temporality and ecology is explored by Jean-Pierre Dupuy in *Pour un catastrophisme éclairé: quand l'impossible est certain* (2003); see also Dupuy's interview, "On peut ruser avec le destin catastrophiste" (2012), but it goes back to Hans Jonas, *The Imperative of Responsibility: In Search of an Ethics for the Technological Age* (1984). It is quite clearly present, as well, in the theology underlying Pope Francis's encyclical *Laudato Sí: On Care for Our Common Home* (2015).

If we can't hope to cure ourselves for good, we might at least gamble on the lesser of two evils. After all, one form of treatment entails "living well with one's ailments," or even simply "living well." If ecology drives us crazy, it's because what we call ecology is in effect an alteration *of the alteration* in our relations with the world. In this respect ecology is both a new form of madness and a new way of struggling against the forms of madness that preceded it. There is no other solution to the problem of treating ourselves without hoping for a cure: we have to get to the bottom of the situation of dereliction in which we all find ourselves, whatever nuances our anxieties may take.[15]

*

The expression "relation to the world" itself demonstrates the extent to which we are, so to speak, *alienated*. The ecological crisis is often presented as the eternally renewed discovery that "man *belongs to nature*" – a seemingly simple expression that is actually very obscure (and not only because "man" is obviously also "woman"). Is it a way of talking about humans who finally understand that they are part of a "natural world" to which they must learn to conform? In the Western tradition, in fact, most definitions of the human stress the extent to which it is *distinguished* from nature. This is what is meant, most often, by the notions of "culture," "society," or "civilization." As a result, every time we attempt to "bring humans closer to nature," we are prevented from doing so by the objection that a human is above all, or is also, a cultural being who has to escape from, or in any case *be distinguished from*, nature.[16] Thus we shall never be able to say too crudely of humans "that they belong to nature." Moreover, if human beings were truly "natural," and only that, they would be deemed no longer human at all but only "material objects" or "pure animals" (to use even more ambiguous expressions).

---

[15] As of now, no one has taken this exploration of the relation to time further than Déborah Danowski and Eduardo Viveiros de Castro in *The Ends of the World* (2016).

[16] I am interested here only in the relation established by modern philosophy between subject and object, on the assumption that the opposition between nature in the sense of wildness – "wildlife" – and artifice has been so thoroughly criticized by historians of the environment that there is no need to go back over it. See the classic study edited by William Cronon, *Uncommon Ground: Rethinking the Human Place in Nature* (1996), and the recent overview offered by Fabien Locher and Gregory Quenet, "L'histoire environnementale: origines, enjeux et perspectives d'un nouveau chantier" (2009). For a particularly striking example of the artificialization of an ecosystem, see Gregory Quenet, *Versailles: une histoire naturelle* (2015).

We understand, then, why every definition of the ecological crisis as a "return of the human to nature" immediately unleashes a sort of panic, since we never know if we are being asked to return to the state of brute beasts or to resume the deep movement of human existence. "But I am not a natural being! I am first of all a cultural being." "Except that, of course, in fact, you are first of all a natural being, how could you forget that?" Enough to drive us crazy, indeed, and without even mentioning the "return to nature" understood as a "return to the Cave Man era," whose pathetic lighting system serves as an argument for any ill-tempered modernist who runs into an ecologist of some standing: "If we listened to you, we'd still be lighting with candles!"

The difficulty lies in the very expression "relation to the world," which presupposes two sorts of domains, that of nature and that of culture, domains that are at once distinct and impossible to separate completely. Don't try to define nature alone, for you'll have to define the term "culture" as well (the human is what escapes nature: a little, a lot, passionately); don't try to define "culture" alone, either, for you'll immediately have to define the term "nature" (the human is what cannot "totally escape" the constraints of nature). Which means that we are not dealing with *domains* but rather with one and the same *concept* divided into two parts, which turn out to be bound together, as it were, by a sturdy rubber band. In the Western tradition, we never speak of the one without speaking of the other: there is no other nature but *this* definition of culture, and no other culture but *this* definition of nature. They were born together, as inseparable as Siamese twins who hug or hit each other without ceasing to belong to the same body.[17]

As this argument is essential for what follows, but always difficult to grasp, I need to go back over it several times. You surely remember the period, not so long ago, before the feminist revolution, when the word "man" was used to speak of "everyone," in an undifferentiated and rather lazy way. In contrast, when the word "woman" was used, it was necessarily a specific term that could designate nothing other than what was then called the "weaker sex," or the "second sex." In the vocabulary of anthropologists, this means that the term "man" is an *un*marked category: it poses no problem and attracts no attention. When the term "woman" is used, attention is drawn to a specific

---

[17]This is the sense in which we have never been modern: we may believe we have been modern as long as we believe it possible to bring two distinct domains into existence, and we stop having been modern as soon as we realize that there are not two; see Bruno Latour, *We Have Never Been Modern* ([1991] 1993).

feature, namely, her sex; this is the feature that makes the category *marked* and thus detached from the unmarked category that serves as its background. Hence the efforts to replace "man" by "human" and to proceed as if this term common to the two halves of the same humanity signified at once woman and man – each with her or his own sex, or in any case her or his own gender, which distinguishes them both equally, as it were.[18]

Well, we could make headway on these questions if we could bring about exactly the same gap with the "nature/culture" pairing, so that "nature" would stop sounding like an unmarked category. (The two pairings are historically linked, moreover, but inversely, since "woman" is often found on the side of nature and "man" on the side of culture.)[19] Thus I would like to bring into existence a place – a purely conceptual place, for the time being, but one that I shall try to flesh out later on[20] – that makes it possible to define culture and nature as equally marked categories. If you recall the wonderfully ingenious devices adopted to avoid the sexist use of language, you understand that it would be very convenient to have an equivalent for this bond between nature and culture. Alas, since there is no accepted term that plays the same role as "human," in order to obtain the same effect of correcting the reader's attention I propose to link the two typographically by referring to Nature/Culture. If the use of "he/she" allows us to avoid taking the male sex as a universal (unmarked) category, similarly we can avoid making nature something universally self-evident against which the marked category of culture would stand out.[21]

Let us take another comparison, this one borrowed from art history and linked more directly to our perception of nature. We are familiar with the very odd habit in Western painting, starting in the fifteenth century, of organizing the viewer's gaze so that it can serve as a

---

[18] See Vinciane Despret and Isabelle Despret, *Les faiseuses d'histoires: que font les femmes à la pensée?* (2011).

[19] This reversal has been subject to a great deal of study since Carolyn Merchant's classic work *The Death of Nature: Women, Ecology and the Scientific Revolution* (1980); Donna Haraway's *Simians, Cyborgs, and Women: The Reinvention of Nature* (1991); and, more recently, Silvia Federici's *Caliban and the Witch: Women, the Body and Primitive Accumulation* (2004). The same inverted pairing can be seen in the trouble women scientists have making their voices heard; see the classic example studied by Evelyn Fox Keller, *A Feeling for the Organism: The Life and Work of Barbara McClintock* (1983).

[20] This is the focus of the last four lectures.

[21] A crucial work by Philippe Descola has made this position much easier to understand: see *Beyond Nature and Culture* ([2005] 2013).

counterpart to a spectacle of objects or landscapes. Viewers must not only remain at a certain distance from what they are looking at, but what they see must be arranged, prepared, aligned so as to be rendered perfectly visible. Between the two, there is the plane of the painting, which occupies the midpoint between the object and the subject. Historians have given a lot of thought to the oddness of this *scopic regime* and the position it assigns to the viewing subject.[22] But we do not pay enough attention to the symmetrical strangeness that gives the object the very odd role of being there only so as to be seen by a subject. Someone who is looking, for example, at a still life (the expression itself is significant) is entirely programmed so as to become the subject in relation to this *type* of object, whereas the objects – for example, oysters, lemons, capons, bowls, bunches of gold-tinged grapes arrayed on the folds of a white tablecloth – have no role other than to be presented to the sight of *this particular type* of gaze.

We can see clearly in this case how absurd it would be to take the subject who is looking as a historical oddity while considering what he/she is looking at – still life!– as something *natural* or, as it were, self-evident. The two cannot be separated or critiqued separately. What has been invented by Western painting is *a pair whose two members are* equally bizarre, not to say exotic, a pairing that has not been observed in any other civilization: the object *for* this subject, the subject *for* this object. Here, then, is proof that there is an operator, an operation, that *distributes* object and subject, exactly as there is a common concept that distributes the respective roles of Nature/Culture by occupying the same place "human" occupies with respect to the marked categories man/woman.

To make the presence of this operator less abstract, I asked an artist to draw it.[23] He chose to put an architect – Le Corbusier, as it happened! – in the obviously virtual position of someone who slipped into the plane of the painting and staged, symmetrically, the two positions, the one as unnatural as the other, of object and subject. The role of the viewer who is presumed to be contemplating a painting in the Western style is so improbable that the artist represented

---

[22] In the wake of Panofsky's classic studies, this quite particular type of attention has been the object of significant historical work; see, for example, Jonathan Crary, *Suspensions of Perception: Attention, Spectacle, and Modern Culture* (1999), and, more recently, Lorraine Daston and Peter Galison, *Objectivity* (2007). (The expression "scopic regime" comes from Christian Metz; see *Psychoanalysis and Cinema: The Imaginary Signifier*, [1975] 1982.)
[23] Samuel Garcia Perez agreed to do the drawings. For the complete gallery, see http://modesofexistence.org.

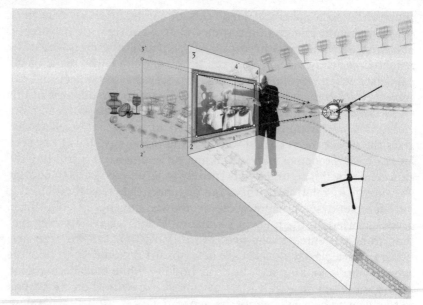

Figure 1.1 Drawing by Samuel Garcia Perez to flesh out the staging operation through which subject and object are visually constructed.

him/her in the form of a tripod to which an enormous single eye is attached![24] But what is not noticed often enough is that the object that serves as counterpart to this eye is just as implausible. To prepare a still life, the artist first has to kill it, as it were, or at least interrupt its movement – hence the lines that trace the trajectory of an object of which the manipulator seizes only a moment, through what is quite appropriately called a "freeze frame."[25] One might say, with very little exaggeration, that there were no more objects in the world before this procedure than there were persons before the invention of photography smiling foolishly in front of a camera while someone yelled "Cheese!".

This schema makes it easier, I hope, to understand why it would be pointless to seek to "reconcile" or "go beyond" the subject and the object without taking into account the operator – represented here by the architect-manipulator – who has *distributed* the roles to these strange characters, some of whom are going to play the role of nature – for a subject – and others the role of consciousness – of

---

[24] The oddness of the cognitive apparatus imposed on such subjects has been well known since the publication of Erwin Panofsky's *Perspective as Symbolic Form* ([1927] 1991).
[25] See Julie Berger Hochstrasser, *Still Life and Trade in the Dutch Golden Age* (2007).

this object. The example is all the more clarifying in that it is in very large part from painting – landscape painting in particular – that we draw the basis for our conceptions of nature.[26] The manipulator actually exists: he/she is a painter. When Westerners are said to be "naturalists," it means that they are fond of painted landscapes, and that Descartes imagined the world as if projected onto the canvas of a still life whose manipulator would be God.[27]

Emphasizing this work of distribution makes it clearer that the expression "belonging to nature" is almost meaningless, since nature is only one element in a complex consisting of at least *three terms*, the second serving as its counterpart, culture, and the third being the one that distributes features between the first two. In this sense, nature does not exist (as a domain); it exists only as *one half of a pair pertaining to one single concept*. We must thus take the Nature/Culture opposition as the *topic* on which to focus our attention and not at all, any longer, as the *resource* that would allow us to get out of our difficulties.[28] To keep this point in mind, I shall adopt the habit of carefully surrounding "nature" with protective quotation marks, as a reminder that we are dealing with a coding system common to both categories. (To speak of the beings, entities, multiplicities, agents that people used to try to stuff into so-called "nature," we shall need an additional term, one that I shall introduce toward the end of this lecture.)

If ecology sets off panic reactions, we now understand why: because it obliges us to experience the full force of the instability of this concept, when it is interpreted as the impossible opposition between two domains that are presumed actually to exist in the real world. Above all, don't try to turn "toward nature." You might just as well try to cross through the plane of the painting to eat the oysters that gleam in the still life. Whatever you do, you will be tripped up, because you will never know whether you're designating the domains or the concept. And it will be worse if you think you can "reconcile"

---

[26] Interestingly, the object of Philippe Descola's recent seminars and ongoing work is precisely to link the question of the invention of nature to the history of painting; this approach can be glimpsed in the catalog of his exhibition at the Musée du Quai Branly, *La fabrique des images* (2010).
[27] On the whole question of "empirical style" and the invention of the theme of copy and model, so contrary to scientific practice, see Bruno Latour, *What Is the Style of Matters of Concern? Two Lectures on Empirical Philosophy* (2008c).
[28] Transforming what is an explanatory resource into an object to be explained (shifting from *resource* to *topic*) amounts to depriving yourself intentionally of an element of metalanguage and making the element instead a basis for study. Instead of having it at your back, you finally have it in front of you.

nature and culture or "go beyond" the opposition through "pacified" relations between the two.[29] Despite the title of a justifiably famous work, we cannot go "beyond nature and culture."[30]

But perhaps it is not entirely impossible to probe *on the near side*. If we are indeed dealing with one and the same concept consisting of two parts, this demonstrates that the parts are held together by a common core that distributes differences between them. If only we could approach this core, this differential, this apparatus, this manipulator, we could imagine how to get around it. Starting with a language that uses the opposition, we would become capable of translating what we want to say into another language that does not use it. This would give us something with which to begin to treat our madness – by inoculating ourselves with a different one, obviously; I have no illusions about this.

*

Now, we begin to spot this common core as soon as we take an interest in expressions such as "acting in keeping with one's nature," or in the classic line about living "according to one's *true* nature." It isn't hard, here, to detect the *normative dimension* of such expressions, since they purport to orient all existence according to a model of life that obliges us to choose between false and true ways of being in the world. In this case, the normative power that one would expect to find rather on the "culture" side turns out to be clearly imputed, on the contrary, to the "nature" side of the twofold concept. This curious imputation is more obvious when we mobilize the theme of "human nature," which one is supposed to "learn to respect" or against which, on the contrary, one is supposed to "learn to struggle."

When we invoke "natural law," we are expressing even more directly the idea that "nature" can be conceived as a set of quasi-legal regulations. In this case, oddly enough, the adjective "natural" becomes a synonym for "moral," "legal," and "respectable." But of course there is never any way to stabilize its meaning or respect the

---

[29] This is the difficulty that many contemporary philosophers run into when they approach the question of nature: they want to go beyond the division even as they continue to maintain it as the only available explanatory resource. This has been the problem from Catherine Larrère, *Les philosophies de l'environnement* (1997), through Dominique Bourg, *Vers une démocratie écologique: le citoyen, le savant et le politique* (2010), to Pierre Charbonnier, *La fin d'un grand partage: de Durkheim à Descola* (2015); the last keeps "the great distribution" in place even though he declares that the end has come.

[30] I am of course referring here to Descola's *Beyond Nature and Culture* (2013).

injunction. As soon as any authority sets out on a campaign to keep acts said to be "against nature" from being committed, protests arise at once: in the name of what do you dare decide which behavioral norms are "natural" and which are "against nature"? Since morality has been the object of vehement disputes for a very long time in our societies, any effort to stabilize an ethical judgment by the invocation of nature will appear as the scarcely concealed disguise of an ideology. The indignation aroused by such invocations is proof enough that "nature," here with its quotation marks, cannot invoke *nature*, without quotes, in order to end a moral controversy.

In other words, on these subjects, as on that of "organic" products or "100% natural" yogurt, we are all fairly likely to be constructivists – not to say relativists. As soon as we are told that a product is "natural," we understand clearly, at worst, that someone is trying to trick us and, at best, that someone has discovered another way of being "artificial." What was possible for Aristotle is no longer possible today: nature cannot unify the polity. It suffices to say that a position has been "naturalized" for us to conclude that the position has to be contested, historicized, or at least contextualized. We are at the point where the moral connotation of the notion of "nature" has been so clearly overturned that the first reflex of every critical tradition consists in fighting naturalization. In fact, as soon as anyone "naturalizes" or "essentializes" a state of affairs, the proposition becomes almost inevitably the assertion of a legal imperative. So much so that, in practice, it is as though common sense had fused the statements *de facto* and *de jure*.

Everyone understands that, if ecology consisted in going back to that sort of appeal to nature and its laws, we would not manage to understand one another any time soon. In today's pluralistic society, a stable meaning for the adjective "natural" is no easier to establish than meanings for "moral," "legal," or "respectable." Here, then, we have a set of cases in which the Nature/Culture theme appears in broad daylight as a distribution of roles, functions, and arguments that cannot be reduced to just one of its two components, despite the claims of those who use it. The more you talk about "staying within the limits of what is natural," the less you will get general agreement.[31]

---

[31] I have heard about militants who are fighting to prevent judges in Lebanon from continuing to use the expression "unnatural acts" to condemn homosexuality, but who are also seeking to introduce the idea of crimes *against nature* to protect rivers against industrial pollution! Such an example highlights the extent to which the appeal to nature can be unstable.

The situation is entirely different with the other family of notions associated with "nature" in the expression "natural world." In this case, it does seem possible actually to distinguish the two parts of the same theme and reach agreement. Or, at least, we thought so before the ecological crises and, more precisely, before the New Climate Regime made the invocation of "nature" as polemical as that of natural law.

And yet, at first glance, the situation ought to be quite different, because the "natural world," as everyone seems to agree, cannot dictate to humans what they must do. Between what is and what must be,[32] there must exist a gulf that cannot be crossed? This is in effect the default position of the ordinary epistemology that is adopted as soon as someone claims to be "turning to nature as it is." No more ideologies: states of affairs speak "for themselves," and one has to take endless precautions not to draw any moral conclusions from them. No prescription may emerge from their description. No passion may be added to the dispassionate presentation of the simple connections of cause to effect. The highly celebrated cloak of "axiological neutrality" is de rigueur in such presentations. Contrary to the previous case, here what is "natural" thus defines not *what is just*, but only what "*is just there, nothing more.*"

It suffices to reflect for only a moment, obviously, to notice that the difference between the two meanings of the word "just" is very slight, and that the default position is very unstable. Every time someone starts to invoke the "natural world," in any sort of argument, the normative dimension will remain present, but in a more convoluted form, since the principal injunction will insist precisely that the "natural world" *will not have* a moral lesson to impart, or even that it *will not allow anyone* to draw any moral lesson whatsoever. Here is a very powerful moral requirement: the one according to which one must *abstain* completely from moral judgment if one wants to take the full measure of the reality of what is![33] One might as well deny Mr Spock and the inhabitants of Vulcan any sense of good and evil. As for the "nothing more," it seems as though that point is not going to be maintained for long! On the contrary, what

---

[32] *Trans.*: The French verb *devoir*, like the English verb "must," can convey either supposition based on evidence ("She must have left...") or imperative obligation ("She must leave!").

[33] Tracing the history of these moral attitudes is the object of Lorraine Daston's systematic work, starting with "The Factual Sensibility: An Essay Review on Artifact and Experiment" (1988), all the way to her book with Fernando Vidal, *The Moral Authority of Nature* (2004).

a long *sequel* of arguments will be rolled out in the process of setting forth the uncontestable necessity of what *is* against the muddled uncertainties of what *must be*!

All the more so in that the simple description is accompanied by an extremely constraining set of injunctions: one "must" learn to respect brute facts; one "must not" draw hasty conclusions either about the way they are ordered or about the lessons that ought to be drawn from them; above all, they "must" be known first of all "in complete objectivity"; and, when they impose themselves, it "must be" in an uncontested and non-controversial way. Here we have a lot of "musts" imposed by something that is supposed to be "just there, nothing more." Such is in fact the paradox of the invocation of "nature": a formidable prescriptive charge conveyed by what is not supposed to possess any prescriptive dimension.[34]

The instability of this second-degree normative dimension is usually summarized in the following expression: "[One must respect] the laws of nature [which] impose themselves on everyone [whatever one may do and whatever one may think]." If the expression were really sufficient, the components in square brackets would not be needed; we would simply have a statement of what is imposed. And yet the normative injunction is indeed implied, since, in practice, those at risk of *not obeying* these laws are always the ones who have to be reminded. This interlocutory situation, most often disputational, sometimes polemical, is found every time someone uses the non-moral existence of the "natural world" to criticize some cultural choice or some human behavior. The pure, brute existence of incontestable facts enters abruptly into the discussion to bring it to an end, thus fully playing the normative role that these facts were not supposed to have – the role of unchallenged arbiter coming precisely from their "purely natural" existence.

Since this simple existence is in such contrast to the desires, needs, ideals, and fantasies of humans, every time someone insists on the facts their insistence brings to light an eminent value that is held to be more cherished than all the others: "*Respect that which* quite simply *is, whether you want to or not!*" The allusion to the arbitrary human will, which one "has to" know how to oppose, brings back at full strength the normative charge that had initially been removed. It is because the always divisive questions of morality have been set aside that agreement will finally be reached: "And you must do this

---

[34] It is to Friedrich Nietzsche, especially in *The Gay Science*, that we owe the analysis of the moral wellsprings of the scientific attitude of objectivity: see *The Gay Science, with a Prelude in Rhymes and an Appendix of Songs* ([1882] 1974).

whether you like it or not!" Here I am simply offering a philosophical comment on the virile gesture of someone who pounds his/her fist on the table to bring a discussion to an end.[35]

The invocation of nature is never satisfied with defining a moral law; it always serves, as well, to *recall to order* those who are straying from it. In the notion of "nature," there is thus always, inevitably, a polemical dimension. The requirement of sticking to the facts is normative to the second degree. Not content to introduce the supreme moral value, this requirement purports, in addition, to be achieving the political ideal par excellence: *the agreement of minds despite disagreements on moral questions.*[36] Clearly, it is hard not to see here once again the contrast between the two parts of the Nature/Culture concept. The two sides of the concept that we are trying to get around are thus indeed present at the same time, exactly as they are in the interminable, constantly renewed quarrels over the force of "natural law." Appearances notwithstanding, the invocation of the "natural world" offers an even stronger prescriptive charge than in the previous case. In all cases, what people are seeking to detect are indeed acts "against nature," but, as soon as someone claims to have found one, the accusation of "naturalizing" a simple set of facts into a legal imperative obliges critics to spring into action. As we can sense quite readily, what is *de facto*, in practice, is also, here again, *de jure*.

*

Oddly, those who first remarked upon this paradox in public were not the ecologists but their most relentless adversaries. In fact, without the immense undermining work undertaken by the climate skeptics against the sciences of the Earth System, we would never have grasped the extent to which the invocation of the "natural world" had ceased to be stable. Thanks to this false quarrel, an argument that had remained the discovery of a small number of historians of science is now becoming visible in broad daylight.[37]

---

[35]The classic article by Malcolm Ashmore, Derek Edwards, and Jonathan Potter remains unequalled: see "The Bottom Line: The Rhetoric of Reality Demonstrations" (1994).

[36]The social history of the sciences, from its beginnings (see, for example, Barry Barnes and Steven Shapin, eds, *Natural Order: Historical Studies of Scientific Culture*, 1979), has explored all possible ways of understanding the political effect of epistemology in the course of controversies.

[37]One can say that all the questions in the realm of "science studies" (see Dominique Pestre, *Introduction aux science studies*, 2006) have become public, in this context, and that the questions raised, for example, by Steven Shapin in *The Scientific Life: A Moral History of a Late Modern Vocation* (2008) are now shared by the researchers

From the 1990s on, as we know, powerful pressure groups have been mobilized to cast doubt on the "facts" (a mix of more and more complex and at the same time more and more robust models and measures) that were beginning to establish a consensus within research communities about the human origin of climate mutations.[38] Despite the distinction between facts and values that is so dear to philosophers and ethicists alike, the heads of the major companies under threat identified the stakes right away. They saw that, if the facts were known (CO_2 emissions are the principal source of climate change), politicians, pressed by the anxiety of the public, would immediately demand that measures be taken. We owe to the astute Republican strategist Frank Luntz, a psychosociologist and unrivalled rhetorician, the celebrated inventor of the expression "climate change" in the place of "global warming,"[39] the best formulation of this profound philosophy: the *description* of the facts is so dangerously close to the *prescription* of a policy that, to put a stop to the challenges addressed to the industrial way of life, one has to cast doubt on the facts themselves:

> Most scientists believe that warming is caused largely by manmade pollutants that require strict regulation. Mr. Luntz seems to acknowledge as much when he says that "the scientific debate is closing against us." His advice, however, is to emphasize that the evidence is not complete. "Should the public come to believe that the scientific issues are settled," he writes, "their views about global warming will change accordingly. Therefore, you need to continue to make the *lack of scientific certainty* a primary issue."[40]

The prescriptive charge of scientific certainties is so powerful that these are what must be attacked first.[41] Hence the development of

---

under attack by the "skeptics." See especially Mike Hulme, *Why We Disagree about Climate Change: Understanding Controversy, Inaction and Opportunity* (2009), and the book edited by Clive Hamilton, Christophe Bonneuil, and François Gemenne, *The Anthropocene and the Global Environmental Crisis: Rethinking Modernity in a New Epoch* (2015).

[38] There is now an abundant literature on the topic, starting with Naomi Oreskes's 2004 article "Beyond the Ivory Tower: The Scientific Consensus on Climate Change," then her 2010 book with Erik M. Conway, *Merchants of Doubt*. See also James Hoggan, *Climate Cover-Up: The Crusade to Deny Global Warming* (2009).

[39] Frank Luntz, *Words That Work* (2005), is cited at length in reporting about "communicators"; see the film by Barak Goodman and Rachel Dretzen, *The Persuaders* (2004).

[40] "Environmental Word Games" (2003), emphasis added.

[41] The use of the epistemological position to destroy the authority of the sciences through the attribution of a sort of auto-immune disease to the scientific institution has struck me ever since the emergence of Mr Luntz; see Bruno Latour, "Why Has Critique Run out of Steam? From Matters of Fact to Matters of Concern" (2004a).

this pseudo-controversy that has so wonderfully succeeded in convincing a large part of the public that climate science remains completely uncertain, and that climatologists are just one lobby among others, the Intergovernmental Panel on Climate Change (IPCC) is just an attempt on the part of mad scientists to dominate the planet, the chemistry of the upper atmosphere is just a plot "against the American way of life," and ecology is just an attack on humanity's inviolable right to modernize itself.[42] All this without managing to shake the consensus of the experts, a consensus whose validity has become more solid every year.[43]

If there were general agreement that CO_2, and thus coal as well as gasoline, was the *cause* of climate change, the industrialists and the financiers have understood perfectly that the description of the facts could never again be kept apart from their moral implications – and from the subsequent development of a policy. The imputation of *responsibility* demands a *response* – especially of course when the cause is "human."[44] If the industrialists and the financiers don't fight energetically, the factual reality will become the equivalent of a legal imperative. To describe is always not only to inform but also to alarm, to move, to set into motion, to call to action, perhaps even to sound the death knell. This has been known, of course; but it still needed to be shown in broad daylight.

Facing the enormity of the first climate threat (the one that emerged from research work), pressure groups were mobilized to respond to an even greater threat, as they saw it – one that stemmed directly from the first: the public was going to hold them responsible, and consequently would impose a profound transformation of the regulatory environment. It hardly needs saying that, in the face of such an emergency, ordinary philosophy of science doesn't carry much weight. You won't intimidate the powerful by pounding on the table; it does

---

[42] The reverberations of this strategy in France have been apparent in the lasting effectiveness with which Claude Allègre, mixing media, politics, and science, has managed right up to the present day to spread the belief that there are "two schools of thought" on this key question. See Edwin Zaccai, François Gemenne, and Jean-Michel Decroly, eds, *Controverses climatiques, sciences et politiques* (2012).
[43] The experts publish overviews "above the fray" (see Catherine Jeandel and Remy Mosseri, *Le climat à découvert: outils et méthodes en recherche climatique*, 2011; Virginie Masson-Delmotte, *Climat: le vrai et le faux*, 2011), but in vain: they are heard only as taking sides, something that is new for them. Even the reports of the IPCC have not succeeded in closing the debate as far as the public is concerned.
[44] We shall come back to the impossibility of distinguishing between facts and values in the next lecture, and also in the fourth, where I shall introduce the notion of the Anthropocene.

no good to say to them: "The facts are there, dear CEOs, whether you like them or not!" The celebrated "axiological neutrality" will be shattered to bits. The lobbyists have set into motion a whole panoply of communicators, paid experts, and even academics above suspicion, to generate a demand on the part of the public for something entirely different, on the strength of quite different facts. As one of them has written, carbon is "innocent" and must be thoroughly scrubbed free of all accusations and all responsibility.[45] No doubt about it: other non-facts will result in other non-policies!

We can grasp the full perversity of the appeals to the "state of the natural world" when we note that the counter-attack has been able to work only because the default position, that of ordinary philosophy of science, continued to look like common sense to everyone: to the public, to politicians, and especially – most astonishingly – to climate experts, those who found themselves so violently and so unfairly attacked because, according to their adversaries, they had crossed the yellow line between facts and values. In fact, if the lobbyists had said, "We do not *believe* in these facts; they do not *suit* us; they lead to sacrifices that we do not *want* to make," or, as President George H. W. Bush said, "Our way of life is not *negotiable*,"[46] everyone would have seen through them. No one can get away with saying, of the "natural world," that one "doesn't want" it, doesn't want to deal with it. Facts, as they say, are presumed to be "stubborn"; that is their way of *prescribing*. One can't negotiate with them or adjust them to suit oneself.

The climate skeptics have thus been clever enough to turn ordinary philosophy of science against their adversaries. They have stuck with the facts alone, by calmly asserting that "the facts *aren't there*, whether you like it or not." And they have started pounding forcefully on the table. The trap is well set: whereas the powerful have it both ways, discerning the prescriptive charge of the facts perfectly well and at the same time strictly limiting the debate to the discussion of only those discoveries whose existence they deny, the others sense that the facts lead to action but don't allow themselves to follow those facts

---

[45] François Gervais, *L'innocence du carbone: l'effet de serre remise en question* (2013). Conversely, P. K. Haff and Erle C. Ellis have proposed that geologists take a solemn vow when they finish their studies, a new form of the Hippocratic oath, given the importance to society of their future responsibilities: see Ruggero Matteucci et al., "A Hippocratic Oath for Geologists?" (2012), which confirms the passage from geochemistry to geophysiology and the transformation of the earth sciences into sciences of intensive care.

[46] In 1992, at the Earth Summit in Rio: "The American way of life is not negotiable." In the sixth lecture we shall trace the theological origin of such an assertion.

across the barrier that their adversaries nevertheless cross nimbly in both directions! Result: the pseudo-skeptics have made mincemeat of their unfortunate opponents. Mr Spock's mechanical voice is not supposed to quaver before the measurements, the alarms, the warnings, and the imputations of responsibility. Yet the climatologists' voice never stopped quavering before discoveries that were all the more awkward in that the experts didn't know how to handle their moral and political charge, even though the implications were quite obvious.[47] What is to be done, indeed, in the face of "inconvenient truths" if you possess only the right of uttering them with a mechanical voice and without adding any recommendation to them?[48] You will remain paralyzed.

This is why for some twenty years now we have been watching the astonishing spectacle of a pitched battle between one party that has perfectly grasped the normative function that invocations of the natural world perform – and for this reason denies the existence of that world – and another party that does not dare unleash the prescriptive force of the facts it has discovered and must limit itself, as if it had its hands tied behind its back, to speaking only of "science."[49] In a superb reversal of the situation, the earth science experts are the ones today who look like over-excited militants of a cause; fanatics, catastrophists, and climate skeptics are the ones assuming the role of stern scientists who at least do not confuse the way the world is going with the way it ought to go! They have even succeeded in appropriating – while reversing its meaning – the fine word "skeptic."[50]

*

In Pierre Daubigny's play *Gaia Global Circus*, which serves as a red thread running through these lectures, Virginia, a climatologist who

---

[47] Oddly enough, the experts' anguish has been made most perceptible in a graphic novel: see Philippe Squarzoni's admirable *Climate Changed: A Personal Journey through the Science* (2014), the best introduction to the New Climate Regime grasped from the standpoint of its *aesthetic* – in the etymological sense of learning to become sensitive.
[48] See Al Gore, *An Inconvenient Truth: The Planetary Emergency of Global Warming and What We Can Do About It* (2006).
[49] Fortunately, more and more scientists are realizing that they must not agree to argue about science with the climate skeptics. See, for example, the blog post by climatologist Mark Maslin, "Why I'll Talk Politics with Climate Change Deniers – but Not Science" (2014). As Aykut and Dahan make clear, the question is no longer – and hasn't been for a long time – a question of knowledge (*Gouverner le climat?*).
[50] This tradition has nothing to do with the polarization of confirmed facts, as we see in Frédéric Brahami, *Le travail du scepticisme: Montaigne, Bayle, Hume* (2001).

sums up the confirmed facts before an audience of bloggers despite the constant interruptions of a paid climate skeptic called Ted, is given a line that would make it possible to get out of the trap in which scientists have let themselves be caught. She proposes to use a means that would amount to modifying the relation between the sciences and politics, and in particular the relation between scientists and the world with which they are trying to enter into resonance. Scientists would have to accept their responsibilities, in Donna Haraway's sense: they would have to become *capable* of responding, would have to acknowledge that they have *"response-ability."*[51]

On stage, pushed to the limit by Ted, who never stops demanding a "democratic" debate, "fair and balanced" in the sense of Fox News, where skeptics would carry the same weight as the "warming sect,"[52] Virginia, like an evolutionist obliged to answer the objections of a creationist, hesitates to take up the challenge. She knows that the trap consists in acting as though there were a dearth of debates, as though the question had not been discussed fully enough. And yet the discussion *has taken place*: successive reports of the IPCC have summarized nearly twenty years of documentation, and the estimated degree of certainty is close to 98 percent – at least concerning the human origin of global warming.[53] On the massive phenomenon against which Ted is trying to turn the audience, the question was settled long before it entered this amphitheater. Virginia would now like to move on to the large number of questions that remain controversial, the most interesting ones in her eyes. Yet, if Ted is going to win, it will not be because he knows the subject better than she does or because he introduces new facts. He is paid to apply the philosophy of Mr Luntz: all he has to do to win is persuade the audience in the room that there is a debate among experts. To agree to respond is to reproduce a televised discussion in which Ms Pro is confronting Mr Con for the maximum pleasure of the audience, which will come away reassured by a demobilizing "what does anyone know?"[54] The very organ of

---

[51] Donna Haraway, *Staying with the Trouble: Making Kin in the Chthulucene* (2016), p. 16.
[52] This is the term, quite well chosen, it must be said, that Ted uses to designate those who "believe" (as if it were a matter of belief!) in warming that is human in origin.
[53] It goes without saying that there remain countless controversies over the consequences to draw from this causality, and about its precise mechanisms, the reliability of the models, the quality of the data, and of course the measures to be taken. The consensus bears only on the vast scope of the phenomenon and its urgency.
[54] The effectiveness of the procedure is ensured, as can be seen in an opinion piece ("À quoi peut encore servir la COP 21?") by the economist and social theorist Jacques Attali in *L'Express* on March 16, 2015: *"First of all,* there is no consensus on the

reason, open debate, becomes in this case the organ of manipulation.⁵⁵ And yet, if Virginia refuses to engage in the exercise that is being imposed, she knows quite well that she'll appear dogmatic – a mortal sin in the era of unlimited commentary on the Web...

But what to do? In the current context, there is no alternative. A scientist has to appear cool, distant, indifferent, and disinterested. For several seconds, in suspense, Virginia explores other solutions, each one more calamitous than the one before. This is when, in a moment of inspiration and panic, she cries out against Ted, whom the spectators are on the verge of driving out of the room: "Go tell your masters that the scientists are on the warpath!"

However, in the next scene she admits sheepishly that she doesn't know what that means. For scientists, in fact, the warpath doesn't exist. The others, the ones who sent Ted to disrupt Virginia's talk, are the ones at war, as they have been for a long time. Neither the honest researchers, like Virginia – before her outburst – nor her well-meaning audience know that they are at war. They think they are still safely behind the Maginot line of rational debate carried out between reasonable people in an enclosed and protected space reserved for questions of lesser importance or with remote applications. As soon as they hear talk about "respecting the facts," they feel obliged to respond politely, since respect for the facts is the basic principle of their method, too. If Virginia hadn't responded so energetically, the trap of negationism would have snapped shut on her.⁵⁶

Except that this negationism does not apply to past facts, facts long since confirmed that are now criticized only by people whose ideology is too clearly apparent – they cannot live in a world in which humans could be capable of committing such crimes. This time what is at stake involves present facts, facts that are reaching us, acts that are being committed, right now. And here the ideology is not so easy to detect, for they are legion, those who would like not to live in a

---

mechanisms involved: for some, the sun is responsible, and there is nothing we can do about it. For others, human activities are responsible, and in particular the emission of greenhouse gases; and there is a lot we can do about it. For still others, finally, on a worldwide basis the temperature has not been increasing for more than ten years; the worst is over and there is no point worrying about it." Isn't that "first of all" admirable?!

⁵⁵ James Hoggan, *I'm Right and You're an Idiot: The Toxic State of Public Discourse and How to Clean it Up* (2016).

⁵⁶ The trap works if one responds empirically but also, on the contrary, if one refuses to respond empirically, as it does in cases where *past* crimes are being denied. See Pierre Vidal-Naquet, *Assassins of Memory: Essays on the Denial of the Holocaust* ([1991] 1992).

On the instability of the (notion of) nature 31

world where humans are capable of such crimes! We are touched in our most intimate being by the hope that humanity will never have such capability. We are constantly at risk of conspiring with our enemies. This is what it really means to find ourselves at war: to have to decide, without any pre-established rules, which side we're going to have to be on.[57]

All the more so in that the negationists, this time, are not marginal types who play at "breaking down the taboos" of the elites; these are the elites themselves, at war against other elites.[58] The phenomena being debated bear upon the near future; they oblige us to rethink the entire past, but above all they entail a frontal attack on the decisions of many pressure groups, and they bear upon questions of direct interest to billions of humans obliged to change their mode of life down to the smallest details of their existence. How can we hope that the scientists will be heard without a fight?

And to complicate the situation further, the scientific disciplines that have come together to develop these facts that have become so sturdy do not come from the prestigious sciences such as particle physics or mathematics; they come from a multitude of earth sciences whose certainties have been achieved not by some earth-shaking, fool-proof demonstration but by the weaving together of thousands of tiny facts, reworked through modeling into a tissue of proofs that draw their robustness from the multiplicity of data, each piece of which remains obviously fragile.[59] Between a tissue of proofs and a tissue of lies, we understand that people who know nothing about the practice of science are quick to confuse the two – especially if it's really in their interest that the data prove false. Poor Virginia. What a dereliction, and what a cry! How could she not be ashamed to feel in her own trembling hand the weight of the tomahawk that

---

[57] I shall come back to this essential principle in the seventh lecture.

[58] The Academy of the Sciences (at least in France) is mobilized in this context, along with major media such as the *Wall Street Journal*, with the signatures of Nobel laureates (see Claude Allègre et al., "No Need to Panic about Global Warming," 2012). Their views are not so easy to sweep away on the same basis as the pompous predictions of those who campaign against vaccinations or who believe in the Hollow Earth.

[59] As Spencer Weart (*The Discovery of Global Warming*, 2003) and Paul N. Edwards (*A Vast Machine: Computer Models, Climate Data, and the Politics of Global Warming*, 2010) have shown, the climate sciences are very different from those that gave rise to the hope, in the twentieth century, that they were establishing the foundation for all the others. Along with the importance taken on by models, the most acceptable explanation for the skepticism on the part of certain scientists is the very variety of these disciplines, which are often close to natural history: many scientists in the twentieth century were expecting an entirely different scientific revolution.

she has just dug up? Ted is driven out, but a new nightmare is now beginning for Virginia.

For her exclamation to be understood, the community of climatologists to which she belongs has to acknowledge that they actually do have a politics. That they can answer back by asking: "Whom do you represent, and for whom are you fighting?" The question in fact makes sense. When climate skeptics denigrate the science of climatologists, whom they accuse of behaving as a lobby, they too are assembled as a group, for which they have defined admissions tests and drawn boundaries, distributing the components of the world in a different way – what one can expect of politics and how science is supposed to function (this is what we shall call, later on, their "cosmogram").[60] Why wouldn't the climatologists do the same thing? There is no reason for them to keep claiming that they are not in the game, as if they were speaking from nowhere and behaving as if they didn't belong to any earthbound population. One would be tempted to offer them some advice: "But finally, instead of believing that you have to make your science meet the impossibly inflated demands of an epistemology that requires you to be disembodied and located nowhere, just say where you are situated."[61]

We would like Virginia to be able to say, finally: "Why aren't you proud of having invented this extraordinary equipment that allows you to give voice to mute things *as if* they were in a position to speak?[62] If your adversaries tell you that you are engaged in politics by taking yourselves as representatives of numerous neglected voices, for heaven's sake answer 'Yes, of course!' If politics consists in representing the voices of the oppressed and the unknown, then we would all be in a much better situation if, instead of pretending that the others are the ones engaged in politics and that you are engaged 'only in science,' you recognized that you were also in fact trying to assemble another political body and to live in a coherent cosmos composed in a different way. If it is entirely correct that you are not speaking in the name of an institution limited by the borders of nation-states and that the basis for your authority rests on a very strange system of election and proofs, this is precisely what makes your *political power to represent* so many new agents so important. That power of

---

[60] The term is borrowed from John Tresch, "Cosmogram" (2005).
[61] Here we see the full importance of the notion of "situated knowledge" developed by Donna Haraway in "A Cyborg Manifesto: Science, Technology, and Socialist-Feminism in the Late Twentieth Century" (1991).
[62] The analysis of this system of scientific and political representation is the focus of my two related works *Pandora's Hope: Essay on the Reality of Science Studies* (1999) and *Politics of Nature: How to Bring the Sciences into Democracy* ([1999] 2004b), which serve as the background for this argument.

representation will be of capital importance in the coming conflicts over the form of the world and the new geopolitics. Don't sell it for a mess of pottage."

Such a confession would not cast the slightest shadow of doubt on the quality, the objectivity, or the solidity of the scientific disciplines, since it is now clear that the network of instruments, the Vast Machine that the climatologists have built, ends up producing knowledge that is robust enough to withstand the *objections*. In any case, on this Earth, the adjective *objective* has no other meaning. There is no other source that can surpass the type of certainties that you have been capable of accumulating. What could it mean to *know* the human origin of climate change *better* than the climatologists do? This thesis was harder to advocate, I acknowledge, in an earlier period, when the apparatus, the groups, the cost, the institutions, and the controversies over the facts were not as visible.[63] But this is no longer the case. Just as no GPS point can be determined without the vast array of satellite equipment that makes zeroing in on it possible, every somewhat solid fact has to be accompanied by a whole suite of instruments, by its assembly of experts engaged in debate, and by its public. One cannot act as though one knew more and more without being caught up oneself in the machinery of knowledge production. To plead against the results of science, there is no Supreme Court, certainly not the Supreme Court of Nature. It is the scientific institution that we have to learn to protect.

*

At the risk of shocking my climatologist friends, I am for my part beginning to think that, philosophically, the billions spent by the climate-skeptic lobbies to create the false controversy over the climate will not have been spent in vain, since we can now see quite clearly to what extent claims about the "natural world" are no more apt to promote agreement than claims about "natural law." Nor, for anyone observing the pseudo-controversies over the climate, does the appeal to the "laws of nature" allow us to reach uncontestable agreement

---

[63] Although certain scientist friends believe that I have stopped being a "relativist" and have started "believing" in the "facts" about the climate, it is on the contrary because I have never thought that "facts" were objects of belief, and because, ever since *Laboratory Life: The Social Construction of Scientific Facts* (with Steve Woolgar, 1979), I have described the *institution* that makes it possible to ensure their validity in place of the epistemology that claimed to defend them, that I feel better armed today to help researchers protect themselves from the attacks of negationists. It is not I who have changed, but those who, finding themselves suddenly attacked, have understood to what an extent their epistemology was protecting them badly.

every time, although that appeal belongs to a distinct historical tradition. What "nature" wants, what it requires, what it allows, is what will bring both closure and provocation, even new venom, to the debates. It's no use contrasting what is and what must be: when we're talking about "nature," we must always learn to reckon with both.

If ecology drives us crazy, it's because it obliges us to plunge head first into the confusion created by reference to a "natural world" that is said to be at once fully endowed and not at all endowed with a normative dimension. "Not at all," since it describes only an order; "fully," since there is no order more sovereign than the order to obey that order. We can understand that the humans whom we are going to ask to define their relations to the world will find themselves in an uncomfortable situation if they understand such a request as meaning: "Can you please spell out the way you belong to nature?" If they answer, they are headed for the confusions pointed out earlier, as they seek to obtain an indisputable peace agreement with notions that are all exceedingly polemical.

In spite of the vast literature on the indispensable chiasmus between facts and values, it is evident that defining the former necessarily bears decisively on the latter. When "nature" is involved, what is a matter of fact is *necessarily* also a matter of law. By pretending to oppose the two, we find ourselves with *two forms of having-to-be, two moralities instead of one*. What is *just there* is fundamentally also always what is *just*. Or, to put it in still another way, to *order* (in the sense of ordering the world) is to *order* (in the sense of giving orders). How could it be otherwise when it means, in addition, evaluating the responsibility of humans mixed with the responsibility of things? Invoking "nature" does not bring peace. If we find it hard to think this, Ted and those who finance him have nevertheless understood it; what is new is that they have finally forced Virginia to understand it as well.

Instability of this sort disrupts all the disciplines, but none more directly than ecology, a term I have been using for a while as though it had an agreed-upon definition. Efforts have certainly been made in the past to distinguish *scientific* ecology from *political* ecology, by assuming that the first is concerned only with the "natural world" and the second only with the moral, ideological, and political consequences that must be drawn – or not – from the first.[64] These efforts

---

[64] See Jean-Paul Deléage, *Histoire de l'écologie: une science de l'homme et de la nature* (1991); Jean-Marc Drouin, *Réinventer la nature* (1991); Florian Charvolin, *L'invention de l'environnement en France: chroniques anthropologiques d'une institutionnalisation* (2003); Pascal Acot, ed., *The European Origins of Scientific*

have only reinforced the confusion, since we now encounter combinations of what is and what must be at every level.

The New Climate Regime does indeed hinge on a renewed form of natural law, or in any case a link to be restored between nature and law, that would enable a revitalization of the expression "laws of nature" – laws whose mode of operation tends to be too hastily simplified.

Clearly, the bad news with which we are bombarded every day about the state of the planet incites us to become aware of a *new instability of nature*. But since we don't manage to evaluate these warnings, or really even to take them into account, they drive us crazy in several ways. At which point we notice the existence of yet another instability, this time in *the very notion of "nature."* The invocation of the "natural world," which was supposed to stabilize, pacify, reassure, and bring minds into agreement, seems to have lost the capacity to achieve these goals since the onset of the false climate quarrel. It had never actually possessed such a capacity, but the goals had nonetheless remained an ideal, as long as we were dealing with questions that were not of planetary importance. This state of dereliction, from which it would be useless to try to escape, stems from the fact that we are caught in the middle of two instabilities. Let us now try to dig down a little deeper, *beneath* the ever-so-equivocal notion of "nature," and thus *before*, or *just short of*, the paired concepts that I have termed Nature/Culture.

Since the madness in question is diagnosed as an alteration in the relation to the world, is it possible to detach the term "world" from its association – an almost automatic one, to be sure – with the term "natural world"? We would have to be able to introduce an opposition, not between nature and culture this time (since the incessant vibrations between the two are what drives us crazy), but between Nature/Culture on one side and, on the other, a term that would include each one of them as a particular case. I propose simply to use the term *world*, or "worlding,"[65] for this more open concept, defining it, in an obviously very speculative fashion, as that which opens to the multiplicity of *existents*, on the one hand, and to the *multiplicity* of ways they have of existing, on the other.[66]

Ecology ([1988] 1998); and, more recently, John R. McNeil, *Something New under the Sun: An Environmental History of the Twentieth-Century World* (2000), which is in part a history of ecology as a science.
[65] Haraway, 2016, p. 16.
[66] The pluralism of the universe, in the sense of William James, *A Pluralistic Universe: Hibbert Lectures at Manchester College on the Present Situation in Philosophy*

Let's be careful: let's not rush into saying that we're already familiar with the list of existents and the way they are related to one another, for example, by saying that there exist two and only two forms – causal relations and symbolic relations – or by claiming that all existents form a Whole that can be encompassed by thought. This would amount to stuffing them all back into the single frame of Nature/Culture, which we are seeking, precisely, to circumvent. No, we have to agree to remain open to the dizzying otherness of existents, the list of which is not closed, and to the multiple ways they have of existing or of relating among themselves, without regrouping them too quickly in some set, whatever it might be – and certainly not in "nature." It is this opening to otherness that William James proposed to call the *pluriverse*.[67]

Only if we place ourselves inside this world will we be able to recognize as one particular arrangement the choice of existents and their ways of connecting that we call Nature/Culture and that has served for a long time to format our collective understanding (at least in the Western tradition).[68] Ecology clearly is not the irruption of nature into the public space but the *end of "nature"* as a concept that would allow us to sum up our relations to the world and pacify them.[69] What makes us ill, justifiably, is the sense that that Old Regime is coming to an end. The concept of "nature" now appears as a truncated, simplified, exaggeratedly moralistic, excessively polemical, and prematurely political version of the otherness of the world to which we must open ourselves if we are not to become collectively mad – *alienated*, let us say. To sum it up rather too quickly: for Westerners and those who have imitated them, "nature" has made the *world* uninhabitable.

This is why, in everything that follows, we are going to try to descend from "nature" down toward the multiplicity of the world,

---

([1909] 2012), offers a good definition. "Nature is but a name for excess," James says. This is also the direction taken by Whitehead: "We are instinctively willing to believe that by due attention, more can be found in nature than that which is observed at first sight. But we will not be content with less" (*The Concept of Nature*, 1920, p. 29). See Didier Debaise's commentary on this statement in *L'appât des possibles: reprise de Whitehead* (2015).

[67]This question of pluralism is at the heart of the AIME project – Bruno Latour, *An Inquiry into Modes of Existence: An Anthropology of the Moderns* (2013b).

[68]Let us recall once again that the Nature/Culture pairing is not a universal – a matter that has been well explored by anthropology; see Descola (2013).

[69]The seeming paradox in the fact that the so-called question of the environment appeared only when the external environment disappeared was what led me to investigate these ecological questions, in the context of a study of the implementation of a new law on water in France. See Bruno Latour, "To Modernize or to Ecologize, That Is the Question" (1998).

but of course, as we proceed, we have to avoid ending up solely with the diversity of cultures. The operation comes down to reopening the two canonical questions: what existents have been *chosen*, and what forms of existence have been *preferred*?

Every time these two questions are answered in a more or less organized way, we can say that we are dealing with *metaphysics*. These are in fact the sorts of questions that philosophers raise as a matter of course. But in the most recent Western tradition the tendency has been to turn rather toward anthropologists when we want to *compare* the various metaphysical schemas that have given different answers to questions about the number and the quality of relations among existents.[70] What matters is that the term "world" remains open enough to preclude premature answers either to the question regarding the set of existents or to the question regarding the forms of existence – thus open enough to allow for proposing other arrangements.

If the notion of "nature" in its two versions – natural law and laws of nature – is so troublesome for those who are trying to find out whether they belong to nature or not, it is because this notion bears the accumulated weight of a large number of previous decisions. You now understand that, if we were to agree to begin with the metaphysics of "nature," we would not discern these decisions at all. Thus we have an interest in going deeper, seeking in other accounts that retrace other cosmologies, other metaphysics, the reason for the particular choices that have led to the current mutation. This choice of method is not an easy one, as I know all too well: the temptation is always to return to the idea of a "natural world" and use a contrastive approach to raise moral, political, or managerial questions about the way to deal with that world; or to dream of a more subjective, more "human," less "reductive" approach to that same "nature"; or to confuse the plurality of cultures with the pluralism of the world. Here, I am simply proposing to *frame* the notion of Nature/Culture, yes, in the literal sense: to relativize it by placing it among other accounts with which it shares, or does not share, certain features. In other words, I propose to make it a question of composition – in all senses of the word.[71]

---

[70] Provided that the anthropologists in question not only define a culture but also venture to inquire into ontological conflicts, as does Eduardo Viveiros de Castro in *Cannibal Metaphysics: For a Post-Structuralist Anthropology* ([2009] 2014) or Eduardo Kohn in *How Forests Think: Toward an Anthropology beyond the Human* (2013).
[71] On the notion of composition, see Bruno Latour, "Steps toward the Writing of a Compositionist Manifesto" (2010c).

The interest of this broadened definition of the term "world" is that one sees very quickly that the concept of "nature" can in no case be taken as one of its synonyms. To speak about "nature," of "man in nature," of "following nature" or "coming back" to it or "obeying" it or "learning to know" it, is *to have already decided on an answer* to the two canonical questions about the set of existents and the choice of forms of existence that connect them.[72] In order not to confuse the two or to take them as synonymous, let us capitalize Nature to remind ourselves that it is a sort of *proper noun*, the name of one *cosmological figure* among many others, and to which we shall soon learn to prefer a different figure, designated by another proper noun, one that will take charge of other existents and other ways of connecting them in an entirely different way, by imposing on them other obligations, other moralities, and other laws.

*

Have we made a little progress? I have proposed something like a first course of treatment that would very cautiously play the ways of being in the world off against one another. Which amounts to raising very old and very banal questions: who, where, when, how, and why? Who are we, we who still call ourselves "humans"? Where do we reside? What type of territory, ground, space, or place are we apt to inhabit, and with whom are we ready to cohabit? In what epoch do we find ourselves – not in terms of the calendar but, rather, in terms of the rhythm, the scansion, the movement of time? How and why have we reached this situation, in which the question of ecology is driving us crazy? What paths have we followed, and what motives lie behind our decisions? Each of these questions has several answers, and this is precisely what disorients us – even more so when the answers become totally incommensurable, as is the case today, with the double instability of nature and of the notion of "nature."

What would happen, for example, if we were to give entirely different answers to the questions that serve to define our relation to the world? Who would we be? Let us say Earthbound rather than Human. Where would we find ourselves? On Earth and not in Nature. And, even more precisely, on land shared with other often bizarre beings whose requirements are multiform. When? After profound transformations, and even catastrophes, or just before the immanence of cataclysms, something that would give the impres-

---

[72] This is why Descola (2013) decided to call "naturalists" those who use the Nature/Culture schema to organize the distribution of existents.

sion of living in an atmosphere of end times – the end of the earlier times, in any case. How would we have reached that point? Precisely through a series of tracking errors during earlier episodes involving Nature. We would have attributed capacities, dimensions, a morality and even a politics to Nature that it was not fashioned to bear. The composition we had chosen would have collapsed. We would find ourselves, literally, *decomposed*.

How could we not be destabilized in realizing that the revolution longed for by progressive minds has perhaps *already come about*? And that it has come not from a presumed change in the "ownership of means of production" but from a stupefying acceleration in the movement of the carbon cycle![73] Even the Engels of the *Dialectics of Nature* could never have imagined how right he was when he asserted that all the agents of the planet would end up being mobilized for real in the intoxicating frenzy of historical action. Even the Hegel of the *Phenomenology of Spirit* could not have envisaged that the advent of the Anthropocene would so radically reverse the direction of his project that humans would be dialectically immersed no longer in the adventures of the Absolute Mind but in those of geohistory. Imagine what he would have said if he had seen that the breath of Spirit is now overcome, surpassed, *aufgehoben*, intoxicated by carbon dioxide!

In an era when commentators are deploring the "lack of a revolutionary spirit" and the "collapse of emancipatory ideals," how can we not be astonished that historians of nature are the ones revealing, under the name of the Great Acceleration whose beginning marks the Anthropocene, that the revolution has already occurred, that the events we have to confront are not situated in the future but in a recent past?[74] The revolutionary activists are brought up short when they realize that, whatever we do today, the threat will remain with us for centuries, millennia, because the relay of so many irreversible actions, committed *by humans*, has been *taken up* by the inertial warming of the sea, the changes in polar albedo, the growing acidity of the oceans, and because it is a matter not of progressive reforms but of catastrophic changes, once the line has been crossed, no longer like the Pillars of Hercules but like tipping points.[75] This is enough to

---

[73] In "The Climate of History: Four Theses" (2009), Dipesh Chakrabarty was one of the first to connect the history of the Marxist tradition with that of carbon. See also, more recently, his "Climate and Capital: On Conjoined Histories" (2014).
[74] See Will Steffen et al., "The Trajectory of the Anthropocene: The Great Acceleration" (2015a).
[75] On the tipping points that have become so important in the history of the earth, see Fred Pearce, *With Speed and Violence: Why Scientists Fear Tipping Points in Climate Change* (2007).

disorient us. At the root of climate skepticism, there is this surprising reversal of the very tenor of progress, of the definition of what is to come and of what it means to belong to a territory. In practice, we are all counter-revolutionaries, trying to minimize the consequences of a revolution that has taken place without us, against us, and, at the same time, through us.

It would be thrilling to live in such an era, if only we could contemplate the tragedy from a distant shore that would have *no history*. But from now on there are no more spectators, because there is no shore that has not been mobilized in the drama of geohistory. Because there are no more tourists, the feeling of the sublime has disappeared along with the safety of the onlookers.[76] It's a shipwreck, to be sure, but there are no more spectators.[77] It looks more like *Life of Pi*: in the lifeboat, there is – there might be – a Bengal tiger! The unfortunate young shipwreck survivor has no more solid shore from which he can enjoy the spectacle of the struggle for survival alongside an untamable wild beast for whom he serves as both tamer and lunch![78] What is coming toward us is what I call Gaia; this is what we have to look at head on if we don't want to be driven crazy for real.

---

[76] See Bruno Latour and Émilie Hache, "Morality or Moralism? An Exercise in Sensitization" (2010).

[77] See Hans Blumenberg, *Shipwreck with Spectator: Paradigm of a Metaphor for Existence* ([1979] 1997).

[78] See Yann Martel, *Life of Pi* (2001) – with the wrinkle that, in the end, there was no tiger...

SECOND LECTURE

How not to (de-)animate nature

Disturbing "truths" • Describing in order to warn • In which we concentrate on agency • On the difficulty of distinguishing between humans and nonhumans • "And yet it moves!" • A new version of natural law • On an unfortunate tendency to confuse cause and creation • Toward a nature that would no longer be a religion?

How are we poor readers supposed to react when we come across a headline like this: "Highest level of CO_2 in the air in 2.5 million years," with an even more disturbing subtitle: "The threshold of 400 ppm of carbonic gas, the principal agent in warming, is about to be crossed"? And the journalist explains:

> A *symbolic threshold* is about to be crossed. For the *first time* since man appeared on earth. And even in the last 2.5 million years. The threshold of 400 parts per million (ppm) of atmospheric carbon dioxide (CO_2) is expected to be reached in May, at the Mauna Loa station in Hawaii, the historic point from which the first measurements in the modern era were taken starting in 1958 by the American David Keeling.[1]

This is an actual fact, the result of a confirmed observation obtained with great difficulty thanks to Keeling's persistence. As he tells the

---

[1] Stéphane Foucart, "Le taux de CO_2 dans l'air au plus haut depuis plus de 2,5 millions d'années" (2013), p. 4, emphasis added.

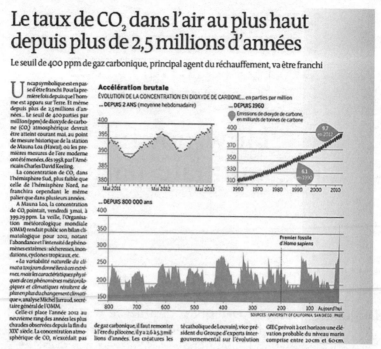

Figure 2.1 Reproduction of an article by Stéphane Foucart in *Le Monde*, May 7, 2013.

story in a book testifying to the daunting challenge of equipping the Earth with sufficiently sensitive instruments, if he succeeded in maintaining his measuring equipment over a long period of time, it was against the skepticism and indifference of the financing agencies and of a number of his colleagues as well.[2] But, at the same time, when a newspaper article mentions lines about to be crossed, symbolic thresholds, and a principal warming agent, the reader can't help but suppose that this piece of news is intended as a *warning*. This is certainly what one of the researchers cited by the journalist asks us to do:

> Crossing the threshold of 400 ppm of CO_2 *carries a powerful symbolic charge*, according to climatologist Michael Mann, director of the Earth System Science Center of the University of Pennsylvania. This comes to remind us to what an extent the *dangerous experiment* that we are carrying out on our planet is *out of control*.[3]

---

[2] Charles David Keeling, "Rewards and Penalties of Recording the Earth" (1998), a stunning example of scientific autosociology.
[3] Foucart, 2013, emphasis added.

Here is one of the hybrid expressions that we identified in the first lecture. To say that a threshold has been crossed and that we are carrying out an out-of-control experiment is to cross the supposedly inviolable gap between pure description and vigorous prescription: we have to *do* something – but we are not told what to do.

Michael Mann, the author of a famous curve in the form of a hockey stick, would be the last to deny that it is a matter of politics as much as of morality.[4] In the history of the sciences, no diagram has been subject to more attacks than this one (there is a simplified version in figure 2.1). The climate skeptics, astute devotees, as we have seen, of a strict distinction between what is and what must be, attacked it so viciously that Mann had to give the book in which he related his adventures a telling subtitle: *The Hockey Stick and the Climate Wars: Dispatches from the Front Lines*. Nothing has improved since 2013, either in the out-of-control and "dangerous experiment that we are carrying out" or in the attacks renewed daily on the "front lines" that are intended to make this inconvenient truth disappear from the face of the Earth. If it is true that "the first victim of war is truth," then the second is certainly axiological neutrality, which is quite unable to resist the unbearable tension between description and prescription that has been created by the New Climate Regime. What Mann discovered, and what we are going to explore in depth throughout these lectures, is that we really are encountering a situation of war – and not only a "climate war."[5] How else can we explain why in 2007 the Intergovernmental Panel on Climate Change (IPCC), itself a diplomatico-scientific body, was awarded the Nobel Peace Prize rather than the prize in physics or chemistry?

The tension is all the stronger in that, as Michael Mann adds with false innocence at the end of the piece in *Le Monde*: "There is a real possibility that with the current levels of CO_2 *we have already crossed* the threshold of a dangerous influence on our climate." Not only do we find ourselves placed at a historic moment without any known precedent ("To find such levels of carbonic gas, we have to go back to the Pliocene, 2.6 to 5.3 million years ago. The creatures nearest to humans that walked the surface of the Earth at the time

---

[4]Michael Mann, *The Hockey Stick and the Climate Wars: Dispatches from the Front Lines* (2013). The link between description and warning is perfectly explicit in the opinion piece Mann published in the *New York Times*, "If You See Something, Say Something" (2014) – impossible to be more explicit.
[5]Harald Welzer, *Climate Wars: Why People Will Be Killed in the Twenty-First Century* ([2008] 2012). The link between climate and war considerably predates today's geo-engineering, as James Rodger Fleming shows in *Fixing the Sky: The Checkered History of Weather and Climate Control* (2010).

were Australopithecenes!"); not only have we crossed a threshold – a term that is at once legal, scientific, moral, and political; not only is humanity responsible for this truly revolutionary transformation (this is implied by the well-known association between CO_2 emissions and the industrial way of life); but in addition we have probably already passed the moment when we could still do something about it.[6] The revolution was started by us, but without us, in a terribly recent past of which we are becoming aware too late! And to make the picture all the more dramatic, the diagram that accompanies the latest series of measurements underlines, with a detail that can be read as black humor, the moment when this history began: "First *Homo sapiens* fossils" – waiting for the last... Between the Australopithecenes and the *Homo oeconomicus* of the "modern era," the reader is treated to a lightning-fast summary: a brief history divided between what has happened to the Earth and what has happened to the humans who, in former times, inhabited it without having much influence on it.

I was not exaggerating, then, in saying that the climate question is driving us crazy. Everything in these reports is dizzying: they offer a sense of the immense complexity of the scientific arrangements capable of establishing reliable measures over such vast distances in time, not to mention the extraordinary layering of disciplines – paleontology, archaeology, geochemistry – capable of converging on models that make it possible to predict at what precise moment we are crossing thresholds.[7] But the most vertiginous experience of all comes when we place the long history of the planet and the short history of humans on the same chart, not in order to stress the *insignificance* of humanity in the face of the Earth's vast history, as we used to do, but, on the contrary, in order to put the burden of unprecedented *geological power* abruptly on that same humanity's shoulders.[8] And

---

[6] While the idea of the Anthropocene may flatter humans because it means they have finally won power over the planet, it is much less agreeable to learn that this power to influence may well have already been lost! See what Wallace Broecker has to say: "The paleoclimate record shouts out to us that, far from being self-stabilizing, the Earth's climate system is an ornery beast which overreacts even to small nudges" ("Ice Cores – Cooling the Tropics," 1995). It is the strangeness of this phenomenon that justifies the title of Timothy Morton's book *Hyperobjects: Philosophy and Ecology after the End of the World* (2013).
[7] For the general public, the best introduction to the everyday work of researchers remains a series of videos accessible at www.thiniceclimate.org.
[8] The common-sense reflex of historians consists in saying that what appears unprecedented to us has already happened many times The interest of the work of researchers focusing on the Anthropocene is precisely that it challenges the argument that there is nothing new under the sun. One example among hundreds: "The early-twentieth-century invention of the Haber–Bosch process, which allows the conversion of

it's not over: after turning the tiny creatures that we thought we were into a giant Atlas, they tell us very calmly at the same time that we're hurtling toward our doom if we do nothing – but that it's probably too late to do anything about it in any case.

How could we not be panic-stricken by such short-circuits, unimaginable earlier, between the rhythm of history and that of *geohistory*, as "full of sound and fury" as the earlier history?[9] We had heard about the acceleration of history, but the idea that this history could also accelerate geological history is what leaves us stupefied. It is not speaking ill of humanity to recall the extent to which we are all ill-equipped – emotionally, intellectually, morally, politically, culturally – to absorb such news. It would be much wiser, and even more rational, to ignore it altogether – if that weren't the surest way of giving in to real delirium!

That there is an enormous difference between responding to a threat under the auspices of politics and responding under the auspices of knowledge is easy to see when we compare the rapid, anxiety-ridden arms race set off by the Cold War with the sluggish pace of negotiations over the climate. Hundreds of billions of dollars were spent on atomic weapons in response to a threat about which the information acquired by spies was, at best, very slim, while the threat created by the anthropic origins of the "climate upheaval" is probably the best documented and the most objectively developed piece of knowledge on which we can rely before moving into action. And yet, in the first case, all the traditional emotions of wartime politics led, in the name of precaution, to the establishment of an arsenal that was *disproportionate* in the extreme; while, in the second case, we are expending a great deal of energy delaying, in the name of the same precaution, the knowledge needed to trigger *barely proportionate* expenses.

It suffices to compare the reception of George Kennan's secret "long telegram" on Soviet strategy in 1946 to that of Sir Nicolas Stern's public report in 2006 on the small sums that would have to be expended by the industrialized countries to avoid most of the

---

atmospheric nitrogen to ammonia for use as fertilizer, has altered the global nitrogen cycle so fundamentally that the nearest suggested geological comparison refers to events about 2.5 billion years ago.... [Human action] has increased ocean water acidity at a rate probably not exceeded in the last 300 million years" (Simon L. Lewis and Mark A. Maslin, "Defining the Anthropocene," 2015, p. 172). On the question of the absence of precedents, to which I shall return in the fourth lecture, see Clive Hamilton and Jacques Grinevald, "Was the Anthropocene Anticipated?" (2015).
[9] The term "geohistory" sums up very nicely Dipesh Chakrabarty's 2009 article "The Climate of History: Four Theses" (2009).

deleterious effects of climate change.¹⁰ In the first case, the clear presence of an enemy, of war and politics, gave the word "precaution" the sense of *rapid action*; in the second, the uncertainty as to the enemy, the war, and the politics gave "precaution" the calming connotation of "*let's wait and see*, we can always sort things out later." A *panic attack* in the first case, resulting in a general mobilization; in the second case, demobilization – and yet we are dealing with the great god Pan in person!¹¹

In the face of such a gap in reaction time, the ecology activists are tempted to accelerate matters by appealing, they think, to the power of conviction of the sciences. "Since we now know for sure what is going on, you have to act. If you don't, you're behaving like criminals." Thus they attribute to the inviolable laws of an indifferent Nature the highly political function of mobilizing the masses, which are indifferent to the threat – while adding a touch of moral indignation. This is a version of what has been called "strategic essentialism."¹² One relies on the notion of incontrovertible certainty to achieve an effect of mobilization that could not be achieved otherwise. The danger of such a tactic is that it bypasses the hard work of politics by attributing to science an incontrovertible certainty that it is far from having – yet without mobilizing anyone at all.

As I showed in *Politics of Nature*, the ecologists have too often repainted in green this same grey Nature that had been conceived in the seventeenth century as a way of making politics if not impotent, then at least subservient to Science – the Nature to which the role of "disinterested third party" had been assigned, capable in the last analysis of serving as arbiter of all the other disputes; the Nature in whose bosom so many scientists still think they have to take refuge in order to protect themselves from the dirty work of politics; the Nature that has inherited, as we shall see later on, all the functions of the all-seeing and all-encompassing God of the old days, and who is just as incapable of bringing its Providence to have any effect whatsoever on the Earth! Ecology can be summed up not as politics

---

¹⁰Compare John Lewis Gaddis, *The Cold War: A New History* (2006), on the rapidity of the response to the Soviet threat, to Nicholas Stern, *The Economics of Climate Change*, to see how slowly mobilization occurs where the climate threat is concerned.
¹¹*Trans.*: The English word "panic" is derived, via the French *panique*, from the Greek *panikos*, referring to Pan, the god of wild nature; Pan was credited with arousing terror in lonely travelers passing through woodlands.
¹²This controversial idea was introduced by Gayatri Chakravorty Spivak. Without believing seriously in the essential character of social identities, a "strategic essentialist" nevertheless uses the notion when it seems expedient in certain struggles, since it is the adversaries' weapon of choice.

taking Nature into account but, rather, as the end of the Nature that served as the consort of politics.[13] This is why we have to choose between a Nature that hides its Politics and a Politics in which the role of Nature is explicit.

*

It is not certain, even so, that the most troubling factor is the hybrid character of these statements, even if they seem very worrying to those who think that a strict separation must be maintained between science and politics. After a moment of surprise, we can readily understand how we should interpret the statements. If data like those in the shape of a hockey stick are no longer objective in the ordinary sense (detached from any prescription), they are indeed objective in the sense that those who prepared them have answered all the *objections* that could be raised against them (this is the only known way in which a statement can be transformed into a fact).[14] The only originality in these *data*[15] is that they concern us so directly that their mere expression sounds like an alarm to those who have to attend to them, a bit like the sound of the instruments that track a patient's heart rate and breathing for attendants in a recovery room.

In practice, the difference between *constative* and *performative* statements (to use the vocabulary of linguists), even though it has been of great concern to philosophers, has always been very slight.[16] If you are on a bus and you see that a passenger is about to sit down on a seat where you have put your baby, the statement that you won't fail to make – "There's a baby on the seat" – will certainly be a constative utterance (as self-evident as "the cat is on the mat"), but you hardly qualify as human if you are not making it also *in order to elicit a*

---

[13] Bruno Latour, *Politics of Nature: How to Bring the Sciences into Democracy* ([1999] 2004b).
[14] The virtues of objectivity have a long history (see Lorraine Daston and Peter Galison, *Objectivity*, 2007), which makes it possible not to confuse the final result – attributed to the known object – with the very complex institution through which the objections have passed one after another. Objectivity is neither a state of the world nor a state of mind; it is the result of a well-maintained public life.
[15] Instead of data (French *données*, or "givens"), we should speak of *obtenus* (elements "obtained"). In English (or in Latin), the term *data* would be more comprehensible if we were to speak of *sublata*.
[16] The vast literature in linguistics, sociolinguistics, and speech act theory has continually whittled away at the distance between description and prescription, a distinction already challenged in J. L. Austin's seminal book *How to Do Things with Words* ([1955] 1962).

reaction from the person to whom it is addressed (this is one of the uses of language we designate with the word "performative"). Don't try to pretend that you are just saying "the baby is there," nothing more. You are not simply stating an objective fact – all the passengers can verify that the baby is indeed on the seat; you are vigorously *objecting* to a behavior that would crush said baby under the bottom of said passenger. "There is a baby on the seat" is thus at one and the same time a constative and a performative utterance. And this is so whether you are making it in a calm, icy, tense, automatic, excited, or screaming tone of voice. The entire success of the good Mr Spock, that famous spokesperson for Reason, lies in the fact that, despite his mechanical voice, he actually tells Captain Kirk what *must be done* in order to take into account what *is*.

Earlier, one could ignore that self-evidence, by imagining that scientists had to remain as *external* to the phenomena they were describing as were those they were addressing. But from now on, if you speak of any part of the Earth to humans, whether it's a question of geology, the climate, living species, the chemistry of the upper atmosphere, carbon, or caribous, we all find ourselves in the same boat – or rather on the same bus. This is why everything scientists say about this thin film of life sounds entirely unlike the indisputable old speech uttered from nowhere to talk about things that did not directly concern either those who were speaking or those who were listening. Only the climate skeptics are still trying to make us believe that objectivity must not lead to any form of action because, in order to sound scientific, one must remain disinterested with respect to what one is saying. But, in seeking to separate science from their interests, the skeptics are actually insisting on sheltering their interests from any objection. And now, it shows! It is on Earth, on the contrary, that people such as Keeling in Mauna Loa are producing utterances that are truly objective and interesting, because they have responded to the objections of their adversaries and, *consequently*, they make it possible to prepare their listeners to take an interest in what is happening to them.[17]

What doubtless explains in part the old idea that description entails no prescription is that these warnings obviously do not spell out *in detail* what has to be done. They are merely ways of putting

---

[17]The return to this so poorly understood notion of disinterestedness characterizes much of the philosophy of science of Isabelle Stengers, from *The Invention of Modern Science* ([1993] 2000) to *La vierge et le neutrino* (2005), and which led her, in *In Catastrophic Times: Resisting the Coming Barbarism* ([2009] 2015), to take up a position facing the intrusion of Gaia.

collective action *under tension*. Which is exactly what one asks of an alarm. Instead of a *difference* in principle between the world of facts and the world of values, a gulf that must never be crossed if one is to remain rational, we see that we have to become accustomed to a *continuous linkage* of actions that *begin* with facts that *are extended* into a warning and that *point* toward decisions – a process that goes in both directions. This double linkage is disallowed by the idea of axiological neutrality, which prematurely cuts off the first link from the preceding ones.[18] This claim of descriptive neutrality made it possible to forget that one never plunges into description except in order to act, and that, before looking into what must be done, we must be impelled to action by a particular type of utterance that touches our hearts in order to set us in motion – yes, to move us. Astonishingly, this type of utterance now comes not only from poets, lovers, politicians, and prophets but also from geochemists, naturalists, modelers, and geologists.

*

How are we to explain that the sciences are multipliers of *agency*[19] even as they purport to speak only of agents that come to be transformed into presumably inert "material beings"? To approach this question, I would like to compare different types of narratives in order to give a sense of the way characters are endowed with a capacity for action, however these characters may be represented in other respects; some of them clearly belong to the repertory of humans, others to that of "beings of nature." I hope to show that what characterizes the so-called scientific ways of expressing oneself is not the fact that scientists' objects of study are *inanimate* but only the fact that our *degree of familiarity* with these objects or "actors" is *very slight*; the inanimate "actors," or *actants*,[20] thus need to be presented

---

[18]The recommendations made to the writers of the IPCC reports insist that what is "policy relevant but not policy descriptive" must be distinguished (IPCC, "Statement on IPCC Principles and Procedure," 2010). See Kari De Pryck, "Le groupe d'experts intergouvernemental sur l'évolution du climat, ou les défis d'un mariage arrangé entre science et politique" (2014).

[19]Even though the word "agency" in English often refers to persons, and most of the time to human entities, I take it, following the insight of semiotics, as a concept that precedes the attributions of humanity and personhood. I have pursued this argument with some obstinacy in "Agency at the Time of the Anthropocene" (2014a, used in part in this lecture) and, more recently, in "How Better to Register the Agency of Things" (2016b).

[20]For an introduction to the notion of actant, see Algirdas Greimas and Joseph Courtés, eds, *Semiotics and Language: An Analytical Dictionary* ([1979] 1982).

at greater length than the characters we call anthropomorphic, with whom we believe we're better acquainted.

I am going to compare three short excerpts: one from a novel, one from a newspaper story, and one from an article on neuroscience. As we listen to them in turn, let us try to be sensitive not to the obviously distinct genres to which they belong, but to the multiplicity of modes of action that they are capable of intermingling. I am asking you, in other words, to suspend the usual reading grid that makes us tend to contrast human and nonhuman actors, for example, subjects and objects; I'd like you to remain attentive to what constitutes their common repertory. It will then become clear that to say of an actor that he/she/it is inert – in the sense of having no agency – or, conversely, that he/she/it is animated – in the sense of "endowed with a soul" – is a *secondary* and *derivative* operation.

One feature of a great novel is that its characters do not conform to repertories of predictable actions; they avoid the clichés we use to simplify our stories as if we were playing "Clue": for example, the Butler, the Detective, the Lost Girl, or the Villain. This is certainly the case in the well-known passage of Tolstoy's *War and Peace* that narrates Marshal Kutuzov's (non-)decision on the eve of the famous Battle of Tarutino on October 12, 1812. The Marshal thinks that launching a battle to defeat Napoleon is pointless:

> The Cossack's report, confirmed by horse patrols who were sent out, was the final proof that events had matured. The tightly coiled spring was released, the clock began to whirr and the chimes to play. Despite all his supposed power, his intellect, his experience, and his knowledge of men, Kutuzov – having taken into consideration the Cossack's report, a note from Bennigsen who sent personal reports to the Emperor, the wishes he supposed the Emperor to hold, and the fact that all the generals expressed the same wish – could no longer check the inevitable movement, and gave the order to do what he regarded as useless and harmful – *gave his approval, that is, to the accomplished fact.*[21]

As readers of the novel surely remember, in what follows this passage Kutuzov does everything he can to postpone the engagement, which he will nevertheless win in the end because he will have managed to remain almost immobile in the face of the advances and counter-advances of Napoleon's Grand Army! If there is one system of commandment in which we believe it possible for the supreme leader to make sure he is obeyed, it is certainly the case of an army at war. Yet, in this battle narrative, exactly the opposite happens: the human subject who should be in full control and able to achieve

---

[21] Leo Tolstoy, *War and Peace* ([1865–6] 1996), p. 879, emphasis added.

his intentions is precisely the one who *is made to act* by *objective* forces that he cannot "check." Certain of these are "natural" – the "events had matured," the "tightly coiled spring" is released; others are clearly human and social – the report of the Cossack scouts, the betrayal of Kutuzov's aide-de-camp, Bennigsen, the wishes of his generals; still others, finally, might be called cognitive – "experience" and "knowledge of men," the wishes imputed to the Emperor. All this obliges Kutuzov to give "the order to do what he [regards] as useless and even harmful," since he can do nothing but give "his approval to the accomplished fact." He ought to have goals; but he is so powerless in his power that he does not even manage to define them.

One can hardly pretend that this is a story dealing exclusively with human actors; we see that a novelist, as soon as he becomes attentive to the ins and outs of the human soul, multiplies the forms of action that make it difficult to say exactly where the *anthropomorphic* aspect of his characters resides. Kutuzov is given his *form* – this is the meaning of the Greek root "morphic" – by forces that have entirely different characteristics. This is what specialists in literary analysis mean when they distinguish figuration from agency: Kutuzov indeed has the figure of a human being, but what makes him act comes to him from elsewhere, from forces Tolstoy spells out in detail.[22]

Someone will object that novelists are paid to probe the depths of the human soul and that it is hardly surprising that they delight in complicating the lives of philosophers who would prefer to see the subjects of the "human world" radically opposed to the objects of the "material world." It is true that, in the example of Kutuzov, there is no agent that can count as a truly credible natural force. Despite the metaphors of the "maturing" situation, of the "spring [that] was released," and the "chimes [that] began to play,"[23] we remain from start to finish, and for our maximum pleasure, within the human comedy.

Let us now take an excerpt from a best-seller with a very modernist title: *The Control of Nature*.[24] John McPhee's book is a series of remarkable stories about the way heroic humans stand up to

[22] The difference between actants and actors is an essential principle of the semiotics inspired by Greimas; see Jacques Fontanille, *The Semiotics of Discourse* ([1998] 2006).
[23] Curiously, throughout the appendix to the novel, Tolstoy uses a technical metaphor for a Providence that acts with such necessity that the characters' freedom of maneuver, though it has been amply deployed throughout the novel, completely disappears. Here is evidence that the discourse of causality can multiply or reduce agency at will without any change in composition. The *attribution* of causes is always a secondary process with respect to the primary process of the *composition* of forces.
[24] John McPhee, *The Control of Nature* (1989).

invincible natural agents – water, landslides, and lava flows. In one chapter, he describes another battle, the one that hydraulic engineers carry out against the tendency not of a hostile army but of a river, the Mississippi, to let itself be captured insidiously by the course of a much smaller and much less well-known river with the wonderful Indian name Atchafalaya. Its course is situated *below* the Mississippi's.

If the Mississippi continues to flow to the east of New Orleans, it is thanks to a rather small and quite fragile work of craftsmanship constructed upstream in a bend in the river, a dam that protects the massive current from being captured by the bed of the Atchafalaya, which is much narrower but several meters lower. If this dam should break (as it threatens to do almost every year, making the whole region tremble), the entire Mississippi, after devastating the Atchafalaya valley and carrying off the town of Morgan City, would come out, through a shortcut of several hundred kilometers, to the *west* of New Orleans, causing massive flooding and destroying a major part of the huge Mississippi delta toward which a quarter of the American economy flows. It is a question no longer of generals, war, treason, wishes, or presumed intentions but of two rivers, and a collective character rather than an individual like Kutuzov, a character that McPhee describes as acting "like a single man": the Army Corps of Engineers. This institution is charged with conducting the battle to "control nature" under the supervision of a commission responsible for infrastructure projects – the River Commission.

Thus here we are truly facing a *natural* actor. But whoever has felt the presence of a stream, a tributary, a river, and especially a river like the Mississippi, will react as Mark Twain did:

> One who knows the Mississippi will promptly aver – not aloud, but to himself – that ten thousand *River Commissions*, with the mines of the world at their back, cannot *tame* that *lawless* stream, cannot curb it or confine it, cannot say to it, "Go here," or "Go there," and make it obey...the Commission might as well *bully* the comets in their courses and undertake to *make* them behave, as try to *bully* the Mississippi into right and reasonable conduct.[25]

A force of nature is obviously just the opposite of an inert actor; every novelist and poet knows this as well as every expert in hydraulics or geomorphology. If the Mississippi possesses anything at all, it is *agency* – such powerful agency that it imposes itself on the agency of all the bureaucrats. But the least one can say is that the Army Corps of Engineers did not follow Mark Twain's intuition. On the contrary, it decided to make the "lawless stream" obey, to "curb"

---

[25] Mark Twain, *Life on the Mississippi* ([1883] 1944), p. 168.

and "confine" it, to "bully" it to the point of keeping it, for two centuries now, from abruptly modifying its meanderings, as it had been doing for millennia, and ordering it to "go here and not there." As the tragedy of Katrina has reminded us,[26] the entire Mississippi basin, completely artificialized, is attempting to protect itself behind the fragile front line of its dikes. The agents we are dealing with here are so mixed that the extent of the technical and legal responsibility of the Corps is a function of both the power of the Mississippi and the level of the Atchafalaya, which stubbornly continues to dig down. The whole business is ultimately concentrated in the little artisanal construction that a slightly stronger than anticipated surge could carry away. And what is the consequence of these exchanges of capacities? A situation of negotiation – almost a contractual relation – between anthropomorphic beings (the Corps of Engineers in particular) and others, which can logically be called *hydromorphs*.

> The Corps was not in a *political or moral* position to *kill* the Atchafalaya. It had to *feed it* water. By the *principles of nature*, the more the Atchafalaya was given, the more it would want to take, because it was the *steeper* stream. The more it was given, the deeper it would make its bed. The difference in level between the Atchafalaya and the Mississippi would continue to increase, magnifying the *conditions* for *capture*. The Corps would have to *deal with* that. The Corps would have to *build* something that could *give* the Atchafalaya a portion of the Mississippi and at the same time *prevent it from taking all*.[27]

Let us note that the expression "by the principles of nature" does not *withdraw* agency from the conflicts between the two rivers featured by McPhee any more than the "accomplished fact" mentioned by Tolstoy is capable of eliminating any will in Kutuzov's decision (as the general in charge, he still has to "give his approval"). Quite to the contrary, there is a will here – that of the competing rivers. But the author represents what it means to "will" quite differently in this case: the connection between a smaller but deeper river and another much bigger but higher one is what supplies the *goals* of the two protagonists, what gives their action a *vector*. It hardly matters that one is evoked as having intentionality or will and the other as simply a force, because it is the *tension that makes the actor*, and not the way actors have been endowed with a more or less plausible set of attitudes.[28]

How can we doubt that the Atchafalaya "wants to capture" the Mississippi? It is a manner of speaking, yes, but one that justifies

---

[26] Hurricane Katrina devastated New Orleans on August 29, 2005.
[27] McPhee, 1989, emphasis added.
[28] See Algirdas Greimas and Jacques Fontanille, *The Semiotics of Passions: From States of Affairs to States of Feeling* ([1991] 1993).

using legal terms, the vocabulary of battle – "give," "supply," "take into account," "prevent" – to give the sense, the direction, the movement of a river that is indeed dangerous. Or rather that has been *made dangerous* by the will of the Corps to bully the Mississippi by introducing a corset of dikes. If this is violence against violence, how can we be surprised that behavioral features shift from one repertory into the other? If you want to avoid anthropomorphisms, the Corps would have had to avoid anthropomorphizing the Mississippi delta! What moralists tend to ignore is something engineers know: on the side of the subject, there is no mastery; on the side of the object, no possible deanimation.[29] As one of the engineers says, "It is not a question of whether or not the Atchafalaya will end up capturing the entire river, but a question of *when*." And he calmly asserts: "Up to now, we have been able only to win some time."[30] "Win some time": there is an expression that Kutuzov would have understood very well!

*

All this is very amusing, you may well say, but journalists are journalists, just tale-tellers, just like novelists, we know how they work; they always feel obliged to *add* a bit of action to what, in its essence, ought to be *deprived* of any form of will, goal, target, or obsession. Even when they take in interest in science and nature, they cannot keep from adding some drama to what contains no drama. Anthropomorphism is the only way they know to tell stories and sell their newspapers. If they had to write "objectively" on the subject of "purely objective natural forces," their stories would be significantly less dramatic. The concatenation of causes and effects – and isn't that, after all, what the material world consists in? – must not lead to any dramatic effects, precisely because – and herein lies its beauty – the consequences are *already there* in the *cause*: there is no suspense, nothing to wait for, no sudden transformation, no metamorphosis, no ambiguity. Time passes *from the past toward the present*. In these stories (which are in fact *not* stories), then, nothing happens, in any case no *adventure*. Isn't this the salient point of rationalism? That no one should create any drama, and no one should tell any more stories.

---

[29] This is the origin of the principle of symmetry introduced into sociology by Michel Callon in "Some Elements of a Sociology of Translation: Domestication of the Scallops and the Fishermen of St Brieux Bay" (1986); it is the basis for actor-network theory. Instead of a distinction between subject and object, one obtains nuances along a gradient in which human and nonhuman figures are mixed.
[30] McPhee, 1989, p. 55.

Such at least is the conventional way in which scientific reports are supposed to be written, or so the experts claim. That convention may be insisted on endlessly in classrooms, but even a superficial reading of the first scientific paper that comes to hand will suffice to call it into question. Let us take for example the beginning of an article published by my former colleagues at the Salk Institute in San Diego.[31]

> The *ability* of the body to *adapt* to *stressful* stimuli and the *role* of stress maladaptation in human diseases has been intensively *investigated*. Corticotropin releasing factor (CRF) (1), a 41-residue peptide, and its three paralogous peptides, urocortin (Ucn) 1, 2, and 3, *play important and diverse roles* in *coordinating* endocrine, autonomic, metabolic, and behavioral responses to stress (2, 3). CRF family peptides and their receptors are also *implicated* in the *modulation* of additional central nervous system functions including appetite, addiction, hearing, and neurogenesis and *act peripherally* within the endocrine, cardiovascular, reproductive, gastrointestinal, and immune systems (4, 5). CRF and related ligands initially *act by binding* to their G protein-coupled receptors (GPCRs).[32]

Once we've taken care of the acronyms (CRF, Ucn, GPCR), which are convenient for the experts but off-putting for neophytes, and once we've replaced the passive forms (a stylistic obligation of the genre) by the actions of the scientists who have "intensively investigated" the question, we confront – here again, here as always – an actor whose agency is the very object of the article: the factor that releases corticotropin. How can we pretend that CRF is inert when it "plays an important role" and is "implicated in the modulation" of a dizzying number of functions? Having a function is its way of having goals, or in any case of being defined as a vector, and thus as an agent.

To be sure, this introduction doesn't lend itself to reading with the same pleasure as *War and Peace*! But there is no doubt that by following CRF we penetrate into the twists and turns of an action that turns out to be even more complex than the intricacies of Kutuzov's decision or the meanderings of the Mississippi. Imagine, moreover, how a Tolstoy of today, clever enough to add CRF to his cast of characters, would have depicted Kutuzov on the eve of a crucial battle.[33]

---

[31] Some context is offered in Bruno Latour and Steve Woolgar, *Laboratory Life: The Social Construction of Scientific Facts* (1979).
[32] Christy Rani R. Grace, Marilyn H. Perrin, Jozsef Gulyas, Michael R. DiGruccio, Jeffrey P. Cantle, Jean E. Rivier, Wylie W. Vale, and Roland Riek, "Structure of the N-Terminal Domain of a Type of B1 G Protein-Coupled Receptor in Complex with a Peptide Ligand" (2007), emphasis added.
[33] This is most likely what the novelist Richard Powers would have done; it is what he has attempted to do for example in *The Echo Maker* (2006), or even more directly in *Gain* (1998), and it is what accounts for the entirely new aspect of his characters; on this point, see Bruno Latour, "The Powers of Facsimiles: A Turing Test on Science and Literature" (2008a).

Is there anything more stressful than a battle situation? The CRF would have spread in his intestine, would have modified his hearing, modulated his response to the microbes; and how could we doubt that Bennigsen, stressed by his betrayal, and soon the whole general staff, not to mention the poor soldiers sent up as cannon fodder, would not all be transformed by the flow of CRF? When it is a matter of understanding what it means to act and to be acted upon, novelists, journalists, and scientists are engaged in one and the same fight, and they steal from one another incessantly.

There is of course a difference between this last example and the two earlier ones, but, as I discovered many years ago in that same laboratory at the Salk Institute, the difference does not arise from the fact that the first two stories deal with "human" agents endowed with goals, while the last one deals with objects of "nature" that have no goals or wills.[34] The only real difference – at least as far as the story is concerned – comes from the fact that the readers of Tolstoy's masterpiece or of McPhee's story can easily endow the characters with a certain consistency on the basis of their past experience, whereas they cannot do the same thing for the case of CRF – unless they are specialists in neurotransmitters, of course. What makes scientific reports so propitious for studying the multiple character of agency is that the character of the agents mobilized cannot be described except through the *actions* by means of which they have to be slowly pinned down.

Unlike generals such as Kutuzov and rivers such as the Mississippi, the *competences* of these agents – that is, what they *are* – are defined only *through their performances* – that is, after observers have succeeded in recording how they *behave*.[35] For a marshal or a river, you can act as though you started from their essence to infer some of their properties. Not for CRF. If you know nothing about it, you will necessarily – whether you are its discoverers or readers of the article cited – begin by exploring what it does. And, since there is no prior knowledge of CRF, since what justifies publishing an article about it is its novelty, every feature has to be produced by a certain experiment, a specific trial, and these have to be listed, line by line.[36] What is CRF? It is what releases corticotropin. What is corticotropin? It is what releases corticostimulin in the pituitary gland. And so on.

---

[34] See Bruno Latour and Paolo Fabbri, "The Rhetoric of Science: Authority and Duty in an Article from the Exact Sciences" (2000), and especially Françoise Bastide, *Una notte con Saturno* (2001).
[35] See the entries "performances" and "competences" in the bible of semiotics: Greimas and Courtés ([1979] 1982).
[36] This is the crucial point of the classic article by Harold Garfinkel, Michael Lynch, and Eric Livingston, "The Work of a Discovering Science Constructed with Materials from the Optically Discovered Pulsar" ([1981] 2011).

Table 2.1

Actants	Actors
Performances	Competences
Names of action	Names of thing
Attributes	Substance
Before	After
Unstable	Stable

If we aren't specialists in this unknown object, we struggle, of course, but the procedure is exactly the same as the one we engage in every day when we consult the Internet for information about a person, place, event, or product that someone has mentioned in passing. We begin with a name that at the outset "means nothing to us"; then we unfold, on screen, a list of situations; later, after we have become familiar with them, we invert the order of things, and we get in the habit of starting from the name to deduce or summarize what it does. In the same way, CRF was initially a list of actions, well before it was, as they say, "characterized." From that moment on, its competences begin to precede and no longer to follow its performances. If we read as much scientific literature as we read novels, CRF would be as familiar to us as Pierre Bezukov and Natasha Rostov – as familiar as endorphins are today, thanks in part to work done at the same Salk Laboratory. In the little chart I have drawn up (table 2.1), the last feature is particularly important: it is through stabilization that a substance acquires its consistency.

I wanted to compare these three examples briefly in order to bring out the gap that separates the common-sense assumption that one can easily distinguish between the objects of the natural world, on the one hand, and the subjects of the human world, on the other, from the extreme difficulty of making this distinction in practice. The actors, with their multiple forms and capacities, never stop exchanging their properties. One sees quite well how the so-called anthropomorphic representations are as unstable as those qualified as *hydro*morphic, *bio*morphic, or *phusi*morphic, since what counts is not the initial snapshot but the *metamorphoses* that Kutuzov, the Atchafalaya, or CRF undergo in the course of the story.[37] Kutuzov does not resemble the traditional human subject ("master of himself and of the universe") any more than the Mississippi or CRF resemble the "objects"

---

[37] I am using terms that are much too crude – *phusis* for nature, *bio* for biology, and so on – simply to point out the importance of the term *morph* to which they are apposed.

of material nature, as we are used to calling them when we want to make them the simple background for human subjects. We must not confuse the perceptions enacted by subjects and objects with what the world is made of. If it is the world that interests us – and no longer "nature" – then we must learn to inhabit what could be called a *metamorphic zone*, borrowing a metaphor from geology, to capture in a single word all the "morphisms" that we are going to have to register in order to follow these transactions.[38]

In the final analysis, the distinction between humans and nonhumans has no more meaning than the Nature/Culture distinction. It would be just as artificial a distinction as putting Kutuzov and the Army Corps of Engineers in one box and the Mississippi and the CRF in another, as though the first were characterized by a form of soul or consciousness or mind and as though the second were, if not inert, then at least lacking in goals and intentions. The distinction between humans and nonhumans and the difference between culture and nature have to be treated the same way: to be sure that we are not using them as resources but rather as objects of study, we have to go a level deeper, to the common concept that distributes the figures into separate parts.[39] To believe that these terms describe anything at all about the real world amounts to taking an abstraction for a description.

When we claim that there is, on one side, a natural world and, on the other, a human world, we are simply proposing to say, after the fact, that an arbitrary portion of the actors will be *stripped of all action* and that another portion, equally arbitrary, will be *endowed with souls* (or consciousness). But these two secondary operations leave perfectly intact the only interesting phenomenon: the exchange of forms of action through the transactions between agencies of multiple origins and forms at the core of the metamorphic zones. This may appear paradoxical, but, to gain in realism, we have to leave aside the pseudo-realism that purports to be drawing the portrait of humans parading against a background of things.

*

---

[38] According to the dictionary, metamorphism is an internal process of the terrestrial globe in which extreme heat or pressure produce a solid structure by altering the texture and mineralogical composition of a rock formation.
[39] This is the same displacement of a term used as an analytical tool transformed into an object of study (the shift from *resource* to *topic*) that I presented in the previous lecture.

Displacing our attention toward this zone common to writers and scientists may allow us to understand differently the idea that the Earth "retroacts" in response to what "we" do to it. Michel Serres had already addressed these delicate questions in the early 1990s, at the very moment when nonchalant humanity had inadvertently crossed the dangerous CO_2 threshold.[40] In a bold and singular book, *The Natural Contract*, Serres proposed, among many innovative ideas, a fictional reformulation of Galileo's famous line: *"Eppur si muove!"*[41] Serres starts with an episode from the potted history of science: after having been forbidden by the Holy Inquisition to teach anything at all in public about the movement of the Earth, Galileo is said to have muttered: "And yet it moves." Serres calls this episode Galileo's *first trial*: a "prophetic" scientist grappling with all the authorities of his time silently reaffirms the objective fact that will eventually destroy those same authorities.

But in our day, according to Serres, we are witnessing *Galileo's second trial*.[42] In the face of all the assembled powers, another equally prophetic scientist (let's say James Lovelock, or Michael Mann, or David Keeling),[43] after being condemned to keep silent by all those who deny the behavior of the Earth, begins to mutter to himself "*Eppur si muove!*," but this time giving it a new and somewhat worrying twist: not "And yet the Earth moves!" but, rather, "And yet the Earth is moved!" in the sense of manifesting an emotional reaction.

> Science won all the rights three centuries ago now, by appealing to the Earth, which responded by moving. So the prophet became king. In our turn, we are appealing to an absent authority, when we cry, like Galileo, but before the court of his successors, former prophets turned kings: "the Earth is moved." The immemorial, fixed Earth, which provided the conditions and foundations of our lives, is moving, the fundamental Earth is trembling.[44]

---

[40] As Foucart says in the article cited at the beginning of this lecture (Foucart 2013): "According to American climatologist James Hansen, the former director of the Goddard Institute for Space Studies (GISS), the concentration of CO_2 that must not be exceeded is around 350 /m. A limit that was reached shortly before 1990."
[41] Michel Serres, *The Natural Contract* (1995) extended in part in *Retour au contrat naturel* (2000).
[42] The situation is all the more piquant in that the figure of Galileo, standing up for what is right all by himself against everyone else, is invoked by the climate skeptics every time they set out to attack the "consensus" of the climatologists.
[43] Serres does not mention Lovelock, but this character, whom we shall meet in the next lecture, is just right for the role.
[44] Serres, 1995, p. 86.

We should not be surprised that a new form of agency ("it is moved," "it reacts") is just as startling for the established powers as the old one ("it moves"). If the Inquisition was shocked by the announcement that the Earth was nothing more than a billiard ball turning endlessly in the vast universe (remember the scene in which Bertolt Brecht showed young monks making fun of Galileo's heliocentrism by turning in pointless circles in a room in the Vatican),[45] the new Inquisition (henceforth economic rather than religious) is shocked to learn that the Earth has become – has become again! – an active, local, limited, sensitive, fragile, trembling, and easily irritated envelope. We would need a new Brecht to show how, in the climate skeptics' talk shows, a whole gang (for example, the Koch brothers, numerous physicists, many intellectuals, a good number of right-wing politicians, and also some pastors, preachers, gurus, and advisors to princes) makes fun of this new as well as very old animated and fragile Earth.

To depict this first new Earth as a body in free fall among all the other bodies in free fall in the universe, Galileo had to strip it from all forms of movement except one, abandoning all the prevailing notions of climate, animation, and metamorphoses. Thus he freed us from the so-called prescientific vision of the Earth as a cesspool, marked with the sign of death and corruption, from which our ancestors, their eyes fixed on the incorruptible spheres of the suns, the stars, and God, had no chance of escaping except by prayer, contemplation, and knowledge. Now, to discover the new Earth, climatologists are again conjuring up the climate and bringing back the animated Earth to a thin film whose fragility recalls the old feeling of living in what was once called the *sublunary zone*.[46] Galileo's Earth could revolve, but it had no "tipping point," no "planetary frontiers," no "critical zones."[47] It had a *movement*, but not a *behavior*. In other words, it was not yet the Earth of the Anthropocene.

---

[45] Bertolt Brecht, *The Life of Galileo* ([1945] 2001).
[46] In the old "pre-Copernican" system, there was a difference in substance between the zone under the Moon (sublunary) and the zone above the Moon (supralunary): the higher one climbed above the corruptible Earth, to the planets and then to the fixed stars, the higher one went in perfection. On the history of this cosmos and its destruction, the classic book by Alexandre Koyré, *From the Closed World to the Infinite Universe* (1957), remains the best introduction, unless one prefers the more novelistic but still very effective version by Arthur Koestler, *The Sleepwalkers: A History of Man's Changing Vision of the Universe* (1959).
[47] It is this agitation on the part of the Earth that makes for the strangeness of books such as Fred Pearce's *With Speed and Violence: Why Scientists Fear Tipping Points in Climate Change* (2007) or Stephen M. Gardiner's *A Perfect Moral Storm: The*

How not to (de-)animate nature 61

Today, through a sort of counter-Copernican revolution, it is the New Climate Regime that compels us to turn our gaze toward the Earth considered once again with all its processes of transformation and metamorphosis, including generation, dissolution, war, pollution, corruption, and death. But, this time, it is useless to try to escape by means of prayer. Here is a dramatic rebound: from the cosmos to the universe, then back again to the cosmos!⁴⁸ Back to the future? Rather, forward to the past! Isn't it precisely radical reversal that the dancer presented in the introduction had marked with her steps? Isn't it embodied in the figure I had glimpsed and given the bizarre name Cosmocolossus?

In establishing a parallel between two trials, two Earths, two climate regimes, Serres's goal is not to move us by asking us to weep for Mother Earth or to go into ecstasy over the fact that she has a soul. It is precisely not a matter of *adding* spirit to what is, alas, deprived of any, in order to make ourselves feel better in a world that would be a little less disenchanted, or, conversely, to make ourselves feel more anxious in a less infinite world. Quite the contrary: Serres directs our attention toward the astonishing *connivance* between formerly distinct agencies – as opposed to one another, as were the old figures of object and subject – that are now so mixed.

For, as of today, the Earth is quaking anew: not because it shifts and moves in its restless, wise orbit, not because it is changing, from its deep plates to its envelope of air, but because it is being *transformed by our doing*. Nature acted as a reference point for ancient law and for modern science *because it had no subject*: objectivity in the legal sense, as in the scientific sense, emanated from *a space without man*, which did not depend on us and on which we depended de jure and de facto. Yet henceforth it *depends so much* on us that it is shaking and that we too are worried by this deviation from expected equilibria. We are disturbing the Earth and making it quake! Now it *has a subject once again*.⁴⁹

Ethical Tragedy of Climate Change (2013). On the controversial question of planetary frontiers, see Johan Rockström, Will Steffen, et al., "Planetary Boundaries: Exploring the Safe Operating Space for Humanity" (2009). On the network of critical zones, see Susan L. Brantley, Martin B. Goldhaber, and K. Vala Ragnarsdottir, "Crossing Disciplines and Scales to Understand the Critical Zone" (2007), and the report by S. A. Banwart, J. Chorover, and J. Gaillardet, *Sustaining Earth's Critical Zone: Basic Science and Interdisciplinary Solutions for Global Challenges* (2013), as well as Bruno Latour, "Some Advantages of the Notion of 'Critical Zone' for Geopolitics: Geochemistry of the Earth's Surface" (2014d).
⁴⁸Émilie Hache tries to capture this unanticipated rebound in the title of her book *De l'univers clos au monde infini* [From the closed universe to the infinite world] (2014), in opposition to Koyré's.
⁴⁹Serres, 1995, p. 86, emphasis added.

Even if his book does not invoke the name "Gaia" and was written before the term "Anthropocene" came into its own, what Serres is registering is this same subversion of the respective positions of subject and object. Since the time of the "scientific revolution," the objectivity of a world without humans had offered solid ground for a sort of uncontested natural law – if not for religion and morality, at least for science and law.[50] In the era of the counter-Copernican revolution, when we turn toward the old solid ground of natural law, what do we find? The traces of our action, visible everywhere! And not in the old way in which the Western Masculine Subject dominated the wild and impetuous world of nature through his courageous, violent, sometimes disproportionate dream of control, in the style of the Army Corps of Engineers. No, this time, just as happens in prescientific and nonmodern myths,[51] we encounter an agent that takes its label, "subject," from the fact that it can be *subjected* to the whims, the bad moods, the emotions, the reactions, and even the revenge of another agent, which also takes its quality as "subject" *from the fact that it is equally subjected to the action of the other.*

Being a subject does not mean acting in an autonomous fashion in relation to an objective context; rather, it means *sharing* agency with other subjects that have also lost their autonomy. It is because we are confronted with these subjects – or rather quasi-subjects – that we have to give up our dreams of control and stop fearing the nightmare of finding ourselves once again prisoners of "nature."[52] As soon as we come close to nonhuman beings, we do not find in them the inertia that would allow us, by contrast, to take ourselves to be agents but, on the contrary, we find agencies that are *no longer without connection* to what we are and what we do. Conversely, on its side (but there are no more "sides"!), the Earth is no longer "objective," in the sense that it can no longer be kept at a distance, considered from the point of view of Sirius and as though it has been emptied of all its humans. Human action is visible everywhere in the construction of knowledge as well as in the generation of the phenomena to which the sciences are called upon to attest. It is impossible, from now on, to play at dialectically opposing subjects and objects. The spring that worked for Kant, Hegel, and Marx is now completely stretched out: there is no longer enough object to oppose to humans, not enough subject

---

[50] I shall come back to this question in the sixth lecture.
[51] See Eduardo Kohn, *How Forests Think: Toward an Anthropology beyond the Human* (2013).
[52] The terms quasi-object and quasi-subject were introduced by Michel Serres in *The Parasite* ([1980] 1982).

to oppose to objects. It is as though, behind the phantasmagoria of dialectics, the metamorphic zone were becoming visible once more. As if, under "nature," the world were reappearing.

*

What is troubling in the hybrid statements proposed by so many researchers about the actions, emotions, movements, and behaviors of the Earth is not their way of establishing continuity between what is and what must be but, rather, the always ambiguous way in which they treat matters of fact. Sometimes it is a question of causal chains that seem to imply no form of action in response to what has been said; sometimes it is just the opposite, with these same researchers unfolding a proliferation of action scenes, some of which inevitably push those who are caught up in the stories to act. This double language is the source of the idea of an infinite distance between description and prescription: if one follows a causal chain in which nothing is supposed to happen – no surprises, in any case – then the gulf separating this chain from the terms used to describe moral, political, or artistic action on the part of humans looks immense. But the situation is entirely different when a scientific description sets forth a profusion of actions, many of which resemble those with which humans are accustomed to being credited: in this case, the distance between the various forms of action that continually engage actors with multiple repertoires looks minuscule. Consequently, the question becomes the following: why do those who describe the Earth's actions sometimes assert that nothing is taking place in these actions but "strict chains of causality" and sometimes that a great deal more is happening? This amounts to asking why, if the Earth is *animated* by countless forms of agents, we have sought to conceptualize it as essentially inert and *inanimate*.

To reach an understanding of what the idea of an Earth that would react, retroactively, to our actions can mean, it becomes clear that one must not simplify in advance the distribution of agency between so-called human and nonhuman actors. What Serres explores in *The Natural Contract* is this congenital weakness of natural law, which consists in saying simultaneously that there is indeed law in nature – the prescriptive dimension that we recognized earlier – and that, nevertheless, law, true law, is found only on the other side, in culture. Hence the seemingly absurd idea of a contract with nature, even though everyone recognizes at the same time that nature orders, because it "dictates" to us what must be done through the intermediary of what is. The failure of efforts to define natural law arises not from the desire to seek an order that makes it possible to legislate but from the tendency to act as though there were *two parallel series*, and only two, one belonging to "nature" and the other to law, and trying to figure out which is the copy of the other.

In dramatizing the idea of a contract with nature, an idea borrowed from Rousseau's equally mythical social contract, Serres explores an entirely different solution: if one can neither keep from drawing an order from nature nor discover that order, it is because, even in our Western tradition, there have never been two parallel series, but always this proliferation of exchanges between figures that I have called the metamorphic zone.

> What language do the things of the world speak, that we might come to an understanding with them, contractually? But, after all, the old social contract, too, was unspoken and unwritten: no one has ever read the original, or even a copy. To be sure, we don't know the world's language, or rather we know only the various animistic, religious, or mathematical versions of it.... In fact, the Earth speaks to us in terms of forces, bonds, and interactions, and that's enough to make a contract.[53]

What difference is there between a force – a physical force – and a bond – a legal bond? Let us not forget that *The Natural Contract* is first of all a book of legal philosophy, and that it seeks to take seriously what the word "laws" means in the expression "laws of nature." The book's title notwithstanding, the natural contract is not a deal between two parties, humanity and nature, two figures that cannot be unified in any case,[54] but rather a series of transactions in which one can see how, all along and in the sciences themselves, the various types of entities mobilized by geohistory have exchanged the various traits that define their agency. *Trait* is precisely the technical term, borrowed from law, geopolitics, science, architecture, and geometry, that Serres uses to designate these transactions between the aforementioned subjects and the aforementioned objects. To make himself clear, he offers the most improbable of examples, that of universal gravity.

> Moreover the word *trait*, in French, like *draft* in English, means both the material bond and the basic stroke of writing: dot and long mark, a binary alphabet. A written contract obligates and ties those who write their name, or an X, below

---

[53] Serres, 1995, p. 39.
[54] This is what we shall see in the lectures to come: neither nature nor humanity can grasp itself as sufficiently unified (and now as sufficiently distinct) to be able to establish a contract between the parties. This is a way of measuring how much the situation has changed between the years in which Serres was writing his book and the period in which we are obliged to confront the Anthropocene.

its clauses.... Now the first great scientific system, Newton's, is linked together by *attraction*: there's the *same word again, the same trait, the same notion*. *The great planetary bodies grasp or comprehend one another and are bound by a law, to be sure, but a law that is the spitting image of a contract, in the primary meaning of a set of cords*. The slightest movement of any one planet has immediate effects on all the others, whose reactions act unhindered on the first. Through this set of constraints, the Earth *comprehends*, in a way, the *point of view* of the other bodies since it must reverberate with the events of the whole system.[55]

Serres is not proposing to *animate* the Earth by claiming that it would benefit from a form of comprehension, sympathy, or sovereignty. Quite the opposite: he proposes to take the force of attraction itself as a *bond* that would allow us to understand what is meant by the *force* of law and the *power* of understanding. To understand is to grasp, to apprehend something; is there a better way to apprehend something than to be subjected without any obstacle to the resounding echoes of all the other bodies? This is not anthropomorphism – in that case, the metaphor would go from the human to the physical – but rather a phusimorphism – the metaphor goes from force to law. Serres means that, in the last analysis, we indeed speak the language of the world, provided that we learn to translate "the animist, religious, or mathematical versions" from one to another. *Translation*, Serres's great project, becomes the way of understanding by what we are *attached* and on what we *depend*.[56] If we become capable of translating, then the laws of nature begin to have a spirit.

We mustn't see this bond between gravity and law as a matter of poetic license. Simon Schaffer has shown in a magnificent article how Newton must have drawn out of his own culture a set of features for the new agent that later imposed itself as "universal attraction."[57] Newton was obsessed by all forms of action at a distance, as much by that of God acting in matter as by that of credit acting in the economy, or the government acting on subjects.[58] A theologian with a whiff of heresy about him, an expert in alchemy as well as optics, he would have seen no point in "strictly distinguishing" between the world of spirits and that of matter. If he had done that, he would never

---

[55] Serres, 1995, pp. 108–109, emphasis added.
[56] Michel Serres, *Hermès III: la traduction* (1974).
[57] Simon Schaffer, "Newtonian Angels" (2011).
[58] "At exactly the same time, Isaac Newton pursued active work on the spiritual agents evident in alchemical processes, on the proper interpretation of angelic messages in the scriptural prophecies and the Apocalypse, started to compose a scholarly genealogy of idolatry and heresy, discussed the material and spiritual effects of cometary motion and solar vortices and drafted a provisional history of the Church" (Schaffer 2011, p. 92). See also Simon Schaffer, *The Information Order of Isaac Newton's Principia Mathematica* (2008).

have been a physicist. Still, it was not to anthropomorphism that he turned to understand how one body manages to act on another, but to angels. His physics is thus first of all *angelo*morphic!

In fact, to avoid Descartes's whirlwinds (another quite astonishing mix of properties and traits), Newton had to discover an agent capable of instantly transporting action at a distance from one body to another. At the time, there was no character available to him who could transport an instantaneous movement without any obstacle – except angels. Through several hundred pages of angelology, Newton gradually managed to trim their wings and transform this new agent into a "force." A "purely objective" force? Of course, because it had answered the objections, but it was still charged, upstream, by millennia of meditations on an "angelic system of instant messaging." As we know quite well, purity would sterilize the sciences: behind the force, the wings of angels are always beating invisibly.

The problem is that the *aspect* of a human subject like Kutuzov or the Army Corps of Engineers is no better known at the outset than the *aspect* of a river, an angel, a factor in hormone release, or a force such as universal gravity. That is why it makes no sense to accuse novelists, scientists, or engineers of committing the sin of "anthropomorphism" when they attribute "agency" to "something that should not have any." Quite to the contrary: if they have to deal with all sorts of contradictory "morphisms," it is because they are trying to explore the form of these *actants*, which are initially unknown and then gradually domesticated by as many figures as are needed in order to approach them. Before these actants are supplied with a style or a genre – that is, before they become widely recognized as *actors* – they must, if I can put it this way, be ground up, kneaded, and cooked in a single vessel.[59] Even the most respectable entities – characters in novels, scientific concepts, technical artifacts, natural phenomena – are all born from the same witch's kettle, for it is literally here, in this metamorphic zone, that all the *tricksters* and all the *shapeshifters*[60] reside.

*

[59] It is this kneading and then the slow decanting that Frédérique Aït-Touati addresses in *Fictions of the Cosmos: Science and Literature in the Seventeenth Century* (2012) on the gradual invention of the difference, now naturalized, between fictional and scientific narratives.
[60] This is Donna Haraway's favorite term for designating the many bifurcations through which agencies exchange their properties in the most unexpected ways; see Donna Haraway, "A Cyborg Manifesto: Science, Technology, and Socialist-Feminism in the Late Twentieth Century" (1991).

The language of the world thus articulates multiple agencies by translating one repertory into another (one morphism into another) in order to incorporate the new actors that are discovered at every step of the way. But, when I say "language of the world," I still need to make it clear whether I am talking about language or about the world! In fact, the arguments in this lecture will seem improbable and even shocking to scientists and the public alike as long as I fail to pin down this small detail. The scientists will probably think that these exchanges of properties among rivers, forces, neurotransmitters, marshals, and engineers are not metamorphoses but simple *metaphors*. "It is the weakness and the limitation of language," they will say, "that force us to talk about CRF as an actor, of the Atchafalaya as a being to which one has to 'give' water, or of gravitational force as an angelic spirit. If we could express ourselves in *truly scientific terms*, we would put away all these metaphors and speak in a way that would be strictly..." There follows a moment of somewhat embarrassed silence. In fact, this is the point at which things get complicated, for, to "speak in strictly scientific terms," according to them, they would obviously have to avoid speaking at all! And we are left to imagine a rather comical scene in which a mute researcher designates a phenomenon that expresses itself silently on its own while imposing itself without any sign or intermediary on a totally passive human being...clearly not a very realistic situation.

Still, the lack of realism does not prevent this scene from serving as the origin of the very distinction, which the public takes as a matter of good sense, between the "material world," on the one hand, and that of "human language," on the other. It is the material world that we have rendered mute in order to avoid answering the questions "Who or what is speaking? Who or what is acting?" It is in order to understand this strange situation that I must introduce, in addition to the zone of transactions that I have called metamorphic, an entirely different operation through which, *in language and by means of language*, some characters are deprived of any form of agency. This operation is going to *deanimate* some of the actors and give the impression that there is a gulf between inanimate material actors and human subjects endowed with soul – or at least with consciousness. The argument may appear convoluted, but I need it to explain through what *effect of language* people have set about constructing scenes in which language would be only one part, the other part being reserved for the mute presence of the inert things over which language has no hold!

It takes just a few moments' reflection, however, to notice that the idea of an inert world is itself *an effect of style*, a particular *genre*, a certain way of muting the agencies that we cannot prevent ourselves from proliferating as soon as we begin to describe any situation whatsoever. Speaking in a mechanical voice is still speaking. Only the tone is different, not the linking of words. Similarly, the idea of a deanimated world is only a way of linking animations *as if* nothing were happening there. But agency is always there, whatever we may do. The idea of a Nature/Culture distinction, like that of human/nonhuman, is nothing like a great philosophical concept, a profound ontology; it is a *secondary stylistic effect*, posterior, derived, through which we purport to *simplify* the distribution of actors by proceeding to designate some as animate and others as inanimate. This second operation succeeds only in deanimating certain protagonists, called "material," by depriving them of their activity, and in *overanimating* certain others, called "human," by crediting them with admirable capacities for action – freedom, consciousness, reflexivity, a moral sense, and so on.[61]

How can one possibly produce the impression that nothing is happening in a narrative in which events, adventures, exchanges of properties, transactions among agencies are multiplied from one moment to the next? It is surely not in scientific literature that this kind of apparent inertia can be found.[62] No, we have simply to *add* to the unfolding of events something that *reverses* its course and thereby annuls its action. How is this possible? By transforming the concatenation of causes and consequences in such a way that all the action is – or at least appears to be – in the cause, and that there is no more agency left in the consequences. Obviously this is impossible; the consequences are always surprising and, in practice, in the history of discovery, as in the narrative of discovery, and even in the teaching of the most solidly established facts, the cause arrives a long time *after* the consequences.[63] For the same reason that ensures that competences emerge long *after* performances have been carefully registered, a strictly *causalist* narrative in which a single character, the

---

[61] What Whitehead called the bifurcation of nature is, as Didier Debaise shows very well in *L'appât des possibles: reprise de Whitehead* (2015), above all a practical operation.
[62] There is now a vast literature dealing with the realm of "science and literature." Especially pertinent to this book is Bruce Clarke, *Neocybernetics and Narrative* (2014). One striking example of the animation of scientific narratives, all the more interesting in that it was written by one of the people responsible for the term Anthropocene, is Jan Zalasiewicz's book *The Planet in a Pebble: A Journey into Earth's Deep History* (2010).
[63] Even if this appears counter-intuitive at first glance, the cause appears first only in the order of exposition; by definition, in the order of discovery it is always necessarily

sole actor, would be in the cause – and furthermore in the primary cause – is obviously impossible. By definition it would be impossible for anyone to produce such a narrative.

And yet it is possible, by using an appropriate philosophical approach, to act *as if* one could reverse the reversal and deduce all the consequences from the cause.[64] By proceeding this way, it is possible to *dedramatize* the dramatic course of time, to the point of acting as though the world flowed from the past toward the present. The hypothesis is implausible, I know perfectly well, but this is how it is possible to give the feeling of a material world subjected to a strict linking of causalities, as opposed to another world – human, symbolic, subjective, cultural, the terms hardly matter – that would then be defined as the empire of freedom. Curiously, the very distinction between the narratives – by implication, dramatic – and the material world – by implication raw, obstinate, inert, objective, and mute – does not coincide with a real distinction; rather, it originates in a very particular, historically limited way[65] to deanimate, through language, the distribution of what will henceforth play the role of agent – by implication, a human – and what will play the role of inert objects – by implication, the material setting of the human world.

The other hypothesis consists in proposing that what I have designated as a zone of common exchange – that is, the metamorphic zone – is *a property of the world itself* and not only a phenomenon of language *about* the world. Even if it is always difficult to keep this in mind, the analysis of meaning – the science of meaning, or semiotics – has never been limited to discourse, language, texts, or fictions. Signification is a property of all agents, in that they never cease to have agency; this is equally true of Kutuzov, the Mississippi, the CRF receptor, and the gravity through which bodies "comprehend" and mutually "influence" one another. For all agents, to act signifies bringing one's existence, one's subsistence, *from the future*

---

second because it is always on the basis of the consequences that one goes back up the chain toward the cause. In other words, there is always, in a causal narrative, an effect of *montage*.

[64] Charles Péguy, in his *Note conjointe sur Monsieur Descartes et la philosophie cartésienne* ([1914] 1992), plays on Descartes's audacity in *deducing* the existence of the heavens from his principles: "And he found not only the heavens. He found *stars, an earth*. I don't know if you are like me. I find it prodigious that he *found an earth*. For finally, if he had not found it…we know perfectly well that he would not have found the heavens, the stars, and an earth if he had not heard of them" (Péguy 1992, p. 1279).

[65] See Simon Schaffer, "Seeing Double: How to Make Up a Phantom Body Politic" (2005), and Stengers, 2000.

toward the present: they act as long as they take the risk of filling the breach of existence – or else they purely and simply disappear. In other words, existence and signification are synonyms.⁶⁶ *As long as they are acting, agents signify.* This is why their signification can be followed, pursued, captured, translated, formulated in language. Which does not mean that "every *thing* in the world is merely a matter of discourse" but, rather, that every possibility of discourse is due to the presence of agents in quest of their existence.

Although the official philosophy of science takes the second movement of deanimation as the only important and rational one, the opposite is true: animation is the essential phenomenon; and deanimation is the superficial, auxiliary, polemical, and often defensive phenomenon.⁶⁷ One of the great enigmas of Western history is not that "there are still people naïve enough to believe in animism," but that many people still hold the rather naïve belief in a supposedly deanimated "material world."⁶⁸ And this is the case at the very moment when scientists are multiplying the agencies in which they – and we – are more and more implicated every day.

*

With this second lecture, I hope to have prepared the ground for what follows. People who assert that the Earth has not only movement but also a way of being moved that makes it react to what we do to it are not all crazies who have invested in the strange idea of adding a soul to something that has none. The most interesting people, in my eyes, like the scientists who are working on the Earth System, are content simply *not to take away from it* the agency that it has. They do not say necessarily that it is "alive" but only that it *is not dead*. Or at least that it is not inert in the very strange form of inertia produced by the idea of a "material world." A world evidently very remote from *materiality*. Between materiality and matter, it seems that we are going to have to choose.

---

⁶⁶This theme is developed more fully in Bruno Latour, *An Inquiry into Modes of Existence: An Anthropology of the Moderns* ([2012] 2013b).
⁶⁷See David Abram, *The Spell of the Sensuous: Perception and Language in a More-than-Human World* (1996).
⁶⁸Hence the new interest in the question of animism, as we see in the work of Philippe Descola or Eduardo Viveiros de Castro, as if deanimation appeared from now on as a bizarre phenomenon that has to be explained anthropologically and no longer as the default position that makes all the others bizarre. See Eduardo Viveiros de Castro, *The Relative Native: Essays on Indigenous Conceptual Worlds* (2016).

To sum up too quickly an argument that I shall take up again later on, we obtain the apparent inertia of the material world as soon as we distribute agency among causes and consequences in such a way as to attribute everything to the causes and nothing to the consequences, except the property of being traversed by the effect without adding anything to it.[69] We gain access to materiality when we reject this secondary operation that eliminates agents and when we leave the consequences with all the *agency* of which they are capable. It is through the causalist narrative that this effect of deanimation is obtained, but always *after the fact*, once agency has been redistributed among the long series of consequences, once this series has been retooled, set up, and traversed *in reverse order*.

Strangely, and I shall come back to the point, this form of causalist narrative closely resembles the *creationist* stories through which one attributes to a first cause, to a creation deemed *ex nihilo*, the whole series of what follows.[70] Even if, in the wake of the scientific revolution, we are accustomed to opposing science and religion, the idea of matter – for it is in the first place an idea – participates in both realms. This is why, in seeking to shed the idea of "nature," we shall also need to shed the theology that is pinned to it – without forgetting the politics that has been mixed up with it! Through the invention, in the course of lengthy battles during the seventeenth century, of the idea of a "material world" in which the power to act of all the entities that constitute the world has been wiped out,[71] a phantom world has been created to speak of the Earth, one that corresponds too often, alas, to what is called the "scientific worldview" and which is also a certain religious view of the nature of causes. Nothing, literally, *happens* any longer, since the agent is taken to be the "simple cause" of its predecessor. All the action has been placed in the antecedent. It hardly matters, then, whether the antecedent is called an omnipotent Creator or omnipotent Causality. The consequence might as well not be there at all; as we might say colloquially, it is there only "as an

---

[69] I have tried in earlier works to make this difference a technical one by emphasizing the opposition between *intermediaries* (which only transport force) and *mediators* (which cause their causes to bifurcate). This is another way of translating Serres's argument on translation.

[70] This is the object of the fifth and sixth lectures, which will plunge us into "natural theology," the theme of the Gifford Lectures on which this book is based.

[71] This link between scientific revolution, political organization, dematerialization of matter, and theology is the subject of the now classic book by Steven Shapin and Simon Schaffer, *Leviathan and the Air-Pump: Hobbes, Boyle, and the Experimental Life: Including a Translation of Thomas Hobbes, Dialogus physicus de natura aeris by Simon Schaffer* (1985).

extra." We can go on stringing episodes one after another; the quality that made them "events" has disappeared.

The great paradox of the "scientific worldview" is that it has succeeded in *withdrawing* the *historicity* of the world for science as well as for politics and religion. And along with historicity, of course, goes the internal *narrativity* that allows us to be in the world – or, as Donna Haraway prefers to put it, to be "with the world."[72] I am saying not that science has "disenchanted" the world by making us lose any connection with the "lived world," but that science has always *sung a quite different song* and has always *lived fully enmeshed in the world*. Perhaps it might be of some use to offer, at last, a view of materiality that is no longer so directly and awkwardly politico-religious and that offers a pathetically inexact vision of the sciences. We could then get away from any and every "religion of nature." We would have a conception of materiality that is finally worldly, secular – yes, non-religious, or, better still, earthbound.

We have known all this, of course, we who for a long time have been studying this curious obsession of the Moderns with deanimating the world in which they have nevertheless been causing unexpected and surprising agents to proliferate. We were well aware that the rationalizing style had no relationship with the sciences as they are practiced. This was even what had allowed me to assert, twenty-five years ago, that "we have never been modern."[73] But everything changes as soon as we read news briefs like the one with which I began this lecture: "The threshold of 400 parts per million (ppm) of atmospheric carbon dioxide (CO_2) is expected to be reached in May." Here, it seems obvious *to everyone*, and not only to historians of science, that we are immersed in a *history* that can no longer be *deanimated*.

And yet we must not count on the approach of catastrophes to make us more aware – quite the contrary. In *The End*, one of the many terrifying books I read while preparing these lectures, the historian Ian Kershaw showed how Germany lost more soldiers and civilians during the final year of the war, when the Germans had given up any hope of victory, than in the previous four years combined. He shows that, in the most cataclysmic situation, when the Reich was doomed, the war was clearly lost and everyone, from generals to housewives,

---

[72] Donna Haraway, *Staying with the Trouble: Making Kin in the Chthulucene* (2016). This theme is developed at length in Bruno Latour and Christophe Leclercq, eds, *Reset Modernity!* (2016).
[73] Bruno Latour, *We Have Never Been Modern* ([1991] 1993).

was completely aware of this, the fighting went on, and the criminal dictatorial system remained almost intact until the final collapse.[74]

It is because the self-evident character of the threat will not make us change that we have to prepare ourselves to remake politics. If there is nothing agreeable, harmonious, or calming about facing ecological problems; if Lovelock can describe Gaia as being "at war" and "taking its revenge" on humans, whom he compares to the British army in June 1940, trapped in the dunes of Dunkirk, in total disarray, forced to leave its weapons lying useless on the beach,[75] it is because geohistory must not be conceived as a great irruption of Nature finally capable of suppressing all our conflicts, but as a *generalized state of war*.

As horrendous as history has been, geohistory will probably be worse, since what had remained quietly in the background up to now – the landscape that had served as the framework for all human conflicts – has just joined the fight. What was a metaphor up to now – that even the stones cried out in pain in the face of the miseries humans had inflicted on them – has become literal. Clive Hamilton asserts that the enemy of action is *hope*, the unalterable hope that everything will get better and that the worst is not always a sure thing.[76] Hamilton maintains that, before undertaking anything at all, we have to purge hope from our desperately optimistic framing of life. It is thus with many scruples that I am putting this series of lectures under Dante's somber warning: "Abandon all hope." Or, in a more modern style, this query by Dougald Hine, cited by Déborah Danowski and Eduardo Viveiro de Castro: "What do you do, after you stop pretending?"[77]

We were already trembling as we observed the acceleration of history, but how are we to behave in the face of the "great acceleration"?[78] Through a complete reversal of the favorite trope of Western philosophy, human societies seem to be resigning themselves to playing the role of witless object, while it is nature that is unexpectedly taking on the role of active subject! Have you noticed

---

[74]Ian Kershaw, *The End: The Defiance and Destruction of Hitler's Germany, 1944–1945* (2011).
[75]James Lovelock, *The Revenge of Gaia: Earth's Climate in Crisis and the Fate of Humanity* (2006), p. 150.
[76]Clive Hamilton, *Requiem for a Species: Why We Resist the Truth about Climate Change* (2010).
[77]Déborah Danowski and Eduardo Viveiros de Castro, *The Ends of the World* (2016), p. 79.
[78]Will Steffen et al., "The Trajectory of the Anthropocene: The Great Acceleration" (2015a).

that we are now attributing to natural history the terms of human history – tipping points, acceleration, crisis, revolution – and that to speak of human history we are using the words inertia, hysteresis, path dependency, as if humans had taken on the aspect of a passive and immutable nature in order to explain why they are doing nothing against the threat? Such is the meaning of the New Climate Regime: the "warming" is such that the old distance between background and foreground has faded away: it is *human* history that appears cold and *natural* history that is taking on a frenzied aspect. The metamorphic zone has become our common place: it is as though we had indeed ceased to be modern, and, this time, collectively.

THIRD LECTURE

Gaia, a (finally secular) figure for nature

> Galileo, Lovelock: two symmetrical discoveries • Gaia, an exceedingly treacherous mythical name for a scientific theory • A parallel with Pasteur's microbes • Lovelock too makes micro-actors proliferate • How to avoid the idea of a system? • Organisms make their own environment, they do not adapt to it • On a slight complication of Darwinism • Space, an offspring of history

Before long, in the history of science as well as in the popular imagination, a second scene of discovery is likely to become as famous as the one in which, during a chilly night in the late fall of 1609, Galileo raised his telescope from the Venice lagoon toward the Moon. It occurred to him at that moment, he said, that all planets are alike. Three centuries later, another discovery reversed the proposition: the Earth is a planet *like no other*! We have to acknowledge that the symmetry is really too perfect: whereas the first scientist discovered how to shift away from the narrow view of the Grand Canal he had from his window toward the infinite universe, the second discovered how to shift from the infinite universe back to the narrow limits of the blue planet. What the first succeeded in doing with an inexpensive telescope, really a child's toy, the second accomplished by pointing an even lighter apparatus toward the sky – by performing a simple thought experiment. We would need a Plutarch to add a new chapter to his *Parallel Lives*, an Arthur Koestler to write an appendix to *The Sleepwalkers*.[1]

It was in the fall of 1965, at the Jet Propulsion Laboratory in Pasadena, in the offices of the department responsible for extraterrestrial life, that James Lovelock, a somewhat eccentric physiologist and engineer – the English still call him a maverick[2] – wrote an article with Dian Hitchcock (a philosopher employed by NASA) on the possibility of detecting life on Mars.[3] The two authors were a bit embarrassed to have to admit to their colleagues, who were busy conceiving of the complex and expensive machinery for the Voyager and later the Viking missions that they anticipated sending to land on Mars with the help of giant rockets, that to answer such a question the best solution would be to stay right where they were, in Pasadena! That they should be content, the authors said, to aim toward the red planet a modest instrument designed to determine whether the atmosphere was in a state of chemical equilibrium or not, and they would have their answer.[4] No need to fly there at great expense to prove the obvious!

It is hard not to be struck by the symmetry between the gestures of Galileo and Lovelock, raising modest instruments toward the sky to make radically opposing observations. When, on the basis of the shaky, haloed, and distorted images of the Moon captured by his telescope, Galileo decided, thanks to his extensive knowledge of perspective drawing,[5] to see shadows projected by the Sun on lunar hills, mountain chains, and valleys, he quickly established a new type of continuity, not to say a new fraternity, between the Earth and its satellite. They were both planets; they were both bodies made of the same homogeneous matter; they both had the same dignity; and they both revolved around another center. Undifferentiated space could henceforth be extended everywhere. The Earth was no longer relegated to the lower depths of a sublunary world surrounded by circles of dignity each more elevated than the one before, from the supralunary planet to the spheres of the fixed stars, distant only by a few degrees from God himself. The Earth henceforth had the same

---

[1] Arthur Koestler, *The Sleepwalkers: A History of Man's Changing Vision of the Universe* (1959).
[2] The London Science Museum, to which he bequeathed all his papers, devoted an exhibit to him titled "Unlocking Lovelock: Scientist, Inventor, Maverick."
[3] James Lovelock and Dian R. Hitchcock, "Life Detection by Atmospheric Analysis" (1967).
[4] The episode has often been related and embellished; see John Gribbin and Mary Gribbin, *James Lovelock: In Search of Gaia* (2009).
[5] See Erwin Panofsky, *Galileo as a Critic of the Arts* (1954).

importance as all the other celestial bodies, without any hierarchy among them; as for God, he could be encountered anywhere in the vast immensities of the world.

Once the first shock had passed, astronomers, writers, polemicists, priests, and pastors as well as libertines could then propel across these new Earths a vast population of fictional characters who had all sorts of adventures and observed the behaviors of all sorts of strange creatures. The new astronomical narratives of Kepler, Cyrano, Descartes, Fontenelle, and Newton became credible with respect to a world that was constantly being extended because it was everywhere homogeneous.[6] And, since infinite space that was everywhere the same had been invented, it was possible to give some substance to the idea "of a point of view from nowhere" that allowed disembodied and interchangeable minds to write laws applicable to the entire cosmos. By leaving aside the secondary qualities – color, odor, texture, but also procreation, aging, and death – and limiting the focus to the primary qualities – extension and movement – one could treat all the planets, all the suns, all the galaxies as so many billiard balls.[7] After all, bodies in free fall are just that; when you've seen one you've seen them all! The infinite extension of the world, like that of knowledge of the world, became possible, since every place was literally the same as every other, except for its coordinates. As the Latin term *res extensa* indicates, the idea of what a thing is could be in effect *extended* everywhere.[8] To return to Alexandre Koyré's celebrated title, Galileo and his successors made it possible for their readers to pass from a closed world to an infinite universe.[9] The spirit of the laws of nature was hovering over the waters.

It was precisely on the basis of these fictional localizations that Lovelock imagined a Martian astronomer who had no need to travel in a flying saucer to decide, simply by reading his equally fictional

---

[6] These are the delegated characters described by Gilles Deleuze and Félix Guattari in *What Is Philosophy?* ([1991] 1994) and made more concrete by Frédérique Aït-Touati in her *Fictions of the Cosmos: Science and Literature in the Seventeenth Century* (2012).

[7] This distinction between primary and secondary qualities, made by Galileo for practical reasons, has taken on increased philosophical weight over time, coming to look like a "bifurcation of nature" between two incommensurable worlds. See Alfred North Whitehead, *The Concept of Nature* (1920).

[8] *Res extensa* is not a domain of the world that could be opposed to another domain, *res cogitans*, but half of a single concept that has organized the transformation of the world into that of Nature/Culture from Descartes on. This theme belongs as much to the history of painting as to the history of science and philosophy; it can be called the idealism of matter (see the first lecture).

[9] Alexandre Koyré, *From the Closed World to the Infinite Universe* (1957).

instrument, that the Earth was a living planet, since its atmosphere did not return to chemical equilibrium.[10] Such is Lovelock's reasoning: if I can decide, from Pasadena, incontrovertibly, that Mars is a dead star, since its atmosphere is in chemical equilibrium, then, similarly, if I were a little green man, I could conclude with certainty that the Earth is a living star, since its atmosphere is in chemical disequilibrium. If this is the case, the terrestrial astronomer concludes in a flash of intuition, something must maintain this situation in place, some agency that has not yet been made visible, one that is absent on Mars as well as on Venus and on the Moon, of course: a force with some sort of agency that allows it to maintain, or recover, throughout billions of years, a state of affairs durable enough to counteract the disturbances introduced by external events – the increasing brightness of the Sun, bombardments by asteroids, volcanic eruptions. But we mustn't rush to give this power an already known name, for example, "life." We must first understand the singularity of this discovery.

While Galileo, raising his eyes from the horizon to the sky, reinforced the similarity between the Earth and all the other bodies in free fall, Lovelock, lowering his eyes from Mars in our direction, in effect *diminished* the similarity between all the other planets and this so peculiar Earth that is ours. It was by taking "the point of view of nowhere" that he showed that there is no "point of view of nowhere"! From his little office in Pasadena, like someone sliding down the roof of a convertible slowly in order to close it and lock it in place, Lovelock brought his reader down to what should be viewed once again as a *sublunary world*. Not that the Earth lacked perfection, quite the contrary; not that it hid the somber site of Hell in its entrails;[11] but because it held – alone? – the privilege of being in disequilibrium, which also meant that it possessed a certain way of being *corruptible* – or, to use the terms of the previous lecture, of being, in one form or another, *animated*.

In any case, it seems capable of actively maintaining a difference between its inside and its outside. It has something like a skin, an envelope. More oddly still, the blue planet suddenly looks like a long string of historical *events*, random, specific, and contingent events,

---

[10] Episode related by Lovelock himself in *Homage to Gaia: The Life of an Independent Scientist* (2000b).

[11] The particularity of the ancient cosmos – I shall come back to this point in the next lecture – was that it had hell at its center, as we see in *The Divine Comedy*. Galileo devoted an astonishing text, moreover, to the dimensions of that hell: see his "Two Lectures to the Florentine Academy on the Shape, Location and Size of Dante's Inferno" ([1558] n.d.).

as though it were the temporary, fragile result of a geohistory.¹² It is as though, three and a half centuries later, Lovelock had taken into account certain features of that same Earth that Galileo *could not* take into account if he were going to consider it simply as a body in free fall amidst all the others:¹³ its color, its odor, its surface, its texture, its genesis, its aging, perhaps its death, this thin film within which we live, in short, its behavior, in addition to its movement. As though the secondary qualities had come back to the foreground. Serres was right: to complete Galileo's Earth, which *moves*, it was necessary to add Lovelock's Earth, which *is moved*.¹⁴

If the first discovery was shocking, the second is no less so. Remember the cliché of the three "narcissistic wounds" made famous by Freud, not without a certain masochism:¹⁵ first Copernicus, then Darwin, and finally Freud himself. As Freud saw it, three times in a row, human arrogance was deeply wounded by scientific discoveries: first, by the Copernican revolution that drove humans out of the center of the cosmos; then, still more deeply, by Darwinian evolution, which made humans a species of naked monkeys; and, finally, by the Freudian unconscious, which expelled human consciousness from its central position. But, to take such discoveries as a series of narcissistic wounds, Freud must have forgotten the enthusiasm with which the so-called Copernican revolution had been greeted.¹⁶ Far from feeling wounded, on the contrary, it seems that those who lived through that revolution felt freed from their bonds after having suffered so long from being relegated to a dead-end ditch with no way out but the supralunary regions, the only site of incorruptible truths. The infinite universe, the millennial evolution, the tortuous unconscious, all of these are liberating: finally we get out of our hole! We are emancipated at last! Brecht, we recall, in his play about Galileo, had celebrated this escape to open territory when he had his young assistant, Andrea, turn the heavy copper circles of an

---

¹²The fragility of the system is another way of emphasizing its historicity. In *The Medea Hypothesis: Is Life on Earth Ultimately Self-Destructive?* (2009), Peter D. Ward shows that nothing protects Gaia against destruction. This is also the theme developed by James Lovelock and Michael Whitfield in "Life Span of the Biosphere" (1982).
¹³See Isabelle Stengers, *The Invention of Modern Science* ([1993] 2000), pp. 83–7. It is in the protocol of the inclined plane that the relation between past and future is inverted; henceforth Galilean time descends from the past cause toward its consequences. The variety of processes associated with the older "phusis" have disappeared.
¹⁴See the second lecture.
¹⁵Sigmund Freud, "A Difficulty in the Path of Psycho-Analysis" ([1917] 1973).
¹⁶Steven Shapin, *The Scientific Revolution* (1996).

old-style astrolabe. Andrea, watching them move, comments: "But we're so shut in."

   GALILEO *drying himself*: Yes, I felt that the first time I saw one of these. We're not the only ones to feel it. Walls and spheres and immobility! For two thousand years people have believed that the sun and all the stars of heaven rotate around mankind.... But now we are breaking out of it, Andrea, at full speed. Because the old days are over and this is a new time.... Because everything is in motion, my friend.... Soon humanity is going to understand its abode, the heavenly body on which it dwells. What is written in the old books is no longer good enough. For where faith has been enthroned for a thousand years doubt now sits.[17]

"Everything is in motion, my friend," indeed, but not in the direction anticipated. We might say, parodying Brecht: "In a place where belief has been enthroned for *three hundred fifty years*, doubt is being installed right now!" "The old days are over," and soon perhaps "humanity is going to understand its abode, the heavenly body on which it dwells," but on the condition of taking in that other "narcissistic wound," more painful still than the ones Freud imagined. What no longer makes any sense is to transport oneself in dreams, without obstacles and without attachments, into the great expanse of space. This time, we humans are not shocked to learn that the Earth no longer occupies the center and that it spins aimlessly around the Sun; no, if we are so profoundly shocked, it is on the contrary because we find ourselves at the center of its little universe, and because we are imprisoned in its minuscule local atmosphere.

   Suddenly we have to pull back on our imaginary voyages; Galileo's expanding universe is as if suspended, its forward motion interrupted. Koyré's title has to be read in the opposite direction from now on: "*Returning* from the infinite universe to the closed and limited cosmos." All those fictional characters you've sent out? Bring them back! Tell Captain Kirk that the USS Enterprise has to return to port. "Out there, you'll find nothing like us; we're alone with our terrible terrestrial history." As for the planet Pandora, it's not in this direction that the next front line against the Na'vi barbarians is going to continue to stretch. Moreover, in the film *Gravity*, Dr Ryan Stone summed up the situation nicely for us: when she finally made it back down onto the muddy earth, she confessed: "I hate space!"[18]

   Yes, unquestionably, "doubt is being installed." We could always spend huge budgets on what used to be called the "conquest of space,"

---

[17] Bertolt Brecht, *The Life of Galileo* ([1945] 2001), Act 1, scene 1, pp. 6–7.
[18] In addition to *Star Trek*, I am referring to two popular films whose mythologies share the preoccupations of the planetologists: James Cameron's *Avatar* (2009) and Alfonso Cuaron's *Gravity* (2013).

but we would succeed at best only in transporting a half-dozen encapsulated astronauts across inconceivable distances, from a living planet toward some dead ones. The place of the action is here below and right now. Dream no longer, mortals! You won't escape into space. You have no dwelling place but this one, this narrow planet. You can compare the celestial bodies to one another, but not by going to see for yourselves. For you, Earth is the place, what is called in Greek a *hapax*, a name that appears only once, and this name pertains to the members of your species, the Earthbound, just as well – or, if you prefer a term with a similar Greco-Latin etymology, *idiot*. "We are idiots; everything that happens to us happens only once, only to us, only here." If Galileo Galilei managed to have a name that brought him into proximity with the mythical name of the Galilean, we have to acknowledge that Lovelock, too, arranged to have a very enigmatic name: "Love locked," "Locket of love," "Love-locks"? In any case, it's his fault; we're locked in here for good, double-bolted.

*

The name "Gaia" is no less surprising than the name "Lovelock." We have all read *Lord of the Flies*, the story of some young British schoolboys marooned on a desert island from which they can no more escape than we can from our blue planet, and on which they slide little by little down the slippery slope that leads to barbarity.[19] It so happens that its author, William Golding, was Lovelock's neighbor in a little Wiltshire village with the delightful name Bowerchalke, and it is to Golding that Lovelock owes his theory.[20] It does no harm to the novelist's reputation to suspect that when, after a few beers in the local pub, he suggested the name "Gaia," he hadn't reread Hesiod for a long time. Had he done so, he would have known that he was putting his friend's theory under a curse from which it would never entirely escape.

For Gaia, Ge, Earth, is not a goddess properly speaking, but a force from the time before the gods. "In Hesiod's theogony," Marcel Detienne writes, "Earth is a great power of beginnings."[21] Prolific, dangerous, savvy, the ancient Gaia emerges in great outpourings of blood, steam, and terror, in the company of Chaos and Eros.

---

[19] William Golding, *Lord of the Flies* (1954).
[20] Lovelock referred to this episode frequently in his autobiography, *Homage to Gaia* (2000b), and in numerous interviews explaining that, in ignorance of mythology, he had first heard "Gyre" instead of "'Gaia." Gyre would have been a good name too, in the end.
[21] Marcel Detienne, *Apollon, le couteau à la main* (2009), p. 165.

> Verily at the first Chaos [the Yawning Gap] came to be, but next wide-bosomed Earth [Gaia], the ever-sure foundations of all, the deathless ones who hold the peaks of snowy Olympus,...and Eros (Love), fairest among the deathless gods... And Earth first bare starry Heaven [Uranus], equal to herself, to cover her on every side... But afterwards she lay with Heaven and bare... Theia [the Divine] and Rhea, Themis [Just-Custom] and Mnemosyne [Memory] and gold-crowned Phoebe [the Luminous] and lovely Tethys. After them was born Cronos the wily, youngest and most terrible of her children, and he hated his lusty sire.[22]

So who is Gaia, the Gaia of mythology? It is impossible to answer this question without doing for her what we learned to do in the previous lecture: first of all, draw up the long list of her attributes in order to find her essence. As we must do for all beings, but especially for the changeable characters that the myths mix together endlessly, we must deduce her competences – what she is – from her performances – what she does.[23] And these performances are multiple, contradictory, hopelessly confused. Gaia has a thousand names. What is certain is that she is not a figure of harmony. There is nothing maternal about her – or else we have to revise completely what we mean by "Mother"! If she needed rituals, these were surely not the nice New Age dances invented later to celebrate the postmodern Gaia.[24]

We can judge for ourselves: Gaia was the first to invent the horrible stratagem that would allow her to get rid of the oppressing weight of her husband Uranus:

> The world would have remained in that state if Gaia, indignant at a reduced existence, had not imagined a perfidious ruse that was going to change the face of things. She created the white metal, steel, and made a sickle of it; she exhorted her children to castrate their father. They all hesitated, trembling, except the youngest, Cronos, the Titan with a bold heart and warped wit.[25]

In Hesiod's narrative, Gaia plays the role of a terrifying power but also that of an astute advisor. Her cunning is manifested first of all in the fact that she never commits abominable crimes herself, but

---

[22] Hesiod, *Theogony* (1914), lines 116–38.
[23] It is through this way of reconstructing bit by bit the semantic field, the rituals, the archaeological testimonies to the existence of the divine characters and concepts, without worrying about their ideal substance, that the great exegetes of the French school have been able to rescue the anthropology of ancient Greece from a sterile academism. What holds true for the ancient Gaia of mythology holds still more true for the scientific Gaia.
[24] See Bron Taylor, *Dark Green Religion: Nature, Spirituality, and the Planetary Future* (2010); Jacques Galinier and Antoinette Molinié, *The Neo-Indians: A Religion for the Third Millennium* ([2006] 2013).
[25] Jean-Pierre Vernant, preface to Hesiod, *Théogonie: la naissance des dieux* (1981), p. 20.

Gaia, a (finally secular) figure for nature 83

always makes use of those in whom she inspires vengeance as intermediaries. She endlessly goads her immense progeniture of monsters and gods into assassinating one another! However, after thrusting family members into frightful conflicts, she then lavishes advice from her divinations (she is said to be *prôtomantis*, the "first prophetess") on the very ones against whom she has plotted – Uranus, Cronos, Zeus – so that they come out on top:

> Three times, Earth gives decisive advice...: she makes herself understood, she indicates by words more than by signs, she also knows how to "say everything explicitly" when she needs to, but always she foresees, she forewarns, she conceives of schemas that orient the course of things in a decisive way.[26]

A chthonic power, dark-skinned, dark-haired and somber, after having incited her son Cronos to use a "sharp-toothed steel sickle" to cut off her husband's genitals, she does not stop there. With Rhea's complicity, Gaia convinces Zeus to fight his own father and defeat him. But then she schemes to mobilize her own youngest son, Typhon – a monster with a hundred serpents' heads – to destroy the empire of Rhea's son Zeus. It is the Olympian who wins, but forever after the poor humans will be victims of Typhon's winds, storms, and cyclones. Gaia, considered from the viewpoint of the Olympian gods, those divine late-comers, is a figure of violence, genesis, and trickery, a figure that is always antecedent and contradictory. If she is bound to order and law, to Themis, this bond is forged in violence and quakes, but especially in duplicity. As Detienne says, she blows hot and cold.

> It was Gaia who conceived of the subterfuge of the stone wrapped in swaddling clothes in the place of [Rhea's] last-born, hidden in the depths of a cavern in Crete, waiting for him to become Zeus. Throughout this entire "archaeology" of the divine world, Gaia demonstrates a capacity for knowing what is going to happen: she appreciates the present in relation to the future that inhabits it, prefiguring in this way the good advice and informed prudence that are going to characterize the action of Themis, at several points in Zeus's career, and especially at the point when Earth, this time doing the asking, will come to complain of the proliferation of the human species and its increasing impiety on her "broad breast."[27]

The figure who complains about the impiety and the excessive weight of humans is surely not pious herself. Moreover, archaeologists have had great difficulty finding her altars, buried as they are in deep caverns, under the ruins of temples erected much later in the names of more acceptable and more celebrated gods.[28]

[26] Detienne, 2009, p. 165.
[27] Ibid., p. 166.
[28] Ibid.

What is true of the mythological character is also true of the theory that bears its name. Yes, there's no doubt about it, there is a curse attached to the Gaia theory. How many times have I been warned not to use this term, and not to admit out loud that I'm interested in Lovelock's work! – so much so that I'm writing an essay on the subject, and to top it off I'm using him as a focal point of the present lecture series! "You can't really take them seriously," I'm told, "these pseudo-scientific ramblings of an independent old inventor who calmly asserts on television that seven-eighths of humanity will soon be wiped out because, like a new Malthus, he claims to have calculated the 'carrying-capacity' of the planet Earth – about 300 million; and he says it's all the same to him, anyway, because he's going to die far above the earth, in a rocket, during a trip into space, thanks to a free ticket offered him as a reward, sponsored by none other than Richard Branson![29] Come on, this mix of science and vaguely spiritualist intuitions can't be the center of a new vision of science, politics, and religion. What a stupid idea to compare him to our great, our magnificent Galileo."

One of the reasons I resisted these warnings is that I am not entirely sure what my detractors would have said had they lived in 1610, as they read the *Sidereus Nuncius* published by an odd, bearded engineer who signed his name Galileo.[30] After all, a mathematician who went on about God, the Earth, the Moon, the Church, the Bible, and human destiny, who compared the Earth and the planets to billiard balls even as he dedicated his work to one of the Medicis with unadulterated flattery, would probably not have been welcomed more favorably at the time.[31] Richard Branson is not the Duke of Medici, certainly, but between the two cosmologies there is an inverse symmetry so striking that I am determined to explore it. In both cases, what is in question is the movement and the behavior of the Earth as well as the destiny of those who inhabit it and who claim to be familiar with it; this is enough incentive to take them both seriously.

If there is a curse that weighs on the Gaia theory, it is the one modernism has brought into the picture by insisting on always treat-

---

[29] The billionaire entrepreneur Richard Branson, founder of the Virgin Group, has recently devoted considerable resources to space tourism. For a presentation on Lovelock, see "Doomsday Pending" (n.d.).

[30] I shall come back to the date 1610 in the sixth lecture. On the reception of this particular text by Galileo, see Mario Biagioli, *Galileo, Courtier: The Practice of Science in the Culture of Absolutism* (1993).

[31] The imbroglio of politics, religion, diplomacy, and academic competition is studied with care in its relation to the nascent science of economics in Mario Biagioli, *Galileo's Instruments of Credit: Telescopes, Images, Secrecy* (2006).

Gaia, a (finally secular) figure for nature   85

ing our relation to the world according to the Nature/Culture schema that I tried to discredit in the first two lectures. This schema is itself in large part heir to the discovery that we might name, to simplify, *Galilean*.[32] Once introduced into physics for reasons that were initially solely practical, the distinction between primary and secondary qualities then began to proliferate in every domain. If it was indispensable for Galileo to remove all behaviors from bodies and retain only their movement, there was no reason to turn this practice into a general philosophy and still less into the politics of an Earth deprived of any possibility of being moved. What was only a convenient expedient for Galileo was transformed into a metaphysical foundation in the hands of Locke, Descartes, and their successors.[33]

It is nevertheless this unwarranted generalization that gave rise to the strange opinion that has made it possible to deanimate one sector of the world, deemed objective and inert, and to overanimate another sector, deemed to be subjective, conscious, and free. It is this strange distribution – which Whitehead called the *bifurcation of nature*[34] – that weighs, four centuries later, on every interpretation of the Gaia theory. It is because Gaia has no place in the Nature/Culture schema – no more than Galileo's Earth in motion had a place in the medieval cosmos – that we have to take some precautions in evaluating it. In a sense, it is Locke against Lovelock! Let's not rush to a negative judgment in the latter's case the way we rush to a favorable one in Galileo's (but always after the fact!). This time, we have to form our own opinions without the benefit of the retrospective judgment of history.

I could easily escape the curse by claiming that the name of a theory is of no importance, and that, after all, serious scientists avoid the name Gaia as much as possible, preferring the euphemism "sciences of the Earth System." But this would be cheating; it would amount to passing from one ambiguous character to another that is even harder to define. "System"? What weird animal is that? A Titan? A Cyclops? Some twisted divinity? By avoiding the real myth, we would land on a false one.[35] Myth and science, as we well know, speak languages

---

[32] This is the meaning that Edmund Husserl gave the term in *The Crisis of European Sciences and Transcendental Phenomenology: An Introduction to Phenomenological Philosophy* (1970).
[33] Didier Debaise offers a good summary of this history in *L'appât des possibles: reprise de Whitehead* (2015).
[34] Whitehead (1920; see especially chapter 2, "Theories of the Bifurcation of Nature") and Isabelle Stengers's indispensable commentaries in *Thinking with Whitehead: A Free and Wild Creation of Concepts* (2011a).
[35] I shall return to the question of the "Earth System" with its two opposite meanings – connection or totality – at the end of the next lecture.

that are only apparently distinct; as soon as we approach the metamorphic zone that we have learned to identify, they begin to exchange their features, so that they can manage to express, to extend, what they want to say. "There is no pure myth other than [that of] science purified of any myth," as Serres put it.[36]

No, we have to do for the scientific theory of Gaia what the magnificent work of the Hellenists has taught us to do for mythological characters such as the ancient Ge. As always, we have to replace what gods, concepts, objects, and things *are* by what they *do*. To launch the Earth into movement in the infinite universe, Galileo had to mix everything together, of course, everything having to do with God, princes, authority, the form of bodies, and even, as we know, the fine Italian style.[37] The same holds true for Lovelock when he seeks to repatriate this same Earth into a finite cosmos. To translate into a more or less comprehensible language the agency responsible for the fact that the Earth has a behavior – that it appears to outside observers to be endowed with a sensitive and perishable envelope – the inventor, too, has to mix everything together, reknead the metaphors so they fit together differently and can be made in the end to say something quite distinct. Lovelock and Galileo both hesitate. Do they contradict themselves? Yes, of course: to pass from nature to the world is always to plunge into metaphysics, to bury the habits of one's discipline – for Galileo, mechanics; for Lovelock, chemistry – in something more active, more open, more corrosive as well.

But Lovelock's problem is new: how to speak about the Earth *without taking it to be an already composed whole*, without adding to it a coherence that it lacks, and yet without deanimating it by representing the organisms that keep the thin film of the critical zones alive as mere inert and passive passengers on a physio-chemical system? His problem is indeed to understand in what respect the Earth is active, but *without endowing it with a soul*; and to understand, too, what is the immediate consequence of the Earth's activity – in what respect can one say that it *retroacts to the collective actions of humans*? Before condemning him, we need to appreciate how unprecedented this problem is, since, to speak of "nature," Lovelock has at his disposal only the metaphysics inherited from Galileo. This "nature," as we now know, is *only half* of a symmetrical definition of culture, subjectivity, and humanity, and we know that for several centuries it has been conveying a whole bundle of morality, politics,

---

[36] See Michel Serres, *Hermès III: la traduction* (1974), p. 259.
[37] Galileo, *Dialogue Concerning the Two Chief World Systems – Ptolemaic and Copernican* ([1632] 1967).

and theology that it has been unable to shed. Lovelock is neither a philosopher nor particularly well read. He is a self-taught inventor. He has to cobble everything together by himself. But what he succeeds in building, in the end, from bits and pieces, is a version of the Earth that comes entirely from *here below*. Let's say that, to study the Earth, one has to come back down to Earth.

As we are going to see, despite the frequent awkwardness of Lovelock's prose, the concept of Gaia plays a much less religious, much less political, much less moral role than the concept of "nature" as it emerged in Galileo's time. The paradox of the figure that we are attempting to confront is that the name of a proteiform, monstrous, shameless, primitive goddess has been given to what is probably the *least religious entity* produced by Western science. If the adjective "secular" signifies "implying no external cause and no spiritual foundation," and thus "belonging wholly to this world," then Lovelock's intuition may be called *wholly secular*. Alas, "secular" invokes only the contrary of "religious"; "profane" has meaning only in relation to "sacred"; as for "pagan," it is a term of exclusion that is meaningful only for missionaries. The English term "worldly" comes closest.[38] If the term is inadequate or lacking, it is indeed because the situation is new.

*

In the rest of this lecture, I should like to insist on two particularly surprising characteristics of Gaia: first, that it is composed of agents that are neither *deanimat*ed nor *overanimat*ed; then, contrary to what Lovelock's detractors claim, that it is made up of agents that are not *prematurely unified* in a single acting totality. Gaia, the outlaw, is the anti-system.[39]

What agency has Lovelock ascribed to living organisms that are capable of playing a role in the local history of the Earth? The best way to understand this is perhaps to set up another parallel, this time between Lovelock and Louis Pasteur. What makes the parallel so seductive is not only the role they each attributed to micro-organisms but the consequences they both drew from this for medicine. After all, one of Lovelock's books is subtitled *The Practical*

[38] Unfortunately, as we shall see in the sixth lecture, the "secular" is like non-alcoholic beer, it is the religious without religion. But Gaia goes further.
[39] As Oliver Morton pointed out to me in a personal communication (June 21, 2015), this is what connects Lovelock to the tradition of A. G. Tansley, "The Use and Abuse of Vegetational Concepts and Terms" (1935). For the inventor of the notion of ecosystem, too, the systematic following of connections did not imply any holism.

Science of Planetary Medicine.[40] Pasteur, after describing how his microbes worked, immediately tried to convince surgeons that with their infected scalpels they were unwittingly killing their patients. Similarly, Lovelock, as soon as he had drawn Gaia's face, tried to persuade humans of their strange fate: they had inadvertently become Gaia's *malady*.[41] As if the challenge, this time, were not to protect humans against microbes but to understand the dangerous retroaction between microbes and humans! If Pasteur's microbes profoundly transformed all the definitions of collective life, to find ourselves in Lovelock's Gaia is to learn to redraw the front lines between friends and enemies. Just as in Pasteur's era, what is at stake in these new sciences is war and peace.[42]

Let us see first of all how the parallel can work. If we recall the long struggles that pitted microbiology, in its early years, against eminent chemists, we will find striking parallels with Lovelock's battles against the geologists to move from geochemistry to what he calls "geophysiology."[43] In each case, attempts to introduce a hitherto unknown agent were accused of overanimating the world while running headlong into metaphysics. In Pasteur's case, and Lovelock's as well, the intuition that there are agents at work in chemical reactions in addition to the usual suspects – those known at the time – was met with great suspicion.[44]

This was certainly the case for the German chemist Justus von Liebig (1803–1873), Pasteur's *bête noire* in the 1850s. After a century of struggle against mysterious agents and vital forces, the chemists had finally established their paradigm as they learned to account for all the phenomena that they could analyze in the laboratories through "strictly chemical processes."[45] This is why they had no patience, at least initially, for the traitor Pasteur, who was after all a chemist

---

[40] James Lovelock, *Gaia: The Practical Science of Planetary Medicine* ([1991] 2000a).
[41] The last chapter of *Gaia* is titled "The People Plague"!
[42] See Bruno Latour, *The Pasteurization of France* ([1984] 1988). See also the superb biography by René Dubos, *Louis Pasteur, Free Lance of Science* (1950), which multiplies the connections with the ecological crisis (Dubos is also the author of one of the first books addressed to a broad public dealing with the Earth as a common and unified world: see Barbara Ward and René Dubos, *Only One Earth: An Unofficial Report commissioned by the Secretary General of the United Nations Conference on the Human Environment* (1972). Needless to say, the link between Pasteur and Margulis is even more direct.
[43] This term appears in the subtitle of the French translation of *Gaia: Gaïa, une médecine pour la planète: Géophysiologie, nouvelle science de la terre*.
[44] See Gerald Geison and James A. Secord, "Pasteur and the Process of Discovery: The Case of Optical Isomerism" (1988).
[45] See Bernadette Bensaude-Vincent and Isabelle Stengers, *A History of Chemistry* ([1992] 1996).

himself, when he claimed to be able to demonstrate, for example, that sugar could not be transformed into alcohol *without* the addition of an unknown agent, yeast, whose presence was indispensable. In the chemists' eyes, this was a return to the vitalism of the past – or even to a suspect spiritualism.

As we saw in the previous lecture, scientific agents, grasped in their nascent state, are first of all a list of actions, well before they are given a name that sums up these actions – often in a language – ancient Greek – that scientists no longer speak. What an agent is capable of doing is deduced from what it has done – a pragmatic principle if ever there was one. In Liebig's hands, "yeast" was only a product derived from fermentation. In Pasteur's laboratory, the same character is called to a more glorious fate. The text is rightly famous:

> If one examines carefully an ordinary lactic fermentation, there are cases where one can find on top of the deposit of the chalk and nitrogenous material *spots of a gray substance* which sometimes form a layer on the surface of the deposit. At other times, this substance is found adhering to the upper sides of the vessel, where it has been carried by effervescence. Under the microscope, when one is not *forewarned...*, it is *hardly possible* to distinguish it from casein, disaggregated gluten, etc.; in short, *nothing indicates that it is a separate material* or that it originated during the fermentation. Its apparent weight always remains very little as compared to that of the nitrogenous material originally necessary for the carrying out of the process. Finally, very often it is so mixed with the mass of casein and chalk *that there would be no reason to suspect its existence. It is it nevertheless this substance that plays the principal role.* I am going to show, first of all, how to isolate it and to prepare it in a pure state.[46]

If the reader, turning the pages of the memoir on fermentation, moves on from "Until now minute researches have been unable to discover the development of organized beings" to "It is nevertheless this substance that plays the principal role,"[47] it is because Pasteur extracts this "principal role" from a set of laboratory tests in which the emerging character is initially revealed by a series of very modest actions: in the beginning, it is nothing more than "spots of a gray substance"; "nothing indicates that it is a separate material." An actor emerges little by little from its actions; a new substance emerges from its attributes. We find ourselves here in the same situation as in the previous lecture: yeast becomes an agent whose properties can then be deduced.[48]

---

[46] Louis Pasteur, cited in James Bryant Conant, *Pasteur's Study of Fermentation* (1952), p. 28, translation adapted, emphasis added.
[47] *Trans.*: The two passages cited here are not included in Conant's translation.
[48] I have tried to establish, based on the English text of this article, as complete a semiotic inventory as possible; the text can be found at www.bruno-latour.fr/node/257/.

If chemists gradually changed their minds, it was not only because of Pasteur's experimental skills but also because he had successfully carried out the same series of experiments in a different context, against the vitalists, whose cause he was nevertheless accused of embracing. In a series of magnificent experiments, Pasteur had demonstrated that those who continued to believe in spontaneous generation, as Félix-Archimède Pouchet did, had "contaminated" their soup by surreptitiously introducing what were soon to be called "microbes."[49] Where Pouchet saw autonomous and spontaneous agency, Pasteur succeeded in showing, on the contrary, that there was only a "culture medium" in which one could, at will, "seed" microorganisms, but that one could also, at will, decide to keep sterile as long as one wished. In Pasteur's hands, the existence of spontaneous generation faded away, to become a simple error in manipulation.

We can see why it is so important never to stabilize the animation with which one endows agencies once and for all: whereas the chemist Liebig, in Pasteur's eyes, had prematurely deanimated his concoctions, Pouchet, the naturalist, had rushed to give his actors comparably excessive generating capabilities. *An excess of reduction in one case; a lack of reduction in the other.* In Pasteur's skilled hands, the anti-Liebig agent was equally anti-Pouchet. By this attack on two fronts, Pasteur, in less than a decade, managed to trace his path between the Charybdis of reductionism and the Scylla of vitalism. He thus established the wholly original existence of an agent that could be reduced neither to "strict chemistry" nor to any of the mysterious "miasmas" that had disoriented medicine for centuries. To the list of agents he had added an element, the microbe, that was to play a crucial role in the rearrangement of all modes of life.

Pasteur's case proves once again that science proceeds not through the simple *expansion* of an already existing "scientific worldview" but through the *revision* of the list of objects that populate the world, something that philosophers normally and rightly call a *metaphysics* and that the anthropologists call a *cosmology*. The reductionism does not consist in limiting oneself to *a few* well-known characters in order to be able to tell the story of *everything*, as Descartes thought he could do in his artful novel on the systems of nature;[50] it consists rather in using a series of tests to bring out the unexpected characters that make up collective bodies. The world always exceeds nature, or, more exactly, world and nature are temporal reference points: nature

---

[49] Bruno Latour, "Pasteur and Pouchet: The Heterogenesis of the History of Science" (1995a) and "Joliot: History and Physics Mixed Together" (1995b).
[50] See Stéphane Van Damme, *Descartes* (2002).

is what is established; the world is what is coming.⁵¹ This is why the word "metaphysics" should not be so shocking for active scientists, but only for those who believe that the task of populating the world has already been accomplished. Metaphysics is the reserve, always to be refurbished, of physics. And, of course, as soon as you have decided which are the human and nonhuman characters that will be called upon, like yeast, to play the "principal roles," politics will start to nose its way in.

*

The parallelism with Pasteur helps show, more charitably, how Lovelock goes about introducing other "organized agents" to which he attributes the "principal role," where his detractors see only passive entities, mere passengers carried along by a nature that does all the work. This time, it is not the indispensable presence of "spots of a gray substance" that unleashes "active fermentation" but a series of chemical instabilities that require the introduction of another agent to even out the balance sheet. When Lovelock tries to sort out the role played by the strange proportion of oxygen and carbon dioxide in the atmosphere, like Pasteur he exploits the effect of surprise. The drama always unfolds in more or less the same way: the Earth *ought to be* like Mars, a dead star. It is not. So what force is capable *of delaying the disappearance of its atmosphere*?⁵²

> Many biologists today seem to think that [the balance of nature] alone explains the levels of the two great metabolic gases – carbon dioxide and oxygen – in the air. *This view is wrong*. The *picture* of the world it gives is like that of a ship with the pumps connected merely to recirculate the bilge water within it, rather than to pump it out. As water leaked in, the ship would soon sink...So what is this "leak" that thus determines the level of carbon dioxide in the atmosphere? In short *it is rock weathering*...Until the 1990s, geochemists maintained that the presence of life has had no effect on this set of reactions. It is *simple chemistry* that determines the level of carbon dioxide in the atmosphere....But I disagreed....*By their growth*, plants pump carbon dioxide from the air into the soil,...proof [being] the observed 10- to 40-fold enrichment of carbon dioxide in the air spaces of the soil.⁵³

Lovelock's prose always reads a little like a detective story, except that the mystery to be solved is not set off by the discovery of a dead

---

⁵¹ The nuance between the two terms was introduced at the end of the first lecture, to open up questions that the notion of nature cannot help but foreclose.
⁵² The connection with the theme of *katekon*, that which delays the catastrophe in the apocalyptic imagination, is not so incongruous after all; we shall come back to it in the seventh lecture.
⁵³ Lovelock, 2000, p. 108, emphasis added.

body; on the contrary, it starts with the mystery of why a character has not been assassinated – at least not yet! Let us subject the situation to a test to see if the normal laws of geochemistry succeed in explaining this continued existence. Every time the test fails, we shall be forced to add something extra, a bit of indeterminacy, to account for this disequilibrium in the chemical balance. Then we shall have to name the invisible protector that ensures the continuity of what ought to have disappeared billions of years ago, as it did on Mars and Venus.

Just as Pasteur threw out challenges to the believers in spontaneous generation, Lovelock challenges the geochemists: "Go ahead, you 'balance of nature' advocates, try to explain the situation on the basis of the normal laws of chemistry." Take water. It should have vanished long ago, just as it did on the other planets. Why is it still here, and so much of it? "The Earth has abundant oceans because it has evolved, *not by geophysics* and geochemistry alone, but as a system in which the organisms are an *integral part*."[54]

Next, let us reproduce this forensic investigation for all the successive ingredients that are thought to populate the Earth. Carbon dioxide ought to be present in much larger quantities in the air? Where does it fall? Into the soil. By the intermediary of what agent? By the action of micro-organisms and vegetation. Now let us look to see whether these micro-organisms are up to the new role assigned to them. Atmospheric nitrogen is not found where it ought to be, in the oceans. It would have increased salinity so much that no organism could have protected its cellular membrane against salt poisoning. Before such a disequilibrium, we have to ask what forces maintain it in the atmosphere:

> If there were no life on Earth the continued action of lightning would eventually remove most of the nitrogen from the air and leave it as nitrate ions dissolved in the ocean...On a lifeless Earth it seems probable that these inorganic forces would partition nitrogen so that most was in the sea and only a little was in the air.[55]

What is *moving* in Lovelock's prose (and even more in that of his sidekick Lynn Margulis (1938–2011)[56] is that every element that we ignorant readers would have seen as part of the *background* of the majestic cycles of nature, against which human history had always

---

[54] Ibid., p. 127, emphasis added.
[55] Ibid., p. 119.
[56] Lynn Margulis and Dorian Sagan, *Microcosmos: Four Billion Years of Evolution from our Microbial Ancestors* ([1986] 1997); see also the chapter titled "Gaia" in Lynn Margulis, *Symbiotic Planet: A New Look at* Evolution (1998).

stood out, becomes active and mobile thanks to the introduction of new invisible characters capable of reversing the order and the hierarchy of the agents. We knew that a substantial part of any mountain formation consists in the debris of living beings, but perhaps the same thing holds true for the cloud layer, manipulated by marine micro-organisms.[57] Even the slow movement of tectonic plates might have been triggered by the weight of sedimentary rocks.[58]

This staging has a cartoonish aspect, as if every time Lovelock touched some part of the décor with his magic wand, suddenly, as in a Disney version of *Sleeping Beauty*, all the servants in the palace, until then passive and inert, awoke from their sleep, yawning, and began to move frenetically about – the dwarves and also the clock, the trees in the garden and also the knobs on the doors. The humblest accessories henceforth play a role, as if there were no more distinctions between the main characters and the extras. Everything that was a simple *intermediary* serving to transport a slim concatenation of causes and consequences becomes a *mediator* adding its own grain of salt to the story.[59] For Lovelock, everything that is located between the top of the upper atmosphere and the bottom of the sedimentary rock formations – what biochemists aptly call the critical zone[60] – turns out to be caught up in the same seething broth. The Earth's behavior is inexplicable without the addition of the work accomplished by living organisms, just as fermentation, for Pasteur, cannot be started without yeast. Just as the action of micro-organisms, in the nineteenth century, agitated beer, wine, vinegar, milk, and epidemics, from now on the incessant action of organisms succeeds in setting in motion air, water, soil, and, proceeding from one thing to another, the entire climate.

It is dizzying. And our vertigo is much more pronounced than the one set off by Galileo when he described the Earth orbiting around the Sun. It took a good deal of imagination, in the seventeenth century, to be frightened by the "eternal silence of these infinite spaces," since in practice, on Earth, no one could detect the slightest

---

[57]Robert J. Charlson, James E. Lovelock, Meinrat O. Andreae, and Stephen G. Warren, "Oceanic Phytoplankton, Atmospheric Sulphur, Cloud Albedo and Climate" (1987); Timothy Lenton, *Earth System Science* (2016), offers a short and up-to-date presentation of many of Lovelock's insights.
[58]Stephan Harding and Lynn Margulis, "Water Gaia: 3.5 Thousand Million Years of Wetness on Planet Earth" (2009).
[59]Introduced in the second lecture, these two terms make it possible to pay attention to the *agency* attributed to the characters in a narrative.
[60]Susan L. Brantley, Martin B. Goldhaber, and K. Vala Ragnarsdottir, "Crossing Disciplines and Scales to Understand the Critical Zone" (2007).

Third Lecture

difference between the heliocentric version and the geocentric version of everyday experience (this is the great disadvantage of the principle of relativity, no one feels it...). But here, with Lovelock, it is very easy to *feel* the extent to which this new form of *geo*-centrism – I ought to say Gaia-centrism – has consequences! This time, we are not at all in the same world, as each of us can smell. The Earth, like the oak vats in a winery in Burgundy at harvest time, gives off a strong whiff of the action of micro-organisms. We bewildered onlookers find ourselves thrust smack in the middle of all this disequilibrium, and it is "the constant commotion of these fragile spaces" that ought to frighten us for real!

*

You'll say: fine, the image of the Earth is from now on fully active; it has indeed been turned into a real cartoon. But hasn't it been overanimated? This the second feature of the Gaia scenography that I would like to address. How did Lovelock manage to retrace the path between the twin pitfalls of reductionism and vitalism? Was he as clever as Pasteur, who managed to profile his micro-organism in such a way that it acted as much against the adherents to spontaneous generation as against chemists like Liebig?

At first glance, Lovelock seems to have managed rather badly, since the most common definition of the Gaia theory is that Gaia acts as *a single, unique coordinating agent*. Gaia would be the planet Earth considered as a living organism. This is often the way Lovelock presented his discovery:

> Gaia is the planetary life *system* that *includes* everything influenced by and influencing the biota. The Gaia system shares with all living organisms the capacity for *homeostasis* – the regulation of the physical and chemical environment at a *level* that is favourable for life.[61]

"System," "homeostasis," "regulation," "favorable levels," these are all quite treacherous terms. Is there then a superior order in addition to living organisms? The reader, however charitable, has a hard time finding a path through the numerous versions proposed by Lovelock. How are we to understand the following statement, where he asserts in the same breath that the Earth is and is not a unified whole? "When I talk of Gaia *as a superorganism*, I do not for a moment have in mind a *goddess* or some *sentient being*. I am expressing my *intuition*

---

[61] Lovelock, 2000, p. 56, emphasis added.

Gaia, a (finally secular) figure for nature 95

that the Earth behaves *as* a self-regulating system, and that the proper science for its study is *physiology*."[62]

But if it isn't a goddess, why call it Gaia? And what difference is there, for a "superorganism," between the status of "sentient being" and that of "self-regulated system"? This is to place a heavy burden on the poor little conjunction *as*, charged all by itself with preventing us from *really taking* Gaia to be a Whole. And yet, if I claim that Lovelock is circling around something as original as Pasteur's anti-Liebig, anti-Pouchet microbe, it is because Lovelock, too, fights to keep anyone from entrusting all the agencies he has detected to a new, higher level, that of the totality.

To understand why he has so much trouble expressing himself, we have to remember that sociology and biology have continually exchanged their metaphors, and that it is therefore extremely difficult to invent a new solution to the problem of organization.[63] All the sciences, natural or social, are haunted by the specter of the "organism," which always becomes, more or less surreptitiously, a *"superorganism"* – that is, a dispatcher to whom the task – or rather the holy mystery – of successfully coordinating the various parts is attributed.[64] Now the problem Lovelock saw very well is that, in the literal sense, in the objects that he studied, *there are neither parts nor a whole*.

As soon as you imagine parts that "fulfill a function" within a whole, you are inevitably bound to imagine, *also, an engineer* who proceeds to make them work together. Only in technological systems,

---

[62] Ibid., emphasis added. Bruce Clarke (personal communication) shows that this passage read differently in the first edition of the book: "When James Hutton in 1785 referred to the Earth as a superorganism, I do not for a moment suppose that he had in mind a goddess or some sentient being. I think he was using the only language then available to him to express his intuition that the Earth behaved as a self-regulating system, and that the proper science for its study was physiology" (1991, p. 57).

[63] Along with many other authors, especially Dario Gamboni ("Composing the Body Politic: Composite Images and Political Representation, 1651–2004," 2005), I have explored this continual criss-crossing (see Bruno Latour and Peter Weibel, *Making Things Public: The Atmospheres of Democracy*, 2005). This exchange of faulty procedures has continued to amaze me ever since my work with Shirley Strum, "Human Social Origins: Oh Please, Tell Us Another Story" (1986).

[64] This refusal to conceptualize organization on two levels is the fundamental tenet of the actor-network theory, which remains as difficult as ever for the social sciences to grasp. Yet this is also true for the biological sciences, which borrow from political theory the same schemas as those of sociology. See Bruno Latour, *Reassembling the Social: An Introduction to Actor-Network Theory* (2005), and a more technical paper by Bruno Latour, Pablo Jensen, Tommaso Venturini, Sebastian Grauwin, and Dominique Boullier, "The Whole Is Always Smaller Than its Parts – A Digital Test of Gabriel Tarde's Monads" (2012).

in fact, can we distinguish between parts and a whole.[65] This is even the definition of a technological act: on the basis of a *blueprint*, you can anticipate the *roles* that will be played by the elements in relation to a goal. One can obviously extend the technological metaphor to a body, a cell, or a molecule by behaving *as if* the functions "obeyed" a diagram. This technomorphism has been of great use to biology, especially in the study of animal societies.[66] But what do we do if we want to talk about the Earth *in its entirety*? The metaphor of the organism – that strange amalgam of social theory, a conception of the State, and machinism – is meaningless on this scale, unless we imagine a General Engineer, a very clumsy disguise for Providence, capable of giving each of these actors agency for the greatest good of all.

Now it is obvious that technological metaphors cannot be applied to the Earth in a lasting way: it was not fabricated; no one maintains it; even if it were a "space ship" – a comparison that Lovelock constantly contests[67] – there would be no pilot. The Earth has a history, but this does not mean that it was conceived. It is because there is no engineer at work, no divine clockmaker, that a *holistic* conception of Gaia cannot be sustained.[68] And as Gaia cannot be compared to a

---

[65] This is the fundamental and still badly understood point developed by Raymond Ruyer, in *Neofinalism* ([1952] 2016). Interestingly, considered in terms of its project and not its result, a technological system cannot be explained by a technological metaphor either – a point at the heart of Gilbert Simondon's enterprise! See *On the Mode of Existence of Technical Objects* ([1958] 2016). On the overall question of the limited capacity of technological metaphors to explain technology, see Bruno Latour, *Aramis, or the Love of Technology* ([1992] 1996).

[66] On the impossibility of using the notions of parts and whole for cells, see Jean-Jacques Kupiec and Pierre Sonigo, *Ni Dieu ni gène* (2000) (taken up again in a more accessible way in Pierre Sonigo and Isabelle Stengers, *L'évolution*, 2003); for monkey societies, see Shirley S. Strum, "Darwin's Monkey: Why Baboons Can't Become Human" (2012); for ants, see Deborah Gordon, *Ants at Work: How an Insect Society Is Organized* (1999).

[67] For example, in *The Revenge of Gaia: Earth's Climate in Crisis and the Fate of Humanity* (2006), p. 17. The technological metaphor of a space ship is all the clumsier in that, when catastrophes have occurred, we have seen the extent to which the unity of the technological system fails to correspond to the practice. See for example Diane Vaughan, *The Challenger Launch Decision: Risky Technology, Culture, and Deviance at NASA* (1996).

[68] This is also the limit of cybernetic interpretations of Gaia, which have simultaneously to pursue the technical metaphor – but then lose the specificity of Lovelock's argument – or slowly modify the metaphor – but then lose any precise connection with cybernetics taken as a science. This is the problem with which Bruce Clarke has been struggling; see *Earth, Life, and System: Evolution and Ecology on a Gaian Planet* (2015).

machine, it cannot be subjected to any sort of *re-engineering*.⁶⁹ As the activists say: "There is no Planet B." You can't fall back on any NASA toward which a crew in difficulty could turn, in a catastrophe, and that could be summoned by radio, by someone shouting: "Houston, we have a problem!"⁷⁰

The whole originality – and it's true, I recognize it – the whole difficulty – of Lovelock's enterprise is that he plunges head first into an impossible question: how to obtain effects of *connection* among agencies without relying on an untenable conception *of the whole*. He sensed that extending the metaphor of organism to the Earth was senseless, and that micro-organisms were nevertheless indeed *conspiring* by sustaining the long-term existence of this critical zone within which all living entities are combined. If he contradicts himself, it is because he is fighting with all his might to avoid the two pitfalls while trying to trace the connections without taking the Totality route. It is through this type of struggle that we recognize the greatness of researchers such as Pasteur and Lovelock.

All the more so in that Lovelock may well have been the first to ask himself such a question. Those whom he is fighting, for their part, have no trouble taking the Earth as a system, *always already unified* in advance: either they view it in its deanimated version – all the parts "passively obey the laws of nature"⁷¹ – or else they view it in its overanimated version – the parts work for the greatest glory of Life, that curious amalgam of soul, spirit, government, and god. The problem Lovelock is confronting escapes them completely: *how to follow the connections without being holistic*? It is in this sense that his version of the Earth System is anti-systematic: "There is only one Gaia but Gaia is not One."⁷²

Like Pasteur, Lovelock had to invent a new way of fine-tuning the agencies that populate the world, but he faced a supplementary difficulty: he had to find a way of creating a composition that encompassed – without unifying them in advance – all living entities

---

⁶⁹A point that needs emphasizing at a time when geo-engineering dreams purport to be getting it back on the right track. See Clive Hamilton, *Earthmasters: The Dawn of Climate Engineering* (2013).

⁷⁰An allusion to the end of Ron Howard's film *Apollo 13* (1995).

⁷¹Those who accuse Lovelock of conceptualizing the Earth as a unified whole fail to say that they too use an extraordinarily powerful unifier, since they have attributed to the laws of nature – in practice, to equations – the task of *compelling obedience* everywhere, on every point. The problem is how to dispense completely with the theme of obedience and mastery – that is, of government (the etymology of cybernetics).

⁷²Philip Conway, "Back Down to Earth: Reassembling Latour's Anthropocenic Geopolitics" (2016).

within the limits of the fragile envelope that he called Gaia. They all react "as" a superorganism but their unity cannot be attributed to any Governor figure. And this is so despite the attractiveness of technological metaphors like that of the thermostat, or of cybernetics, although Lovelock continued to play with these figures of speech (I shall come back to this point in the next lecture). How did he handle the problem? By abandoning the idea of parts! This was his central intuition; this, then, is what we need to understand.[73]

*

If, as a geophysicist, Lovelock was fighting against the geochemists, he was fighting just as much against the Darwinians, for whom organisms settle for "adapting themselves to" their *own* environments. For Lovelock, organisms, taken as the point of departure for a biochemical reaction, do not develop "in" an environment; rather, each one *bends* the environment around itself, as it were, the better to develop. In this sense, every organism intentionally manipulates what surrounds it "in its own interest" – the whole problem, of course, lies in defining that interest.[74]

This is the sense in which there cannot be, strictly speaking, any parts. No agent on Earth is simply superimposed on another like a brick juxtaposed to another brick. On a dead planet, the components would be placed *partes extra partes*; not on Earth. Each agency modifies its neighbors, however slightly, so as to make its own survival slightly less improbable. This is where the difference between geochemistry and geobiology lies. It means not that Gaia possesses some sort of "great sensitive soul," but that the concept of Gaia captures the distributed intentionality of all the agents, each of which modifies its surroundings for its own purposes.

Up to here, we have nothing really extraordinary. It is only if we push this idea to its limits, as the obstinate Lovelock does, that it becomes truly fertile. All historians acknowledge that humans have adjusted their environment to suit their needs: the nature in which they live is artificial through and through. Lovelock – an inventor,

---

[73] This problem depends in turn on another more fundamental hypothesis, a philosophical hypothesis advanced by Whitehead about the *penetrability* of entities; the same hypothesis is what lends interest to the notion of monad as renewed by Gabriel Tarde, in *Monadology and Sociology* ([1893] 2012), and actor-network theory.

[74] "Interest" here is taken in its etymological sense as what is situated "in between," between two entities – while keeping in mind that intentionality, will, desire, need, function, and force are only different figures for what is arrayed along a gradient expressing the same power to act, as I showed in the second lecture.

Gaia, a (finally secular) figure for nature

it must be remembered – does nothing more than extend this capacity for transformation to every agent, however small. Beavers, birds, ants, and termites are not the only ones who bend the environment around them to make it more favorable; so too do trees, mushrooms, algae, bacteria, and viruses. Is there a risk of anthropomorphism here? Of course; this is even what makes the reasoning so clever: the capacity of humans to rearrange everything around themselves is a *general property of living things*. On this Earth, no one is passive; the consequences *select*, so to speak, the causes that will act on them.

On this point, we have to increase our attention to the distribution of agency. What happens, in fact, if you extend intentionality to all agents?[75] Paradoxically, such an extension quickly wipes out all traces of anthropomorphism, since it introduces, at every level, the possibility of non-intentional retroactions. In fact, what is true for an actor taken as the starting point of the analysis is *equally true for all of the actor's neighbors*. If A modifies B, C, D, and X to benefit A's own survival, it is just as true that B, C, D, and X modify A in return. Animation is immediately propagated at all points.[76] Suppose that, as a good Darwinian, you take interest or profit as the final cause of every organism engaged in a struggle for its own survival: what can "final cause" mean if it is no longer "final," but *interrupted* at each point by the interposition of the just as robust intentions and interests of the *other organisms*?

The more you extend the notion of intentionality to all the actors, the less intentionality you will detect in the whole, even if you can observe more and more positive or negative retroactions, each having as little intentionality as the others![77] It seems that the moralists have

---

[75] The term "semiotics" is used, for example, by the naturalist Jakob von Uexküll, in *A Foray into the Worlds of Animals and Humans* ([1940] 2010), to describe living systems. For him, as for Lovelock, it is a question not of adding meaning to something that would be "strictly material," but of *not withdrawing meaning* from the intersecting mutual interests of living organisms in order, precisely, to make them comprehensible. This is the very method used by Vinciane Despret, *Penser comme un rat* (2009), as well as in *What Would Animals Say if We Asked the Right Questions?* ([2012] 2016), and a key method for Raymond Ruyer, *Neofinalism* ([1952] 2016).
[76] This was continually reinforced by Lovelock's collaboration with Lynn Margulis, as underlined in Bruce Clarke, "Gaia Is Not an Organism: Scenes from the Early Scientific Collaboration between Lynn Margulis and James Lovelock" (2012).
[77] In *Staying with the Trouble: Making Kin in the Chthulucene* (2016), Donna Haraway offers a good summary of Lynn Margulis's solution, "What happens when the best biologies of the twenty-first century cannot do their job with bounded individuals plus contexts, when organisms plus environments, or genes plus whatever they need, no longer sustain the overflowing richness of biological knowledges, if they ever did? What happens when organisms plus environments can hardly be remembered for the same reasons that even Western-indebted people can no longer

never seriously assessed the consequences of the Golden Rule: if we all "do unto others what we would want others to do unto us," the result is neither cooperation nor selfishness but the chaotic history we know very well because we are living in it![78] You can follow the undulations produced by a stone tossed into a pond, but not the waves produced by hundreds of cormorants plunging in all at once to catch fish. With Gaia, Lovelock is asking us to believe not in a single Providence, but in as many Providences as there are organisms on Earth. By generalizing Providence to each agent, he insures that the interests and profits of each actor will be *countered* by numerous other programs. The very idea of Providence is blurred, pixelated, and finally fades away. The simple result of such a distribution of final causes is not the emergence of a supreme Final Cause, but a fine *muddle*. This muddle is Gaia.

Here, too, the parallel with Pasteur is striking, since the latter's discovery was not so much the existence of microbes as the complex interactions of microbes with the terrain that they influence and that influence their development in return.[79] It was only because Pasteur had succeeded in showing that he could cause variations in the virulence of diseases by passing microbes through various species – rabbits, chickens, dogs, and horses – that he was finally able to convince doctors to recognize the role of microbes in the development of diseases.[80] Here, too, reductionism is defined not by the deanimated nature of the agent introduced into history but by the *number* of other agents that take part in the action.

Properly speaking, for Lovelock, and even more clearly for Lynn Margulis, there is *no longer any environment* to which one might adapt. Since all living agents follow their own intentions all along, modifying their neighbors as much as possible, there is no way to distinguish between the environment to which the organism is adapting and the point at which its own action begins. As Timothy Lenton,

---

figure themselves as individuals and societies of individuals in human-only histories?" (pp. 30–1). And later she "evokes the name of Gaia in the way James Lovelock and Lynn Margulis did, to name complex nonlinear couplings between processes that compose and sustain entwined but nonadditive subsystems as a partially cohering systemic whole" (p. 60). Amusingly, she concludes those passages by saying that this is why we should reject the word "Anthropocene," whereas I conclude that we should keep it precisely to "stay with the trouble"!

[78] Or, in John Dewey's lovely expression: "There is no mystery about the fact of the association" (*The Public and its Problems*, 1927, p. 23).
[79] This is exactly the point that allowed Dubos to relate Pasteur's microbiology to ecology, in his *Louis Pasteur* (1950).
[80] See Latour, 1988.

one of Lovelock's collaborators, emphasizes in a review article: "Gaia theory aims to be consistent with evolutionary biology and views the evolution of organisms and their material environment as so closely *coupled* that they form a *single, indivisible process*. Organisms possess environment-altering traits because the benefit that these traits confer (to the fitness of the organisms) outweighs the cost in energy to the individual."[81]

But let's be careful here: "single, indivisible" applies to the process of coupling, not to the results! Here we can see the particular charm of Lovelock's prose, and Margulis's. The inside and outside of all borders are subverted. Not because everything is connected in a "great chain of being"; not because there is some global plan that orders the concatenation of agents; but because the interaction between a neighbor who is actively manipulating his neighbors and all the others who are manipulating the first one defines what could be called *waves of action*, which respect no borders and, even more importantly, never respect any fixed scale.[82] These overlapping waves are the true actors that one ought to follow *all along*, wherever they lead, without being limited by the internal border of an isolated agent considered as an individual "within" an environment "to which" it would adapt.[83] The term is awkward, it is not Lovelock's, and yet these waves of action are the real brush strokes with which he seeks to depict Gaia's face.

*

Up to this point, Lovelock's argument is fully compatible with the Darwinian narratives, since each agent works for itself without being asked to abandon its own interest "for the benefit of a higher whole," which it would obviously be expected to do if there were a giant Dispatcher distributing functions to all parties. Without praise of

---

[81] Timothy Lenton, "Gaia and Natural Selection: A Review Article" (1998).

[82] There is no generally accepted term for it, but the phenomenon is recognizable in Tarde's use of the term "monad," in Ruyer's "absolute domains of survey" (Ruyer 2016, pp. 90–123), and in C. H. Waddington's "chreode" (*Biological Processes in Living Systems: Towards a Theoretical Biology*, [1972] 2012, vol. 4); it has been the object of numerous efforts on the part of researchers to get out of the customary paradigm common to sociology and biology that grasps entities only as parts of a whole – *partes extra partes*. See for example Deborah Gordon, "The Ecology of Collective Behavior" (2014).

[83] This is the argument of "symbiogenesis" in Margulis (1998), and again in Scott F. Gilbert and David Epel, *Ecological Developmental Biology: Integrating Epigenetics, Medicine, and Evolution* (2009).

sacrosanct self-interest, no Darwinism is thinkable.[84] But Lovelock begins to add something to the usual argument at the point where he asks what it really means for an agent to "calculate its interests."

The evolutionists have subjected Lovelock to a great deal of criticism with the counter-argument, at first glance unanswerable, that no one can tell how the organism Earth could manage to survive among a population of planets each struggling for its own survival – the standard format for evolutionary narratives.[85] They have thus indignantly rejected the idea of a "living planet." But this is because they attributed to Lovelock the idea of a *unified* planet, a superorganism, an idea that Lovelock in fact constantly combated. For him, there is no need whatsoever for the standard format in order to detect the ordinary action of evolution. The difficulty with which his opponents charge him is thus wholly imaginary. It depends entirely on the primal scene of evolutionism, which rests, on the one hand, on the idea that one can *assign limits* to the organism whose chances of survival one is claiming to calculate and, on the other hand, on the function of *ultimate arbiter* assigned to the environment in which selection occurs. Now, for Lovelock, there is no limit to the organism that would make its survival "calculable," and no independent arbiter, either, since he tries to do without both concepts, that of the isolated organism calculating its own interests and that of the inert whole to which it would adapt. Far from yielding to the critique of the neo-Darwinians, Lovelock inverts their paradigm: if there is a vestige of Providence, it is rather among the Darwinians that it is more likely to be found.[86]

Even if he was prepared to engage in the obligatory exercise of showing, thanks to the Daisy model,[87] that organisms in conflict

---

[84] The issue of calculating self-interest will come up again in the eighth lecture, but at that point it will serve to delimit the sovereignty of states.

[85] We have learned from the marvelous story-teller Stephen Jay Gould (*Wonderful Life: The Burgess Shale and the Nature of History*, 1989) that evolution is always first and foremost a form of narrative.

[86] The evolution, if I may call it that, of Edward O. Wilson, who shifted from the idea of a superorganism to socio-biology and then from there back to a superorganism (Bert Hölldobler and Edward O. Wilson, *The Superorganism: The Beauty, Elegance, and Strangeness of Insect Societies*, 2008), provides good evidence of the total failure of what is called "kin selection"; that notion first appeared as a biological principle, before it was understood that what was at stake was only the extension of economization to living beings. Biology has never fully extracted itself from Providence; like economics, it always needs the miracle of coordination. The Invisible Hand is always that of God.

[87] This model – a fairly simple one in the beginning, but later increasingly complicated – was used to demonstrate that homeostasis between two distinct organisms in competition was possible. The value of the demonstration was more metaphorical than explanatory, but Lovelock considered it very important. See Stephen H. Schneider,

could obtain homeostatic effects without a pre-established plan (which was fairly obvious), Lovelock was indeed attacking the way the biologists understand adaptation to an environment. This limit is quite clearly that of the *economic theory* used as a model for biology, a theory thanks to which one could distinguish between the *outside* and the *inside* of an agent. According to this theory, one always has to choose between the selfish individual and the integrated system – a dilemma that biologists have borrowed from the social sciences.[88] But what is so implausible in the idea of the "selfish gene" is not that genes are selfish – each agent pursues its own interest up to its sad end – but that one can calculate an agent's "viability" by *externalizing* all the other actors in what would constitute, for a given actor, its "environment." In other words, the problem with the selfish gene is the definition of the *self*.[89] This does not mean that it is necessary to mobilize a superorganism to which the actors would be compelled to sacrifice their well-being; it means only that life is more chaotic than the economists and the Darwinians had imagined, since every selfish goal is submerged by the selfish goals of all the others. Narratives based on natural selection offer a much too idyllic picture of natural history. The comparison to the muddle of Gaia reveals the merciless struggle for life for what it is: a domesticated and rationalized form of natural religion.[90]

The reason Darwin's secular intuition has been so often caricatured in a thinly disguised version of Providence is that the neo-Darwinians have pretended to forget that, if such a calculation does function in the human economy, it is by virtue of the continuous pressure of *accounting procedures* whose goal is to make functional – the technical term is to *perform* – the distinction between what a given agent

---

James R. Miller, Eileen Crist, and Penelope J. Boston, *Scientists Debate Gaia* (2008), and also the Wikipedia entry "Daisyworld" for references, including many films.

[88] Ever since Bernard Mandeville's *The Fable of the Bees: Private Vices, Publick Benefits* ([1714] 1962), there have been endless borrowings in efforts to "naturalize" a very particular version of economics; see Karl Polanyi, *The Great Transformation: The Political and Economic Origins of Our Time* ([1944] 2001).

[89] An allusion to the title of Richard Dawkins's well-known book *The Selfish Gene* (1976). The original difficulty is in the idea of individual: see Scott Gilbert, Jan Sapp, and Alfred Tauber, "A Symbiotic View of Life: We Have Never Been Individuals" (2012).

[90] It is not the reductionism that is shocking in the neo-Darwinian narratives, but the lack of reductionism and the constant appeal to the balance of nature and to the well-being of organisms. Behind natural selection, the benevolent hand of the Creator is recognizable in Darwin and in his successors as well. See Dov Ospovat, *The Development of Darwin's Theory: Natural History, Natural Theology, and Natural Selection, 1838–1859* (1995).

must literally *take into account* and what that agent must decide not to take into account.[91] Without these accounting procedures, it would be impossible to calculate profit and even more so to detach profit from its so-called environment. As soon as Darwinism is extended to all living beings, and thus to that which each one does to all the others on which it depends, calculating optimization becomes simply impossible.[92] Neither internalization nor externalization has any meaning. What one obtains instead are opportunities, chances, feedback loops, noise, and, yes, history. If there is no selfish gene, it is because the self literally has no limit!

*

The evolutionists, in other words, have been in a hurry to treat Gaia as a whole without even trying to understand what Lovelock was exploring. In this way they revealed their entrenched attachment to the classical opposition between the individual and the totality, the actor and the system, a political, sociological, and religious obsession, but with hardly any relation to what can be expected of living beings in the world. The critics of neo-Darwinism were beginning to suspect this: the economy of nature is not the same as that of humans. In the next lecture I shall look more closely at the evolutionists' approach, but to conclude this one I would like to point out another consequence of Lovelock's: if he does without the idea of parts to explain an organism, *he also does without the idea of a whole* that would account for differences in scale.

As soon as we abandon the borders between the outside and the inside of an agent, by following these waves of action we begin to *modify the scale* of the phenomena considered. It is not that we would change levels and make a crude leap from the individual to the "system"; it's simply that we have to *abandon* both viewpoints as being equally inoperative. This is where Margulis plays such an important role. Actually, the connection between the two writers

---

[91] This is the principle of analysis of the *economization* of the collectives pursued by Michel Callon, ed., *The Laws of the Markets* (1998b), Donald MacKenzie, *Material Markets* (2009), and many of their colleagues. See Michel Callon, ed., *Sociologie des agencements marchands* (2013); for the link with politics, see Dominique Pestre, "Néolibéralisme et gouvernement: retour sur une catégorie et ses usages" (2014).

[92] The implausibility of calculating through redistribution between the inside and the outside is a source of the renaissance of the notion of "commons," in Elinor Ostrom, *Governing the Commons: The Evolution of Institutions for Collective Action* (1990).

ought to have alerted the critics, since Margulis upset the understanding of minuscule organisms as surely as Lovelock upset those of the Earth.[93] This is indeed evidence that the very notions of organism, scale, parts and whole, were what they were both attacking. Together, they were trying to get along entirely without the notion of levels layered on top of one another.

One example of such a wave of action has taken on an emblematic character in Lovelock's saga: the gradual appearance of oxygen at the end of the Archean age. Is the oxygen we breathe pertaining to a larger layer than our individual level? Are we "in" the atmosphere? Not really, since this dangerous poison is itself the unforeseen consequence of the action of micro-organisms that have given to other actors – from which we descend – the opportunity to develop. In other words, we *are* the atmosphere. Oxygen is a relative newcomer, a massive case of pollution that was grasped by new forms of life as a golden opportunity, after it had annihilated billions of earlier forms of life:

> Oxygen is poisonous, it is mutagenic and probably carcinogenic, and it thus sets a limit to life spans. But its presence also opens abundant *new opportunities* for organisms. At the end of the Archean, the appearance of a little free oxygen would have *worked wonders* for those early ecosystems.... Oxygen would have *changed* the environmental chemistry. The oxidation of atmospheric nitrogen to nitrates would have increased, as would the weathering of many rocks, particularly on the land surfaces. This would have *made available nutrients* that were previously scarce, and so *allowed an increase* in the abundance of life.[94]

If we live now in an atmosphere dominated by oxygen, this is not the result of a preordained feedback loop. It is because the organisms that transformed this mortal poison into a powerful accelerator of their metabolism *have multiplied*. Oxygen is there not simply as a component of the environment but as the *extended consequence* of an event continued to our day by the proliferation of organisms. In the same way, it is only since the invention of photosynthesis that the Sun has come to play a role in the development of life. Both phenomena are the consequences of historical events that will last no longer than the creatures that sustain them. And, as Lovelock's passage shows, each event opens up "new perspectives" for other creatures.

---

[93] Margulis does this by showing to what extent the cellular organism itself, far from being an indivisible atom, is on the contrary the result of a vast composition of organisms recruited during a very long history; see Margulis and Sagan (1997). Without Margulis, it is probable that the Gaia hypothesis would not have been able to combat the cybernetic metaphor effectively.

[94] Lovelock, 2000a, p. 114.

The crucial point is that scale does not intervene in passing from a local level to a higher point of view. If oxygen had not spread, it would have remained a dangerous pollutant *in the neighborhood* of the archeobacteria. The scale, in this case, was engendered by the very success of the living forms able to benefit from its sudden abundance. If there is a climate for life, it is not because there exists a *res extensa within* which all creatures reside passively. The climate is the historical result of reciprocal connections, which interfere with one another, among all creatures as they grow. It spreads, diminishes, or dies with them.[95] "Nature," in the classical conception, had levels, strata; it was possible to pass from one to another according to a continuous well-ordered process of "zooming."[96] Gaia subverts the levels. There is nothing inert, nothing benevolent, nothing external in Gaia. If climate and life have evolved together, space is not a frame, not even a context: *space is the offspring of time*. Exactly the opposite of what Galileo had begun to unfurl: extending space to everything in order to place each actor within it, *partes extra partes*. For Lovelock, such a space no longer has any sort of meaning: the space in which we live, that of the critical zone, is the very space toward which we are conspiring; it extends as far as we do; we last as long as those entities that make us breathe.

It is in this sense that Gaia is not an organism, and that we cannot apply to it any technological or religious model. It may have an order, but it has no hierarchy; it is not ordered by levels; it is not disordered, either. All the effects of scale result from the expansion of some particularly opportunistic agent grabbing opportunities to develop as they arise: this is what makes Lovelock's Gaia totally secular. If

---

[95] In his fine chapter on Tarde, Pierre Montebello shows that the same argument holds for the extension and "success" of monads. "[Tarde] conceived of the success of an invention as a contamination capable of winning *little by little* the confines of an immense territory. This is what has happened with matter, since triumphant atoms have *managed to spread* their power of attraction over all the nebulae. They have *shaped this physical milieu* that extends in the infinity of space, broken the primitive equilibrium of things, imposed the law of attraction everywhere. The physical stratum has resulted from a political domination, from the *supremacy of a desire* over the entire set of monads.... Here, the image of the political supplants that of the theological" (*L'autre métaphysique: essai sur Ravaisson, Tarde, Nietzsche et Bergson*, 2003, p. 152, emphasis added).

[96] The ordering of entities according to their dimensions within a *res extensa* does not correspond to any real experience, even though it ended up being conflated with the scientific image of the world thanks to films such as *The Powers of Ten* (see Philip Morrison, Phylis Morrison, and the Office of Charles and Ray Eames, *Powers of Ten: A Book about the Relative Size of Things in the Universe and the Effect of Adding Another Zero* (1982), and its critique in Bruno Latour and Christophe Leclercq, eds, *Reset Modernity!*, 2016).

it is an opera, it depends upon constant observation that has neither a score nor an ending, and it is never performed twice on the same stage. If there is no frame, no goal, no direction, we have to consider Gaia as the name of the process by which variable and contingent occurrences have made *later* events more probable. In this sense, Gaia is a creature no more of chance than of necessity. Which means that it closely resembles what we have come to regard *as history itself.*

*

Have we finally sketched Gaia's face? No, of course not. I hope at least that I've said enough to convince you that seeking "Man's place in Nature" – to fall back on an outmoded expression – is not at all the same task as learning to participate in the geohistory of the planet. By bringing into the foreground what was formerly confined to the background, we are not hoping to live at last "in harmony with nature." There is no harmony in that contingent cascade of unforeseen events, nor is there any "nature" – at least not in this sublunary realm of ours. By the same token, learning how to situate human action in this geohistory does not amount, either, to "naturalizing" humans. No unity, no universality, no unchallengeability, no indestructibility can be invoked to simplify the geohistory in which humans find themselves immersed.

The drama is that the intrusion of Gaia is happening at a moment when the figure of the human has never appeared so ill-adapted to take it into account. Whereas we ought to have as many definitions of humanity as there are ways of belonging to the world, this is the very moment when we have finally succeeded in universalizing over the whole surface of the Earth the same economizing and calculating humanoid. Under the name of *globalization*, the culture of this strange GMO – whose Latin name is *Homo oeconomicus* – has spread everywhere. At the very moment when we have a desperate need for other forms of homodiversity! Bad luck, truly: we have to confront the world with humans reduced to a very small number of intellectual competences, endowed with brains capable of making simple calculations of capitalization and consumption, to whom we attribute a very small number of desires and who have finally been persuaded to view themselves as individuals, in the atomic sense of the word.[97] At the very moment when we should be remaking politics,

---

[97] This is what justifies Naomi Klein's use of the word "capitalism" as the form most foreign to the habitation of the planet in *This Changes Everything: Capitalism vs. the Climate* (2015).

we have at our disposal only the pathetic resources of "management" and "governance." Never has a more provincial definition of humanity been transformed into a universal standard of behavior.[98] At the very moment when we ought to be loosening the grip of the first Nature, the second Nature of Economics is imposing its iron cage more strictly than ever.

This disconnect between the old definitions of humanity and what humans must now confront is probably at the origin of the troubling impression that history, or rather historicity, has changed sides. As long as modernism maintained its grip, the "humans" were happy to live divided between, on one side, the "realm of necessity" – the linking of causes and consequences – and, on the other, the "realm of freedom" – the creations of law, morality, liberty, and art. They were exchanging the constraining necessity of Nature for the proliferation of cultures. "*Mono*naturalism," on the one hand, "*multi*culturalism" on the other.[99] Now, the geohistorical event that I am seeking to define has turned this division completely upside down. The power of invention and surprise has shifted from the humans to the nonhumans, as Fredric Jameson notes in a famous quip: "Nowadays it seems easier to imagine the end of the world than to imagine the end of capitalism!"[100]

Can you recall how much energy the social sciences have expended to fight the dangers of biological reductionism and naturalization? Today, it seems difficult to tell whether we gain more freedom of movement if we turn toward nature or toward culture. What is certain is that the glaciers seem to be shrinking more quickly, the ice is melting more rapidly, species are disappearing at a faster pace than the majestic processes of politics, consciousness, and sensibility are progressing. Shelley would be hard put to sing his song today:

> The *everlasting universe* of things
> Flows through the mind, and rolls its rapid waves,
> Now dark – now glittering – now reflecting gloom –
> Now lending splendour, where from secret springs

---

[98] To such an extent that the idea of "commons" looks like a bizarre novelty today! On the history of this actually tragic loss of bearings, see a remarkable article by Fabien Locher, "Les pâturages de la guerre froide: Garrett Hardin et la 'tragédie des communs'" (2013).

[99] Bruno Latour, "The Recall of Modernity – Anthropological Approaches" (2007b).

[100] Jameson's exact wording: "Someone once said that it is easier to imagine the end of the world than to imagine the end of capitalism. We can now revise that and witness the attempt to imagine capitalism by way of imagining the end of the world" (Fredric Jameson, "Future City," 2003, p. 76).

> The source of human thoughts its tribute brings
> Of waters – with a sound but half its own,
> Such as a feeble brook will oft assume,
> In the wild woods, among the mountains lone,
> Where waterfalls around it *leap for ever*,
> Where woods and winds contend, and a vast river
> Over its rocks ceaselessly bursts and raves.[101]

"The everlasting universe of things"? We mustn't count on this any longer! We have stopped believing that waterfalls will "leap forever" and that "a vast river over its rocks" will "ceaselessly burst and rave." If there is still a chiasmus to nourish the blend of "melancholy" and "splendour" that accompanies the feeling of the sublime, it is not because we see poor ephemeral humans bustling about on the stage of an everlasting nature, but because we are compelled to see humans obstinately deaf and dispassionately seated, immobile, while the past setting of their past intrigues is disappearing at a frightening pace! Sublime or tragic, I don't know, but one thing is sure: it is no longer a spectacle that we can appreciate from a distance. We are part of it.

Oddly, the question henceforth is whether humans can rediscover a sense of the history that has been taken away from them by what they had viewed up to now as a mere frame deprived of any capacity to react. The bifurcation of Nature that Whitehead had so criticized finds itself overturned in the most unexpected way; the "primary qualities" are from now on characterized by sensitivity, activity, reactivity, and uncertainty, while the "secondary qualities" are characterized by indifference, insensitivity, and torpor. To such an extent that Whitehead's celebrated remark could be reversed: "so that the course of human history is conceived as being merely the fortunes of matter in its adventure through space."[102]

You might complain that this geohistorical account is marked by an excessive dose of anthropo*morphism*. I hope so! Certainly not in the old sense in which it would "project human values onto an inert

---

[101] Percy Bysshe Shelley, "Mont Blanc – Lines Written in the Vales of Chamouni" (1817). These lines were written during the famous 1816 sojourn in which Mary Shelley wrote *Frankenstein*. It is amusing to note that, if this most famous couple wrote prolifically during their stay, it was also because the eruption of the Mount Tambora volcano that year had ruined the summer vacation period, as recalled in Gillen D'Arcy Wood's delightful *Tambora: The Eruption that Changed the World* (2015).
[102] The original remark referred to the course of "nature" rather than of human history (Whitehead 1920). Let us recall that, between matter and materiality, we are obliged to choose.

world of mute objects" but, on the contrary, in the sense that it "gives humans a shape," or, as one can say in English, that it is beginning to *morph* humans into a more realistic image. One could complain about the dangers of anthropomorphism only in the era when humans strutting on stage were playing roles quite distinct from their surroundings. The roles of all the previous characters in the play are in the process of being redistributed. In any case, how could we avoid the traps of anthropomorphism, if it is true that we are living from now on in the era of the Anthropocene!

FOURTH LECTURE

The Anthropocene and the destruction of (the image of) the Globe

> The Anthropocene: an innovation • *Mente et Malleo* • A debatable term for an uncertain epoch • An ideal opportunity to disaggregate the figures of Man and Nature • Sloterdijk, or the theological origin of the image of the Sphere • Confusion between Science and the Globe • Tyrrell against Lovelock • Feedback loops do not draw a Globe • Finally, a different principle of composition • *Melancholia*, or the end of the Globe

I suppose that not too many of us were waiting impatiently, during the first six months of 2012, for the conclusions of the 34th International Geological Congress that was to take place in Brisbane during the summer. I confess that before then I had not been in the habit of following the work of this eminent academic body – even though their somewhat Nietzschean motto *Mente et Malleo* (By Thought and Hammer) would have suited my own profession very well! If I paid attention in 2012, it was because, like everyone else, I was eager for a clear decision about the epoch in which we are living from the International Commission on Stratigraphy, or, more precisely, the Subcommission on Quaternary Stratigraphy, a working group headed by Dr Jan Zalasiewicz of the University of Leicester.

Defining a historical epoch, and doing so officially, is no small matter! Were they going to declare that the Earth had officially entered into a new epoch, or not?¹ And, if the answer was yes, what was the precise date of entry? The stakes are enormous: for the first time in geohistory, someone was going to make the solemn declaration that the most important force shaping the Earth was that of humanity *taken as a whole and as a single unit*. Hence the name proposed, the Anthropocene (*cene* for "new," *anthropos* for "human"). The Zeitgeist determined by a subcommission? You see why I found the suspense unbearable!²

As I was expecting something solemn, I was a little disappointed when I read the summary report on the Brisbane meeting:

> The "Anthropocene" is currently being considered by the Working Group as a potential geological *epoch*, i.e. at the same hierarchical level as the Pleistocene and Holocene epochs, with the implication that it is within the Quaternary Period, but that the *Holocene has terminated*.³

"Potential" isn't very decisive. On the other hand, to declare that we are no longer living in the Holocene is more radical, since it has been precisely during these eleven thousand years of relative stability between two glaciations that human beings, or, more accurately, civilizations, have been able to develop.⁴ As long as we remained in the Holocene, the Earth remained stable and in the background, indifferent to our histories. It was business as usual, as it were. In contrast,

---

¹See Christophe Bonneuil and Jean-Baptiste Fressoz, eds, *The Shock of the Anthropocene: The Earth, History, and Us* (2016). (I am using the term "epoch" in a non-technical sense. Geologists distinguish time by segments in decreasing order: eons, eras, periods, epochs, and ages.)
²The crucial importance of the Anthropocene is that it attributes practical – that is to say, stratigraphic – truth to the notion of epoch as studied by a historian (but not a geohistorian), Hans Blumenberg, in *The Legitimacy of the Modern Age* ([1976] 1983). The Middle Ages were not viewed as "middle" by anyone living at the time, nor was Antiquity understood as "antique." But when the modern age was defined, explicitly, as the *modern* age, no one knew that it would end up being precisely defined by a subcommission on stratigraphy. In titling his book *The Archaeology of Knowledge* ([1966] 1972), Michel Foucault hadn't anticipated that the archaeological concept would be taken literally! This is another example of the great universal law of history according to which the figurative tends to become literal.
³Subcommission on Quaternary Stratigraphy, "What Is the 'Anthropocene'? – Current Definition and Status" (2011, emphasis added).
⁴The choice of beginning date – from very remote (since the appearance of *Homo faber*) to quite recent (since the industrial revolution) or very recent (since the Second World War) – correlates with profound political and moral differences. The more remote the date, the less the current forms of capitalism are at issue and thus the more responsibilities are diluted. It amounts to settling for saying that, "where there is humanity, there is human influence."

The Anthropocene and the destruction of the Globe   113

if "the Holocene has terminated," this is proof that we have entered into a new period of instability: the Earth is becoming sensitive to our actions and we humans are becoming, to some extent, geology! As we can see, a decision like this requires careful reflection. If stratigraphy has revolutionized the history of the Earth, it is in part thanks to the care with which geologists treat issues of nomenclature. It is thus out of the question that just anyone may be allowed to determine haphazardly the name of the first stratum of rock he or she comes across. The report goes on:

> Broadly, to be accepted as a formal term the "Anthropocene" needs to be (a) scientifically justified (i.e. the "geological *signal*" currently being produced in strata now forming must be sufficiently *large, clear and distinctive*) and (b) *useful* as a formal term to the scientific community. In terms of (b), the currently informal term "Anthropocene" has already proven to be very useful to the *global change research community* and thus will continue to be used, but it remains to be determined whether *formalisation* within the Geological Time Scale would make it more useful or broaden its usefulness to other scientific communities, such as the geological community.[5]

To advance a proposal for naming a geological epoch through the bureaucracy of the International Geological Society is as tortuous as getting a law passed through the committees of a parliament or promoting the beatification of a saint through Vatican diplomacy. And, even if the stratigraphers agree to give humanity a decisive role, they still have to reach agreement on the date and on the marker that will allow all specialists throughout the world to recognize it in the rocks:

> The beginning of the "Anthropocene" is most generally considered to be at c. 1800 CE, around the beginning of the Industrial Revolution in Europe (Crutzen's original suggestion);[6] other potential candidates for time boundaries have been suggested, at both earlier dates (within or even before the Holocene) or later (e.g. at the *start of the nuclear age*).[7] A formal "Anthropocene" might be defined either with reference

---

[5] Subcommission on Quaternary Stratigraphy, 2011, emphasis added.
[6] An article by Paul J. Crutzen and Eugene F. Stoermer, "The 'Anthropocene'" (2000), triggered a major literary movement and led to the creation of several specialized journals: *Anthropocene*, the *Anthropocene Review*, *Elementa: Science of the Anthropocene*, and so on.
[7] A recent article confirms the date of July 16, 1945, the date of the first nuclear explosion, without taking a position on the underlying principle; it simply emphasizes the convenience of being able to identify the geological transition, everywhere in the world, thanks to the signature left by the newly introduced artificial radioactivity. See Jan Zalasiewicz, Mike Walker, Phil Gibbard, and John Lowe, "When Did the Anthropocene Begin? A Mid-Twentieth Century Boundary is Stratigraphically Optimal" (2015).

to a particular point within a stratal section, that is, a Global Stratigraphic Section and Point (GSSP), colloquially known as a *"golden spike"*; or, by a designated time boundary (a Global Standard Stratigraphic Age).[8]

A flood of technical questions that still do not allow us to find out whether or not the Holocene is over and whether the New Climate Regime identified in the earlier lectures has a correlate in the rocks. For I had forgotten that geologists are in the habit of taking their time and speaking of millions and billions of years. It took them nearly a half-century, for example, to decide on the Quaternary Era! That is why, indifferent to the pressure coming from secular voices like mine that were eager to know for certain whether the news was official or not, they calmly noted in their conclusion that they had had to defer their final vote for at least four years! "The Working Group has applied for funding to allow further discussion and networking, and is working to reach a consensus regarding formalisation by, it is hoped, the 2016 International Geological Congress."[9]

Note the nonchalant expression "working to reach a consensus" – as well as the irritating habit researchers have of always requesting more funds.[10] You can understand my disappointment: it is as though we had all the time in the world to decide on the date that attributes to humans responsibility for having become a geological force!

While the decision is pending, the papers published by Zalasiewicz's working group offer to anyone willing to read them a fascinating example of the redistribution of agency that we are following in these lectures. Here we have it, the metamorphic zone I've been trying to designate: all human activities turn out to be transformed, in part, into geological forms; everything that we used to call bedrock is beginning to be humanized – or, in any case, to bear traces of a tempestuously remodeled humanity! It is no longer a question of landscapes, of the occupation of land, or of local impact. From now on,

---

[8] Subcommission on Quaternary Stratigraphy, 2011, emphasis added.
[9] Unfortunately, four years later, in September 2016, almost exactly the same scene took place in South Africa, where the same working group, during the same International Geological Congress, even though it had accumulated much better arguments and data, was not able to reach a conclusion acceptable to the other stratigraphic commissions in charge of the decision. For the new data, see Colin N. Waters, Jan Zalaciewicz, et al. "The Anthropocene Is Functionally and Stratigraphically Distinct from the Holocene" (2016).
[10] See the fascinating project carried out by the Haus der Kulturen der Welt (HKW) in Berlin, "The Anthropocene Curriculum" (n.d.), which includes videos contributed by the project's principal authors. See also the many interviews on the Portail des humanités environnementales, www.humanitesenvironnementales.fr/fr/les-ressources/les-grands-entretiens.

the comparison is made on the scale of terrestrial phenomena. With its increase in energy expenditure, human civilization now "runs," so to speak, at seventeen terawatts, twenty-four hours a day, which ends up making it comparable to the expenditure of energy of volcanos or tsunamis – obviously more violent, but over short periods of time. Certain calculations even end up comparing the power of human transformation to that of plate tectonics.[11]

It is as though the stratigraphers, transporting themselves into the future through an effort of imagination, were undertaking a thought experiment that allowed them to deduce retrospectively, from the rock layers that are beginning to accumulate, what the so-called human epoch had been like.[12] In the rocks, in fact, everything can be seen: the modification, by dams, of the sedimentation of rivers; changes in ocean acidity; the introduction of previously unknown chemical products; the composite ruins of vast infrastructures unlike anything that came before; changes in the rhythm and nature of erosion; variations in the nitrogen cycle; the continual growth of atmospheric CO_2, not to mention the sudden disappearance of living species during what biologists are resigned to calling the "sixth extinction."[13] Everything can be identified all the more legibly in sediments because, as of July 16, 1945, the clear radioactive signals left by atomic explosions offer a serious candidate for the famous "golden spike," easy to detect throughout the world, and they may well allow the geologists to reach consensus.

Each item on the list, and this is what is most fascinating, could have been found throughout the nineteenth and twentieth centuries in narratives boasting of the fabulous exploits of Mankind transforming the Earth the better to master it. With just one difference: the tone is no longer triumphal; there is no longer any question of "mastering" nature. Instead, the focus is on searching the sedimentary ruins for traces of earlier humans who had been *turned to stone*. As in a new master–slave dialectic, features of both, human and stone, end up melding. *Anthropo*morphism of the critical zones, *petro*morphism of

---

[11] In *Eating the Sun* (2007), Oliver Morton estimates the energy of human civilization at a given moment to be 17 TW. If the entire planet lived in the American manner, an expenditure of 90 TW would be required. The energy released by tectonic plates (heat and movement) is estimated, in comparison, to be 40 TW, while primary energy – of biological origin, on earth and in the oceans – is estimated at 130 TW. All this remains negligible, obviously, compared to the 130,000 TW of energy available on Earth through the action of the sun alone.

[12] Jan Zalasiewicz's book *The Earth after Us: What Legacy Will Humans Leave in the Rocks?* (2008) describes this imaginary scene with panache.

[13] Zalasiewicz et al., 2015.

humans. In any case, we have a fusion of geohistorical forces in what truly resembles a witch's cauldron.

This would be amusing if it were not so dramatic, but what gives the members of the subcommission the most pause is the mix of time scales they have to confront. Remember how we were taught in school to stand in awe before the *slow rhythm* of geological time? At a moment when we could hardly imagine even reaching the age of twenty, our teachers bent over backwards to find good pedagogical devices that could abolish the indefinite distance that separated us from the era of the dinosaurs or the epoch of Lucy.[14] And now, suddenly, in a complete reversal, we see geologists stunned by the rapid rhythm of geo-human history, a rhythm that forces them to place their "golden spike" in a segment of two hundred or even just sixty years (depending on whether they choose a recent or very recent temporal border marker to delineate the emergence of the Anthropocene). The formula "geological time" is now used for an event that has come and gone more quickly than the Soviet Union! As though the distinction between history and geohistory has suddenly disappeared, with carbon and nitrogen cycles taking on as much importance on the cosmic scale as the last glaciations or the Manhattan Project.[15]

Let's allow the specialists in stratigraphy to proceed at their own pace, and wait patiently for them to make a decision. Given the importance of what is at stake, we cannot hold it against them if they ask for a little more time in order to adjust the acceleration of time, even if it means adopting the pace of a representative of the academic bureaucracy!

*

What makes the Anthropocene an excellent marker, a "golden spike" clearly detectable beyond the frontier of stratigraphy, is that the name of this geohistorical period may become the most pertinent philosophical, religious, anthropological, and – as we shall soon see – political concept for beginning to turn away for good from the notions of "Modern" and "modernity."

---

[14]Thus reproducing the long history of the extension of time by geologists, archaeologists, exegetes, and other learned scholars during the eighteenth and nineteenth centuries, a story told by Martin Rudwick in *Earth's Deep History: How it Was Discovered and Why it Matters* (2014).
[15]This crossing of historicities that had been totally incompatible before is what first attracted Dipesh Chakrabarty's attention; see his "The Climate of History: Four Theses" (2009).

I find it enticing that this oxymoron linking geology and humanity should be the product of the cogitations of serious geologists who, until recently, had been totally indifferent to the ins and outs of research in the human and social sciences. No postmoderrn philosopher, no anthropologist, no liberal theologian, no political thinker would have dared measure the influence of humans *on the same scale* as rivers, volcanos, erosion, and biochemistry. What "social constructivist," determined to show that scientific facts, power relations, or inequalities between the sexes are "only" historical episodes manufactured by humans, would have dared say the same thing about the chemical composition of the *atmosphere*? What literary critic would have extended the principles for deconstructing texts to the sedimentary strata revealing in all the deltas of the planet the irrefutable traces of *erosion* caused by humans?[16]

At the very moment when it was becoming fashionable to speak of the "post-human" in the blasé tones of those who know that the time of the human is "outdated," the "Anthropos" has come back – and with a vengeance – owing to the thankless empirical work of researchers whose lack of culture intellectuals like to mock by calling them mere "naturalists." Despite all their sophistication, the various fields of the humanities, obsessed as they have been with defending the "human dimension" against the "illegitimate encroachment" of science and the risks of excessive "naturalization," could not detect what the *historians of nature* have to be credited with bringing to light.[17] By giving a totally new *dimension* to the very notion of "human dimension," these historians are proposing the most radical term of all for putting an end to anthropocentrism as well as to the old forms of naturalism; they are thus completely reconstituting the role of human agents. The magazine *The Economist* was quite right to use this slogan on its cover in 2011: "Welcome to the Anthropocene!"[18]

In light of this conceptual advance, it is only fair to pay respectful homage to all geoscientists. Their profession well deserves its motto "*Mente et Malleo*," since it is thanks to the intelligent handling of this hammer that we have begun to realize that our most precious values, when adroitly tapped, emit a rather hollow sound! I am no

---

[16] The amount of erosion of human origin has been deemed comparable to erosion produced by natural forces! See J. R. Ford, S. J. Price, A. H. Cooper, and C. N. Waters, "An Assessment of Lithostratigraphy for Anthropogenic Deposits" (2014).
[17] The ancient and venerable term "natural history," which served as a label for countless "naturalists" for centuries, from Pliny through Buffon to Darwin, takes on a quite different meaning as soon as we stress the word "history" and relate it to human history. Scientists have indeed become the historians of nature.
[18] "Welcome to the Anthropocene!" (2011), p. 13.

longer astonished that Deleuze and Guattari, astute connoisseurs of the "philosopher with a hammer," were prescient enough to draw up a "geology of morality."[19]

It goes without saying that this disruption in the very definitions of the best established categories was immediately misunderstood – and for the same reason that Lovelock's efforts to extract his Gaia from the old idea of "nature" have been drowned in sarcasm. The Nature/Culture format is so powerful that people have rushed to interpret the Anthropocene as the simple superposition – or even the dialectical reconciliation – of "nature" and "humanity," each one taken as a whole; or even as a vast conspiracy on the part of scientists to "naturalize" humanity by transforming it into a stone statue; or, conversely, as an undue politicization of science.[20] It seems more interesting to me to seek to welcome this innovation coming from scientists rather than to bury it at once with yet another critique of naturalization that would increase our risk of losing the opportunity to understand the New Climate Regime.

As it happened, the major science journal *Nature*, four years after *The Economist*, put the Anthropocene on its cover as well.[21] One of the drawings featured in the accompanying article offers a great opportunity to find out whether we are capable of putting new wine in old bottles. The illustration uses the familiar principle of representation known as the "Arcimboldo effect,"[22] in which the earth sciences provide themes used to redraw a still recognizable face.

This image can be used as a personality test: in it do you see the petrification of a human face or, on the contrary, an anthropization of Nature? At first glance, it has the look of a hybrid. Yet, if we look more closely, nothing connects in the highly muddled distribution of features: are we seeing mummy wrappings, scarification, war paintings, tattoos, soil stratifications, or, rather, a blend of the *Carte du Tendre* (a seventeenth-century French allegorical map)

---

[19] See the well-known chapter titled "10,000 B.C.: The Geology of Morals (Who Does the Earth Think It Is?)," in Gilles Deleuze and Félix Guattari, *A Thousand Plateaus: Capitalism and Schizophrenia* ([1980] 1987).
[20] If the label is ultimately rejected, it will probably be because of the excess of interest on the part of intellectuals, philosophers, artists, and activists in a term that the geologists, by definition, have not managed to keep for themselves, owing to the *Anthropos* that they themselves have introduced. I am not aware of any artists or activists mobilizing in favor of the Proterozoic!
[21] *Nature*, March 11, 2015.
[22] Pontus Hultén, ed., *The Arcimboldo Effect: Transformations of the Face from the 16th to the 20th Century* (1987), published on the occasion of the exhibition "The Arcimboldo Effect" at the Palazzo Grassi, Venice.

The Anthropocene and the destruction of the Globe 119

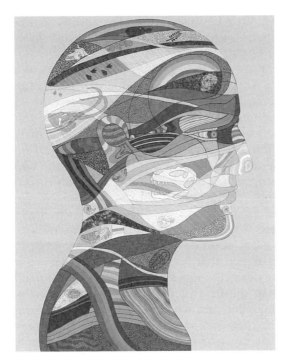

Figure 4.1 Drawing by Jessica Fortner to illustrate an article on the Anthropocene in *Nature*, March 11, 2015.

and a geological inventory designed to shape a colossal stone giant who, like the commendatore in Mozart's *Don Giovanni*, is getting ready to twist our arms to invite us to a deadly new banquet? The journal *Nature* demonstrates rather effectively that it has missed the point, since its cover story is titled "The Human Epoch," whereas the illustration clearly announces, with fanfare, the disappearance of the human! For my part, I see it rather as evidence of the attraction that this zone holds for journalists and illustrators, this metamorphic zone that we have learned to recognize and that is leading us, little by little, beneath and beyond the superficial characterizations, to a radically new distribution of the forms granted to humans, societies, nonhumans, and divinities.

*

Even if the competent institutions of the International Geological Association do not end up voting to adopt "Anthropocene" as the official label for the epoch in which we find ourselves, it is still worth taking advantage of the occasion to continue the work of *disaggregating*, little by little, all the ingredients that contributed to the joint characterization of people and things under the Old Climate Regime.

One thing is certain: the old role of "nature" has to be completely redefined. The Anthropocene directs our attention toward much more than the "reconciliation" of nature and society into a larger system that would be unified by one or the other. In order to bring about such a dialectical reconciliation, we would have to have accepted the dividing line between the social and the natural – the Dr Jekyll and Mr Hyde of modern history (I'll let you decide which is which). But the Anthropocene does not "go beyond" this division: it circumvents it entirely. The geohistorical forces *ceased to be the same* as the geological forces as soon as they fused at multiple points with human actions. Where we were dealing earlier with a "natural" phenomenon, at every point now we meet the "Anthropos" – at least in the sublunary region that is ours – and, wherever we follow human footprints, we discover modes of relating to things that had formerly been located in the field of nature. For example, if we follow the nitrogen cycle, where are we going to place the biography of Franz Haber and the chemistry of plant bacteria?[23] If we draw the carbon cycle, who can say when Joseph Black comes on stage and when the chemists drop out of the game?[24] Even following the course of rivers, you're going to find human influence everywhere.[25] And if, in Hawaii, you come across rocks made partly of lava and partly of a new substance, plastic, how are you going to draw the line between man and nature?[26]

For each of these aforementioned objects of the natural world, cycles like these oblige us rather to feel the effect of a finger running along a Moebius strip. We are gradually forced to *redistribute* entirely what had formerly been called natural and what had been called social or symbolic. Do you remember the gap between "physical" and "human" geography, thought to be unbridgeable, or the one between "physical" and "cultural" anthropology? The distinction between the social sciences and the natural sciences is totally blurred. Neither nature nor society can enter intact into the Anthropocene, waiting to be peaceably "reconciled." What happened to the landscape, for

---

[23] Bernadette Bensaude-Vincent and Isabelle Stengers, *A History of Chemistry* ([1992] 1996).
[24] David Archer, *The Global Carbon Cycle* (2010a).
[25] Mark Williams, Jan Zalasiewicz, Neil Davies, Ilaria Mazzini, Jean-Philippe Goiran, and Stephanie Kane, "Humans as the Third Evolutionary Stage of Biosphere Engineering of Rivers" (2014).
[26] Angus Chen, "Rocks Made of Plastic Found on Hawaiian Beach" (2014).

earlier generations, is now happening to the whole Earth: its gradual artificialization is making the notion of "nature" as obsolete as that of "wilderness."[27]

But the disaggregation is more radical still on the side of the aforementioned humans. Here we encounter the full irony of giving the traditional face of the Anthropos such a new characterization.[28] It would be absurd in fact to think that there is a collective being, human society, that is the new *agent* of geohistory, as the proletariat was thought to be in an earlier epoch. In the face of the old nature – itself reconstituted – there is literally *no one* about whom one can say that he or she is *responsible*. Why? Because there is no way to *unify* the Anthropos as an actor endowed with some sort of moral or political consistency, to the point of charging it with being a character capable of acting on this new global stage.[29] No business-as-usual anthropomorphic character can participate in the Anthropocene: this is where the whole interest of the notion lies.

Speaking of the "anthropic origin" of global warming is meaningless, in fact, if by "anthropic" we mean something like "the human species." Who can claim to speak for the human in general without arousing a thousand protests at once? Indignant voices will be raised to say that they do not hold themselves responsible in any way for these actions on the geological scale – and they will be right![30] The Indian nations deep in the Amazonian forest have nothing to do with the "anthropic origin" of climate change – at least so long as politicians running for election haven't given them chain saws. The same can be said of the poor residents in Bombay's shantytowns, who can only dream of having a carbon footprint more significant than the one left by the soot from their makeshift stoves.[31] No more than the

---

[27] See William Cronon, ed., *Uncommon Ground: Rethinking the Human Place in Nature* (1996), and Bronislaw Szerszynski, "The End of the End of Nature: The Anthropocene and the Fate of the Human" (2012).
[28] As we can see in an extraordinary book by Anna L. Tsing, *The Mushroom at the End of the World: On the Possibility of Life in Capitalist Ruins* (2015), about a mushroom!
[29] This is Chakrabarty's argument: "There is no 'humanity' that can act as a self-aware agent. The fact that the crisis of climate change will be routed through all our 'anthropological differences' can only mean that, however anthropogenic the current global warming may be in its origins, there is no corresponding 'humanity' that in its oneness can act as a political agent" ("Postcolonial Studies and the Challenge of Climate Change," 2012, p. 15).
[30] This is what makes Donna Haraway reject the term (*Staying with the Trouble: Making Kin in the Chthulucene*, 2016), but to stay with the trouble it's better to stay with the word.
[31] The role of soot in global warming seems to have been neglected. See Jeff Tollefson, "Soot a Major Contributing Factor to Climate Change" (2013).

worker forced to travel long distances by car because she hasn't been able to find affordable housing near the factory where she works: who would dare shame her on account of her carbon footprint?

This is why the Anthropocene, despite its name, is not an immoderate extension of anthropo*centrism*, as if we could boast of having really been changed into Supermen of sorts, flying about in red and blue costumes. It is rather the human as a unified agent, as a simple virtual political entity, as a universal concept, that has to be decomposed into several distinct *peoples*, endowed with contradictory interests, competing territories, and brought together by the warring agents – not to say warring divinities. The Anthropos of the Anthropocene? It is Babel *after* the fall of the huge tower. Finally, humans are not universifiable. Finally, they are not off the ground! Finally, they are not outside of terrestrial history!

*

What keeps us from taking advantage of this disaggregation of the traditional figures is a mental image that had remained intact throughout the whole history of philosophy, the idea of a *Sphere* that could allow anyone to "think globally and to bear on his or her shoulders the entire weight of the *Globe* – that strange Western obsession, which is the real "white man's burden." In other words, we have to put an end to what could be called "Atlas's curse." Let us recall that Atlas was one of the Titans, one of the numerous monsters that were born from the blood of those whom Gaia had planned to assassinate (I mean the mythological Gaia whom we encountered in the preceding lecture, the one whose provocative portrait was drawn by Hesiod, the goddess who was more ancient than all the Olympians).[32]

To remove some of this excess weight from our shoulders, we have to indulge in a little *spherology*, the fascinating project invented out of whole cloth by Peter Sloterdijk in his massive three-volume study of the envelopes that are indispensable to the perpetuation of life.[33] Sloterdijk borrowed von Uexküll's notion of Umwelt[34] and extended it to all spheres, all enclosures, all the envelopes that agents have had to invent to differentiate between their inside and their outside. To accept such an extension, one has to consider all the philosophical and scientific questions thus raised as being part of a very broad

---

[32] See the third lecture.
[33] Peter Sloterdijk, *Globes: Macrospherology* ([1999] 2014).
[34] Jakob von Uexküll, *A Foray into the Worlds of Animals and Humans* ([1940] 2010).

The Anthropocene and the destruction of the Globe 123

definition of *immunology*, viewed by Sloterdijk neither as a human science nor as a natural science but, rather, as the first *anthropocenic* discipline!

Sloterdijk is a thinker who takes metaphors seriously and fully tests how well they measure up to reality – for hundreds of pages, if necessary. His immunological challenge is to detect how an entity, whatever it may be, protects itself from destruction by building a sort of well-controlled internal milieu that allows it to create a protective membrane around itself. He asks this question at every level with stubborn determination. Even when he maliciously catches his master Heidegger up short for failing to answer questions such as the following: When you say that the *Dasein* is "thrown into" the world, "into" what is it actually thrown? What is the composition of the air it breathes there? How is the temperature controlled? What sort of materials constitute the walls that keep the *Dasein* from suffocating? In short, what is the *climate* in its air-conditioning system? As Sloterdijk sees it, these are exactly the awkward but essential questions that philosophers and scientists of all tendencies and all species have never agreed to answer with adequate precision.

For Sloterdijk, the complete singularity of Western philosophy, science, theology, and politics lies in the fact that they have infused all the virtues into the figure of a Globe – with a capital G – without paying the slightest attention to the way in which that Globe might be built, tended, maintained, and inhabited. The Globe is supposed to include everything that is true and beautiful, even if this is an architectonic impossibility that will collapse as soon as you think seriously about how and through what it *holds up* and especially *how it is traversed*.

Sloterdijk raises a set of very simple, very humble architectural questions, just as material as those the geologists raise with their hammers: Where are you residing when you say that you have a "global view" of the universe? How are you protected from annihilation? What do you see? What air are you breathing? How do you keep warm, how do you dress, how do you eat? And if you cannot satisfy these fundamental needs of life, how can you keep on claiming to speak of the true and the beautiful, as if you occupied some higher rung on a moral ladder? If you don't specify their air-conditioning system, the values that you are trying to defend are probably already dead, like plants that have been kept inside a greenhouse overexposed to the sun. In Sloterdijk's hands, even more than in Lovelock's, the notions of homeostasis and climate control take on a highly metaphysical dimension. This is what's called taking the atmosphere seriously! It's also the New Climate Regime.

As soon as elementary questions such as these come up, it becomes highly unlikely that one can see anything whatsoever from nowhere. No one has ever lived *in* the infinite universe. And no one has ever even lived "*in* Nature." Those people who frighten themselves by wandering around the infinite universe are always gazing upon a small globe with a surface area of two or three square meters in the warmth of their terrestrial offices under the comfortable light of a lamp.[35] Instead of saying that "the eternal silence of these infinite spaces terrifies me," Pascal should have reassured himself: "The murmur of the instruments confined within these limited spaces soothes me as it informs me." When the epistemologists claim that we can live "in Nature," what they are really doing is carrying out what for Sloterdijk amounts to a criminal act of destruction: breaking through all the protective envelopes necessary for the immunological function of life (and life, for him, is just as much politics as is it biology and sociology).

Every thought, every concept, every project that fails to take into account the necessity of the fragile envelopes that make existence possible amounts to a *contradiction in terms*. Or, rather, a contradiction in architecture and design: it will not have the atmospheric, climatic conditions that could make it viable. Trying to live in such a utopia would be like trying to save all your precious data in the Cloud – *without first investing* in computer clusters and refrigeration towers.[36] If you want to keep using the words "rational" and "rationalists," go ahead, but then also do the work of conceiving of the fully furnished spaces in which the presumed inhabitants can breathe, survive, equip themselves, and reproduce. The uncontrolled materialism of the air-conditioned system is another form of idealism.

Thus from page to page Sloterdijk rematerializes in a new way what it means to be *in* space, *on* this Earth, offering us the first philosophy that responds directly to the requirement of the Anthropocene that we bring ourselves back down to Earth. What interests me in particular is that, in the middle of his second volume, the author devotes some

---

[35] See the fascinating catalog of the exhibit devoted to the *Whole Earth Catalog* by Diedrich Diederichsen and Anselm Frank, eds, *The Whole Earth Catalog: California and the Disappearance of the Outside* (2013). On the implausibility of the Globe as a figure of the Earth, see Kenneth Olwig's research in "The Earth Is Not a Globe: Landscape versus the 'Globalist' Agenda" (2011). On the history of the recent form of the Globe, see Sebastian-Vincent Grevsmühl, *La terre vue d'en haut: l'invention de l'environnement global* (2014); its subtitle is in perfect harmony with Sloterdijk's argument.

[36] See the fascinating site that attempts to map the material infrastructure of what is called the virtual: http://newcloudatlas.org.

hundred pages to a meditation that he titles "*Deus sive Sphaera*," "God, that is, the Sphere." The point is delicate but, as we shall see later on, it allows us to remove the principal difficulty common to the sciences and the humanities when they approach the superorganism question.

The little crack that Sloterdijk is the first to point out, I believe, results from the unresolved bifocalism of the Christian imagery left over from the pre-Copernican epoch, the one we have already encountered with Galileo.[37] What looks like a simple technical defect in design in fact destabilizes the entire architecture of Western cosmology. Despite the practical impossibility of drawing the two types of globes together, theologians have striven to bring them into coincidence: one theocentric, the other geocentric. When God is placed at the center, the Earth must inevitably be relegated to the periphery and revolve around Him. At first glance, this doesn't seem too awkward, because our planet is assigned a modest role, rightly *peripheral*. But the problem becomes more complicated as soon as one puts the Earth at the center, with Hell located in the middle, under the sublunary world: then *it is God* who is removed to the periphery. This positioning is not so readily accessible: God, for rational theology, cannot be peripheral! How, Sloterdijk asks, can you construct an entire cosmology with two contradictory centers, one that revolves around God while the other revolves around the Earth?

For two millennia, Sloterdijk tells us, this little flaw in construction seems to have posed no problem for theologians, artists, or mystics:

> The bifocalism of the "world picture" had to be kept latent, and...there could be no explicit dialogue about the contradictions between the geocentric and theocentric locations of projection within the illusory bubble sphere of the Perennial Philosophy.[38]

This philosophy is eternal, perhaps, but it is entirely empty within its sphere of nonexistence. The curse of the Globe is so powerful that theologians have designed a cosmic god in the form of two wobbly spheres without worrying about its architectonic implausibility. From Dante to Nicholas of Cusa, from Robert Fludd to Athanasius Kircher, right up to modern illustrators such as Gustave Doré, the disconnect remains both patent and constantly denied. Although visually impossible, the gentle emanation of God's grace toward the human Earth was never called into question, even if no one could literally *draw* its mystic rays by continuous lines across the cleft that divided the two

[37] See the third lecture.
[38] Sloterdijk, 2014, pp. 448–9.

systems. This is why it has become so awkward to relate any history of the planet – and still less any geohistory: as soon as philosophy believes it is thinking globally, it becomes incapable of conceiving of time as well as of space.

*

You could protest that we have no reason to attribute any importance to this flaw in the construction of Christian theology. After all, coherence is not the strong suit of religious minds, and one more chink in their operation has little chance of being noted. But what fascinates me in this discovery is that exactly the same incoherence is upheld by the architecture through which *rationality* has been constructed.

What Sloterdijk has detected in Christian imagery has been detected just as clearly by the history of science in scientific texts. There is nothing surprising about this: it is the same problem repeated all over again, appearing first in the history of religion, then in the history of science, owing to the *translatio imperii* of which there are so many examples, and to which I shall return later on. It is as impossible to *situate* the Earth as it is to *stabilize* the center around which the other entity is presumed to revolve. Let us recall how precarious the "Copernican revolution" that Kant claimed to have introduced into philosophy has always been: how could he have made us believe that making the Object revolve around the human Subject could count as an abandonment of anthropocentrism? The metaphor is so badly adjusted that it has thrust every definition of the "human in nature" into oscillations that make one's head spin – and in some cases induce nausea. To return to the first meaning of the word "revolution," it is as though there had never been a stable center around which the Earth could *revolve*.

When it is a question of science as it is practiced, science in action, all of a sudden researchers have to begin to talk about their laboratory lives. The same scientists who used to levitate from nowhere are brought back into terrestrial bodies of flesh and blood in narrowly situated places. When physicists celebrate the great heroes of science, they don't hesitate to mount a plaque on a wall with a text, for example, like the one I spotted in Cambridge and found particularly delectable: "Here in 1897 at the old Cavendish Laboratory J. J. THOMSON discovered the electron subsequently recognized as the first fundamental particle of physics and the basis of chemical bonding electronics and computing."[39]

---

[39] A wall plaque in Free School Lane.

It is hard to discover a more *situated* piece of knowledge than this one. It starts from one precisely determined place, Free School Lane (which has become the temple of the history of science),[40] with electrons that are firmly in the hands of a great scientist, and then it extends to the whole world, since electrons are at the core of all chemical bonds and all computers! But a minute later, these same physicists will have no qualms about explaining to you how the mind of Stephen Hawking wanders through the cosmos in intimate dialogue with the Creator, naïvely ignoring the fact that Hawking's mind benefits not only from a brain but also from a "collective body" composed of a huge network of computers, chairs, instruments, nurses, aides, and voice synthesizers that are necessary for the progressive unfolding of his equations.[41] This bifocal conception of science does not allow the "view from nowhere" to be reconciled with these very particular places: classrooms, offices, laboratory benches, computer centers, meeting rooms, expeditions and field stations, the sites where scientists have to place themselves when they actually have to *obtain* data or really *write* their articles.

The two images of the world in Christian theology are just as irreconcilable as the images that would be represented, for example, by the physics of the electron that is present *everywhere* in the world even as it is safely housed *in* J. J. Thomson's Cavendish Laboratory. But this irreconcilability is denied by scientists and philosophers just as much as by theologians and mystics. Paraphrasing Sloterdijk, I could say: "The 'illusory sphere' of *philosophia perennis* maintains in latency the contradictions between Nature – centered on the cosmos – and that other Nature known by the sciences centered on the laboratory. This contradiction makes any explicit dialogue between the two visions just as impossible as reconciliation between the geocentric and theocentric 'pictures of the world' of medieval cosmology."

Following Sloterdijk's examination of the architecture of Reason, we realize that the Globe is not that of which the world is made but, rather, a Platonic obsession *transferred* into Christian theology and then *deposited* in political epistemology to put a face – but an impossible one – on the dream of total and complete knowledge.[42] A strange fatality is at work here. Every time you think about knowledge in a

---

[40] This is in fact where Simon Schaffer and his colleagues have their offices; historians of science have ended up occupying, after a time, the offices of scientists, who themselves have moved on, following their increasingly cumbersome instruments.
[41] Hélène Mialet, *Hawking Incorporated: Stephen Hawking and the Anthropology of the Knowing Subject* (2012).
[42] On the constitution of this "political epistemology," see Bruno Latour, *Pandora's Hope: Essay on the Reality of Science Studies* (1999).

weightless space – and this is where the epistemologists dream about dwelling – it inevitably takes the form of a transparent sphere that could be inspected by a fleshless body from a place that is nowhere. But once we restore the gravitational field, knowledge immediately loses this mystical spherical form inherited from Platonic philosophy and Christian theology.[43] The data flow in again in their original form as fragments, waiting to be put together in a narrative.

By virtue of this bifocalism, the two portraits of Atlas are equally implausible, the Atlas who is supposed to be holding the world on his shoulders (without being able to look at it, as Sloterdijk remarks), but also the one invented by Mercator, the perfect emblem of the scientific revolution – an Atlas who is supposed to be holding the entire cosmos in his hands, as if it were a soccer ball.[44] By fusing the image of the scientist with the much older metaphor of the hand of God, Mercator gave it a human form, that of an authentic Superman capable of holding everything in his palm. But if the globe is actually held in the hand of some human of average height, then, inevitably, it is a map, a model, a *globe* in the very modest and very local sense of the little instrument in papier maché that many of you, I'm quite sure, like to spin with your fingertips.[45]

Building a globe always amounts to reactivating a theological theme – even when it is a matter of lofty pedagogical sites, a panorama, a geodesic dome, an amusement park invented by compilers of information to give the encyclopedic knowledge they have accumulated a popular form. This was easy to see when Patrick Geddes, the director of the Outlook Tower in Edinburgh,[46] had to give the funeral oration for his friend, the very famous Élisée Reclus, the anarchist geographer who had asked him for help drawing the plans for the giant globe that he intended to build for the Universal Exposition in Paris in 1900 at a scale of 1:100,000. Had it been built so as to cast its immense shadow on the right bank of the Seine, the structure

---

[43] Readers of *Tintin* will recognize in this metaphor the adventure of Captain Haddock in *Explorers on the Moon*: when the Duponts accidentally make the artificial gravity of their rocket disappear, whisky turns into little balls floating about the cabin; see Hergé, *The Adventures of Tintin: Explorers on the Moon* ([1953] 1976).
[44] Frontispiece of the first atlas of the world, attributed to Mercator, figure 4.2.
[45] There is an immense literature on the uses of the globe, but here are two fairly recent works: Franco Farinelli, *De la raison cartographique* (2009), and the very useful survey by Jerry Brotton, *A History of the World in Twelve Maps* (2012).
[46] This tower, a sort of Palace of Discovery and a geodesic dome, is one of the most visited sites in Edinburgh; it is located just a few hundred meters from the room where the Gifford Lectures are given. I thank Pierre Chabard for introducing me to Geddes, an incredible character in his own right: Pierre Chabard, "L'Outlook Tower, anamorphose du monde" (2001).

The Anthropocene and the destruction of the Globe 129

Figure 4.2 Frontispiece of Mercator's *Atlas sive Cosmographicae Meditationes de Fabrica Mundi et Fabricati Figura*, 2nd edn, 1609 (author's personal collection).

would have been almost as tall as the Eiffel Tower and would have cost five times as much.

> This was no mere *scientific model* in its institute, but the image, and shrine, and *temple of the Earth-Mother*, and its expositor no longer a modern professor in his chair, but an arch-*Druid* at sacrifice within his circle of mighty stones, an Eastern Mage, initiator to *cosmic mysteries*....the unity of the world now the basis and symbol of the brotherhood of man upon it; sciences and art, geography and labour uniting into a *reign of peace* and goodwill.[47]

All the words count here, in this relation between the macrocosm and the microcosm, not only the strange displacement from "scientific model" to "temple of the Earth-Mother," but also from "professor" to "arch-Druid," from geography to prophecy through the

---

[47] Patrick Geddes, "A Great Geographer: Elisée Reclus, 1830–1905" (1905), p. 550, emphasis added.

intermediary of poetry. And how strange it is for us, a century later, to hear a celebration of "the brotherhood of man" and "the unity of the world" thanks to the construction of a reduced model, a miniature facsimile, an Atlas of iron and plaster. One thing is certain: today, as yesterday, the same question arises: how can one escape from the excessive burden of the Globe?

To put an end to the fatality of the Globe – what I have called Atlas's curse[48] – we have to stick to the history of the sciences or to Sloterdijk's spherology, while noting that "global" is an adjective that can of course describe the form of a local device apt to be inspected by a group of humans who are looking at it, but *never the world itself in which everything is presumed to be included*. However large the size of the galaxies may be, the map of the galaxies dispersed since the Big Bang is no larger than the *screen* on which the data flows from the Hubble telescope are pixelated and colored. Contrary to the formula "think globally, act locally," no one has ever been able to think Nature globally – still less Gaia. The global, when it is not the attentive analysis of a *reduced model*, is never anything but a tissue of globabble.

*

Whether we are dealing with the idea of the Anthropocene, the theory of Gaia, the notion of a historical actor such as Humanity, or Nature taken as a whole, the danger is always the same: the figure of the Globe authorizes a premature leap to a higher level *by confusing the figures of connection with those of totality*. This perilous slippage is not only the preoccupation of philosophers,[49] politicians, military thinkers,[50] or theologians;[51] it also obsesses the scientists who wish to understand the Anthropocene. I can't resist the temptation of demonstrating this for you with an exemplary case that will allow us to measure, once again, the slope that writers such as Lovelock

---

[48] Not to be confused with the attempt to portray a shrugging Atlas, as in Ayn Rand's infamous novel! [*Trans.*: *Atlas Shrugged* (New York: Random House, 1957).]

[49] This is particularly striking in the case of Michael Ruse, *The Gaia Hypothesis: Science on a Pagan Planet* (2013), in which the author does not seem to suspect for an instant that Lovelock is attempting to compose Gaia and not deducing its form on the basis of a pre-existing Globe.

[50] Grevsmühl pursues the archaeology of this obsession in *La terre vue d'en haut* (2014).

[51] Christophe Boureux, *Dieu est aussi jardinier* (2014), starts from the principle that there is a totality with a common (divine) origin and that its initial composition poses no particular problems.

and Zalasiewicz have to climb when they seek to explain the Earth's retroactive relations to human actions.

There are books that are admirable owing to the perseverance with which they misunderstand their object. This lack of comprehension is visible in the very title of *On Gaia: A Critical Investigation of the Relationship between Life and Earth*.[52] What makes the case of Toby Tyrrell – a professor of Earth System Science at the University of Southampton – so remarkable is that he claims to be producing a legitimate and "strictly scientific" refutation of the Gaia theory. Now Tyrrell cannot present Lovelock's hypothesis without at once turning Gaia into something superior that *encircles* the Earth. Amusingly, and without the slightest awareness of this on the author's part, all the theological phantoms that Patrick Geddes attributed to Élisée Reclus immediately reappear in Tyrrell's account!

Each chapter summarizes quite pedagogically the results of the disciplines traversed by the Gaia theory, and each ends with the conclusion that one cannot discern the existence of a totality that would ensure the stability of the system.[53] The author's thesis is that Lovelock is necessarily wrong because nothing makes it possible to ensure that Gaia *protects* Life on Earth, whereas it *ought to devote itself to this* if it really has the virtues of the Providence that, in Tyrrell's reading, Lovelock seems to promote. We've come back to a problem we encountered in the previous lecture: from beginning to end, Tyrrell imputes to Lovelock the idea that Gaia is a *higher* system than the life forms it manipulates. Not for a moment does he notice that Lovelock's innovation consists precisely in not letting himself get caught in the trap of that habitual trope concerning the Whole and its parts.

Even though the argument is technical, it is worth following the way an ancestral political theme – an amalgam of the fable of the bees and of divine Providence[54] – comes to take over, in parasitical fashion, the prose of a researcher who would otherwise have very respectable reasons to oppose the Gaia theory – if only it were Lovelock's![55] The paradox is that he begins by granting the main thesis:

---

[52] Toby Tyrrell, *On Gaia: A Critical Investigation of the Relationship between Life and Earth* (2013).
[53] A more developed version of this section is to be found in Bruno Latour, "Why Gaia Is Not a God of Totality" (2016c).
[54] Bernard Mandeville's book *The Fable of the Bees: Private Vices, Publick Benefits* ([1714] 1962), with its eloquent subtitle, is one of the many ancestors of the animal models that make it possible to explain the emergence of the optimum – in fact, the Market – on the basis of clashes between individual interests.
[55] Tyrrell rightly worries about the fact that, if Gaia were conceived as a lovable and benevolent Providence, humans would allow themselves to assault it, confident that

Lovelock claimed that life *does modify* the environment. Life is not *simply a passive passenger* living *within* an environment set by physical and geological processes over which it has *no control*. The biota have not lived *within* the Earth's environment and processed it but also, it is suggested, *have shaped it* over time.... There is no doubt that Lovelock is correct, and few now disagree.[56]

But then he asserts, toward the end of the book:

> For these reasons it can be concluded that the long and uninterrupted duration of life-tolerant conditions does not prove the existence of *an all-powerful thermostat*, and does not prove the existence of Gaia.[57]

We are familiar with the obsessive determination of theologians to prove the existence of an *all-powerful* God, but why on earth attribute to Lovelock the idea that he is seeking to prove "the existence of an All-Powerful Thermostat"?! There is no doubt about it: Tyrrell let himself get carried away by the scheme of the Globe. To be sure, as we have seen, Lovelock does talk about a control system, but he goes on to be immediately suspicious of the perilous connotations that the technological metaphor would bring with it. Let us stress here all the risks that would ensue for a scientific author who remained insensitive to the tropisms of prose. It is here, however, where the nuances required for speaking of agency are best revealed. As Lovelock says, in fact:

> I describe Gaia *as* a control system for the Earth – a self-regulating system *something like* the familiar thermostat of a domestic iron or oven. I am an inventor. I find it *easy* to invent a self-regulating device by *first* imagining it as a *mental picture*.... In many ways Gaia, like an invention, is difficult to describe.[58]

For Lovelock, Gaia possesses no omnipotence; it is a "mental picture," a convenience ("easy to invent"), a comparison ("something like") made in an effort to conceptualize, in the manner of an inventor – inventors being better gifted, according to him, at understanding than scientists how things really work[59] – something that he recognizes from the outset as "difficult to describe." Tyrrell

---

Gaia would forgive their straying. On the contrary: "Because the Earth's climate system has *transpired*, as opposed to *evolved*, there is no reason to expect it to be particularly *robust or fail-safe*" (Tyrrell 2013, p. 216, emphasis added). On this everyone agrees, and Lovelock first of all.

[56]Ibid., p. 113, emphasis added.
[57]Ibid., p. 198, emphasis added.
[58]James Lovelock, *Gaia: The Practical Science of Planetary Medicine* ([1991] 2000a, p. 11).
[59]In interviews, Lovelock often emphasized the fact that he was above all an inventor of very sensitive instruments (in particular the famous electron capture detector, or

The Anthropocene and the destruction of the Globe 133

remains completely insensitive to all these linguistic hesitations. Yet it is precisely through these hesitations that a difference arises between a naïvely theological vision – although Tyrrell claims it is "scientific" – and the secular, terrestrial, innovative version of a Lovelock seeking to capture, in the shifting of his convoluted prose, something that is seeking its path, like life on earth itself: something that produces order downstream yet that does not depend on a pre-established order upstream. The Gaia theory comes from an *inventor* talking about an *invention* that is difficult to describe.

> The nearest I can reach is to say that Gaia is an *evolving system*, a system made up from all living things and their surface environment, the oceans, the atmosphere, and crustal rocks, the two parts *tightly coupled and indivisible*. It is an "emergent domain" – a system that has emerged from the *reciprocal* evolution of organisms and their environment over the eons of life on Earth. In this system, the self-regulation of climate and chemical composition are *entirely automatic. Self-regulation emerges as the system evolves.* No foresight, planning or *teleology*...are involved.[60]

It would be hard to be clearer about the absence of Providence. And yet Tyrrell remains deaf to such subtleties. Whereas Lovelock's entire effort consists in avoiding as much as possible the two-level distinction – one for the connections, the other for the regulatory totality – his adversary plunges headlong into the worst cybernetic metaphor there is:

> The Gaia hypothesis is nothing if not daring and provocative. It proposes *planetary regulation* by and for the biota, where the "biota" is the collection of all life. It suggests that life has *conspired in the regulation* of the global environment, so as to keep conditions favorable.[61]

Where the one hesitates, the other does no such thing, even as he believes he can give the first, through this absence of hesitation, a lesson in the scientific method! If planetary regulation existed, the Gaia hypothesis would hardly be "daring and provocative"; in any case, it would not deserve to be published: God the Creator, the one who has always had the form of a Sphere, was there first! Lovelock is trying not to separate the two levels that Tyrrell is imposing here as self-evident from the outset:

ECD), and that it was thanks to such inventions that he became sensitive to the animation of the Earth, since he could detect the presence of chemical elements (when he began his research into pollutants) over very long distances.
[60] Lovelock, 2000a, p. 11, emphasis added.
[61] Tyrrell, 2013, p. 3, emphasis added.

Lovelock suggests that *life has had a hand on the tiller* of environmental control. And the intervention of life in the regulation of the planet has been such as *to promote stability* and keep conditions favorable to life.[62]

The error of interpretation is flagrant, for it is precisely because there is no tiller, and thus no helmsman, no master, no captain, no engineer, no God, that Gaia is an invention that all the subtleties of science must tend to explain. But the strangest thing of all is that Tyrrell objects to Gaia only because he wants to entrust the tiller to a different helmsman, a different captain, a different providential God: Evolution! Whereas Lovelock tries to couple the environment and evolution by definitively blurring the distinction between the two, since organisms also make up their environment, in part, Tyrrell thinks it possible to *oppose* Gaia and Evolution:

> In fact the snug fit between organisms and habitats is more a testament to the *overwhelming*, transforming power of evolution to mold organisms than to the *power* of organisms to make their environment more comfortable.[63]

Here is a nice case of inversion in the figures of Totality: All-Powerful Evolution is supposed to be fully natural, Gaia dangerously providential... Tyrrell does not notice for a second that these two figures can be precisely interchanged. Whereas he thinks he is writing scientifically, we find ourselves here in full Theogony: the "powers" of Evolution struggling for supremacy against the "powers" of Gaia! Or rather in full Theodicy, since it is a matter of finding out what best protects against Evil on Earth: is it the All-Powerful Thermostat or Darwinian evolution that best privileges those who are faithful to it? Tyrrell goes so far as to order Lovelock to make an effort, as Leibniz did, to prove that his God is innocent of the disorders He has introduced here below.[64] The objection is amusing, coming from an author who uses the neo-Darwininan model without the slightest hesitation, a model itself borrowed from the Invisible Hand of the Market!

Am I splitting hairs by accusing Professor Tyrrell of being a theologian in disguise? Yes, of course, for everything depends in fact on the thread that the narrative prose allows us either to follow or to cut off. To be sure, Lovelock is neither a philosopher, nor a poet, nor

---

[62] Ibid., p. 4, emphasis added.
[63] Ibid., p. 48, emphasis added.
[64] Hence this astonishing passage: "to my mind this paradox of nitrogen starvation while being bathed in nitrogen is one of the strongest arguments against the Gaian idea that the biosphere is kept comfortable for the benefit of the life inhabiting it" (ibid., p. 111). It's as though we were reading Voltaire making fun of the proofs of the existence of God drawn from the harmony of nature!

a novelist, nor a historian, but he is fighting against something that resists thought. If he captures the narrative capacity of geohistory, it is because he hesitates and because *he starts over*. Tyrrell swallows metaphors so easily that he can criticize one only by relying on another, whereas Lovelock mistrusts metaphors; he handles them with precaution as the only way to avoid them, little by little.

> At first we explained the Gaia hypothesis in words *such as* "Life, or the biosphere, regulates or maintains the climate and the atmospheric composition at an optimum for itself." This definition was *imprecise*, it is true; but neither Lynn Margulis nor I ever proposed that planetary self-regulation is *purposeful*....In the arguments over Gaia quite often the metaphor not the science was attacked. Metaphor was seen as a pejorative, something inexact and therefore unscientific. In truth, real science is riddled with metaphor.[65]

It is unfair of me to go after a naturalist when the adherents to the social sciences, as I know perfectly well, do no better, and they leap without a moment's hesitation to the global level of society as soon as they have to explain any sort of connection. When they talk about "society as a whole," "the social context," "globalization," they are drawing a figure with their hands that has never been bigger than an ordinary pumpkin! But the fact is that the problem is the same whether we are talking about Nature, Earth, the Global, Capitalism, or God. Each time, we are presupposing the existence of a superorganism.[66] The passage through connections is immediately replaced by a relation between parts and the Whole, and the latter is said – without much thought – to be necessarily *superior* to the sum of its parts – whereas it is always necessarily *inferior* to its parts.[67] Superior does not mean more encompassing; it means *more connected*. One is never as provincial as when one claims to have a "global view."[68]

---

[65] Lovelock, 2000a, p. 11, emphasis added.
[66] Bruno Latour, *Reassembling the Social: An Introduction to Actor-Network Theory* (2005). It is fascinating to see that the problem is exactly the same at every scale, whether it is a matter of ants, as in Deborah Gordon's *Ant Encounters: Interaction Networks and Colony Behavior* (2010) or Gaia. This is the problem that Tarde placed at the heart of the social sciences and that has been swallowed up by the idea of distinct levels going from the individual to the collective; see Gabriel Tarde, *Social Laws: An Outline of Society* (1899).
[67] Bruno Latour, Pablo Jensen, Tommaso Venturini, Sebastian Grauwin, and Dominique Boullier, "The Whole Is Always Smaller Than its Parts – A Digital Test of Gabriel Tarde's Monads" (2012).
[68] There is a confusion between the cartographic globe, which is a way to register as many differences as possible through the simple device of Cartesian coordinates, and the globe of so-called globalization, which is the extension everywhere of as small a set of standard formats as possible.

Scale is not obtained by successive embeddings of spheres of different sizes – as in the case of Russian dolls – but by the capacity to establish more or less numerous relationships, and especially reciprocal ones. The hard lesson of actor-network theory, according to which there is no reason to confuse a *well-connected* locality with the utopia of the Globe, holds true for all associations of living beings.

The reason the relocalization of the global has become so important is that the Earth itself can no longer be grasped globally by anyone. This is precisely the lesson of the Anthropocene. As soon as one unifies it in a terraqueous sphere, one reduces geohistory to the limits of the old format of medieval theology, transported into the nineteenth-century epistemology of Nature, then again poured back into the mold of the twentieth-century military-industrial complex[69] – even if one is a professor of Earth System Science at the University of Southampton. Despite the unanimous enthusiasm that it has aroused, the highly celebrated "blue planet" has poisoned thought in a lasting way. It is a composite image that blends the ancient cosmology of the Greek gods, the old medieval form given to the Christian God, and NASA's complex network for data acquisition, before being projected within the diffracted panorama of the media.[70] What is certain is that the inhabitants of Gaia are not those who view the blue planet as a Globe.

Even so, it must be possible, today, to pull ourselves away from the fascination that the image of the Sphere has held for us since Plato: the spherical form rounds off knowledge in a continuous, complete, transparent, omnipresent volume that masks the extraordinarily difficult task of assembling the data points coming from all instruments and all disciplines. A sphere has no history, no beginning, no end, no holes, no discontinuities of any sort. It is not merely an idea but the very ideal of ideas. Those who pride themselves on thinking globally will never get away from the curse of Atlas: *Orbis terrarum sive Sphaera sive Deus, sive Natura*.

*

To put it in still other terms, he who looks at the Earth as a Globe always sees himself as a God. If the Sphere is what one wishes to

---

[69] We must never forget that environmental preoccupations are first and foremost military, and that total war *through* the modifications of the climate precedes by several dozen years the war *against* the mutations of the climate. On this point, see Ronald E. Doel, "Constituting the Postwar Earth Sciences: The Military's Influence on the Environmental Sciences in the USA after 1945" (2003).
[70] As Grevsmühl shows, the canonical image is in fact a composition made one pixel at a time and is in no way, in a technical sense, a "global" image (Grevsmühl 2014).

contemplate passively when one is tired of history, how can one manage to trace the connections of the Earth without depicting a sphere? By a movement that turns back on itself, in the form of a *loop*. This is the only way to draw a path between agents without resorting to the notions of parts and a Whole that only the presence of an all-powerful Engineer – Providence, Evolution, or Thermostat – could have set up. This is the only way to become secular in science as well as in theology. But let's not hurry to identify this movement, which in the previous lecture I called *waves of action*, with feedback loops in the cybernetic sense: we would revert at once to the model with a rudder, a helmsman, and a world government![71]

Let's begin with the strange reflexive loop that historians of the environment have recently insisted upon: to speak of ecology now is to repeat almost word for word what was said in 1970, in 1950, or even in 1855 or in 1760 to protest against the damage inflicted on nature by industrialization.[72] This theme has been looping back and forth since the very beginnings of the industrial revolution.[73] This does not mean, however, that historians are giving in to their harmless little vice of unearthing, for each novelty, a host of more or less unknown predecessors. It is as though all ecologist writers were led to discover that there is "something new under the sun"; but, because they shape their views in terms that take up earlier ideas quite faithfully, they nevertheless leave us with the impression that, over the long run, there is nothing new under the sun at all.[74] This is hardly astonishing, since it is always to the vocabulary of the sempiternal Globe that we entrust our hopes as well as our anxieties. When we appeal to the blue planet, we cannot help but go around in circles!

---

[71] See Andy Pickering, *The Cybernetic Brain: Sketches of Another Future* (2011).
[72] The argument made by Bonneuil and Fressoz in *The Shock of the Anthropocene* is hard to refute: our predecessors have never stopped deploring the same catastrophe in the same terms, have kept on warning us of the same threats. See Stephen Toulmin, *Cosmopolis: The Hidden Agenda of Modernity* (1990); Barry Commoner, *The Closing Circle* (1971); Barbara Ward and René Dubos, *Only One Earth: An Unofficial Report commissioned by the Secretary General of the United Nations Conference on the Human Environment* (1972); Donella H. Meadows, Dennis L. Meadows, Jørgen Randers, and William W. Behrens III, *The Limits to Growth: A Report for the Club of Rome's Project on the Predicament of Mankind* (1972); Élodie Vieille-Blanchard, "Les limites à la croissance dans un monde global: modélisations, prospectives, réfutations" (2011); but we can go as far back as Eugène Huzar, *La fin du monde par la science* ([1855] 2008), or the anti-vaccination campaigns in 1760.
[73] Jean-Baptiste Fressoz, *L'apocalypse joyeuse: une histoire du risque technologique* (2012).
[74] See the contribution of Clive Hamilton and Jacques Grinevald, "Was the Anthropocene Anticipated?" (2015), and the book by John R. McNeill, *Something New under the Sun: An Environmental History of the Twentieth-Century World* (2000).

If the historians are right to criticize those who claim, with the same enthusiasm every time, that we have just entered into a radically different period,[75] they are mistaken not to see that this repetition is part of a phenomenon that must be accounted for: by definition, geohistory can never be conceptualized in the form of a Sphere whose encompassing form has been discovered *once and for all*. This is why it is just a history and not a "nature." History, for its part, surprises us and obliges us to start all over again every time. The impression of repeating the same thing comes from the form of the Globe with which everyone tries to depict what is happening to it that is new. In contrast, the discovery, shattering every time, of a dramatic new connection between previously unknown agents, and on increasingly more distant scales, and at an increasingly frenetic pace – yes, this is truly new. The Anthropocene, because it dissolves the very thought of the Globe viewed from afar, brings history back to the center of attention.[76] In this sense, despite the critique of historians, there actually has been, since 1860, since 1945, since 1970, something new under the sun.[77] If the feedback loops are similar in form, their contents, rhythms, and extensions are different in each case. This is what I mean by Gaia's *insistence*!

The notions of globe and global thinking include the immense danger of unifying too quickly what first needs to be *composed*. This is above all a material problem: we have to draw a circle before we can generate a sphere. It is also an empirical problem: only because Magellan's boat came back were his contemporaries able to fix in their minds the image of a spherical earth with which they were already

---

[75] I plead guilty, obviously, with the slight exception that, as we have never been modern, and as we have always suspected that we have never been modern, there have never in fact been sharp breaks to which we could hold, even if the Moderns, for reasons that we shall encounter in the sixth lecture, can only live propped up by a radical break.

[76] This return of history is quite well marked by the multiplication of alternatives proposed for the Anthropocene: the "Anglocene" (the combined carbon emission of England and the United States still remains higher than that of the developing countries); the "capitalocene" (Jason Moore, *Capitalism in the Web of Life: Ecology and the Accumulation of Capital*, 2015), not to mention the delicious "Chthulucene" proposed by Donna Haraway in *Staying with the Trouble: Making Kin in the Chthulucene* (2016).

[77] For the moment, the most serious alternative is that of the "Plantatiocene," proposed by Tsing in *The Mushroom at the End of the World* (2015), to describe a pre-industrial regime of land-appropriation that marks the beginning of the "great Columbian exchange" (Charles C. Mann, *1493: Uncovering the New World Columbus Created*, 2011), an ideal golden spike for the beginning of the Great Divergence analyzed by Richard Grove in *Green Imperialism: Colonial Expansion, Tropical Island Edens, and the Origins of Environmentalism, 1600–1860* (1995).

familiar. But it is also a moral problem: it is only when you feel the repercussions of your own action that you understand to what extent you are *responsible* for it. As Sloterdijk has noted, it is only when humans see pollution falling back on them that they begin really to feel that the Earth is in fact round.[78] Or, rather, the roundness of the Earth, known – but always superficially – from the earliest Antiquity, takes on more and more verisimilitude as the number of circles with which it can be surrounded gradually increases. Thus the loop that is required to draw any sphere is pragmatic in John Dewey's sense: you have to feel the consequences of your action before you are able to represent to yourself what you have really done and become aware of the tenor of the world that has resisted your action.[79]

This is why it is so important to move from the Globe to the quasi-feedback loops that tirelessly design it in a way that is broader and denser each time. Without Charles Keeling's observatory in Mauna Loa and the instruments that detect the carbon dioxide cycle, we *would know* less,[80] by which I mean that we would *feel* less strongly that the Earth can be made rounder by our own actions. And, before that, we had to feel the hole in the ozone layer thanks to the campaign with Dobson's instruments,[81] as we had to learn to feel the possibility of nuclear winter thanks to the new models of atmospheric circulation advanced, during the epoch of a virtual nuclear holocaust, by Carl Sagan and his colleagues.[82]

What is at stake in the Anthropocene is this order of understanding. It is not that the little human mind should be suddenly teleported into a global sphere that, in any case, would be much too vast for its small scale. It is rather that we have to slip into, envelop ourselves within, a large number of loops, so that, gradually, step by step, knowledge of the place in which we live and of the requirements of our atmospheric condition can gain greater pertinence and be experienced as urgent. The slow operation that consists in being enveloped in sensor circuits in the form of loops: this is what is meant by "being of this Earth."

---

[78] Peter Sloterdijk, *In the World Interior of Capital: For a Philosophical Theory of Globalization* ([2005] 2013).
[79] John Dewey, *Logic: The Theory of Inquiry* (1938).
[80] See Charles David Keeling, "Rewards and Penalties of Recording the Earth" (1998) (cf. the first and second lectures).
[81] See Grevsmühl, 2014, chapter 7.
[82] See Paul N. Edwards, "Entangled Histories: Climate Science and Nuclear Weapons Research" (2012), and Matthias Dörries, "The Politics of Atmospheric Sciences: 'Nuclear Winter' and Global Climate Change" (2011), on the link between nuclear war and the New Climate Regime. Oliver Morton offers a remarkable summary of those successive events in *The Planet Remade: How Geoengineering Could Change the World* (2015).

But we all have to learn this for ourselves, anew each time. And it has nothing to do with being a human-in-Nature or a human-on-the-Globe. It is rather a slow, gradual fusion of cognitive, emotional, and aesthetic virtues thanks to which the loops are made more and more visible. After each passage through a loop, we become *more sensitive* and *more reactive* to the fragile envelopes that we inhabit.[83]

How many supplementary loops do we have to trace around the Earth before "knowledge" is receptive enough for this shapeless Anthropos to become a real agent of history and an ever-so-slightly credible political actor? It is useless to claim that we *already* knew this and that others have said it before. How many loops have some of you had to follow before giving up smoking? It is possible that you *always knew* that cigarettes caused cancer, but there's a long way to go between that "knowledge" and really stopping smoking. "To know and not to act is not to know." Before weighing what it is to know that one must not smoke, doesn't one need to anticipate the pain in one's flesh, the pain that shocking images on some cigarette packages try to prefigure? In this case, too, there have to be complex institutions and well-equipped bureaucracies for you to reach the point of feeling in advance the effects of your actions on yourself. Similarly, how many loops do you have to go through really to *feel* the roundness of the Earth? How many supplementary institutions, how many bureaucracies do you call for, you personally, to make yourself capable of responding to a phenomenon, at first glance so remote, as the chemical composition of the atmosphere? Especially if others are working for their part to *make* you insensitive by deliberately producing ignorance?[84] (It is no accident that the same lobbies that are financing the climate skeptics have worked so long to conceal the connections between cigarettes and your lungs.)[85]

But there is another, more convincing, ultimate reason why we should be extremely suspicious of any global vision: Gaia is not a Sphere at all. Gaia occupies only a small membrane, hardly more than a few kilometers thick, the delicate envelope of the critical zones. Thus it is not global in the sense that it would work as a system starting from a control booth occupied by some Supreme Distributor, surveying and dominating the whole. Gaia is not a cybernetic machine controlled by

---

[83] See David Abram, *The Spell of the Sensuous: Perception and Language in a More-than-Human World* (1996).
[84] See Robert N. Proctor, *Golden Holocaust: Origins of the Cigarette Catastrophe and the Case for Abolition* (2011).
[85] See the testimony of Al Gore, *The Assault on Reason* (2007), and a more detailed account by James Hoggan, *Climate Cover-Up: The Crusade to Deny Global Warming* (2009).

feedback loops but a series of historical events, each of which extends itself a little further – or not. Understanding the entanglements of the contradictory and conflictual connections is not a job that can be accomplished by leaping up to a higher "global" level to see them act like a single whole; one can only make their potential paths cross with as many instruments as possible in order to have a chance to detect the ways in which these agencies are connected among themselves. Once again, the global, the natural, and the universal operate like so many dangerous poisons that obscure the difficulty of putting in place the networks of equipment by means of which the consequences of action would become visible to all the agencies.

This is what it means to live in the Anthropocene: "sensitivity" is a term that is applied to all the actors capable of spreading their sensors a little farther and making others feel that the consequences of their actions are going to fall back on them, come to haunt them. When the dictionary defines "sensitive" as "something that detects or reacts rapidly to small changes, signals, or influences," the adjective applies to Gaia as well as to the Anthropos – but only if it is equipped with enough sensors to feel the retroactions. Isabelle Stengers often says of Gaia that it is a power that has become "touchy."[86] Nature, the Nature of yesteryear, may well have been indifferent, dominating, a cruel stepmother, but She surely wasn't touchy! On the contrary, her complete lack of sensitivity was the source of thousands of poems, and it was what allowed her, in contrast, to unleash in us the sensation of the sublime: we humans were what She was not – sensitive, responsible, and highly moral.

Gaia, on the other hand, seems to be excessively sensitive to our actions, and it seems to react extremely rapidly to what it feels and detects. No immunology – in Sloterdijk's expansive sense – is possible unless we learn to become sensitive in turn to these multiple, controversial, mutually entangled loops. Those who are not capable of "detecting and responding rapidly to small changes" are doomed. And those who for whatever reason interrupt, eradicate, neglect, diminish, weaken, deny, obscure, discriminate against, or disconnect these loops are not merely insensitive or unreceptive. As we shall see in the following lectures, they are probably, if not criminals, in any case our enemies. This is why it makes sense to call "negationist" those who, denying both our own sensitivity and Gaia's, declare with confidence that the Earth cannot under any circumstances react to our actions.

*

[86] Isabelle Stengers, *In Catastrophic Times: Resisting the Coming Barbarism* ([2009] 2015).

To follow the loops in order to avoid totalizing is obviously also to approach politics. With the concept of Anthropocene, the two great unifying principles – Nature and the Human – become more and more implausible. And it is not the intrusion of Gaia that is going to pull together and unify what is coming apart before our eyes. It is useless to hope that the urgency of the threat is so great and its expansion so "global" that the Earth will act mysteriously as a unifying magnet to turn all the scattered peoples into a single political actor occupied in reconstructing the Babel Tower of Nature. Gaia is not a kindly figure of unification. It is "nature" that was universal, stratified, incontrovertible, systematic, deanimated, global, and indifferent to our fate. But not Gaia, which is only the name proposed for all the intermingled and unpredictable consequences of the agents, each of which is pursuing its own interest by manipulating its own environment.

The multicellular organisms that produce oxygen and the humans who emit carbon dioxide will multiply *or not* according to their success, and they will win exactly the dimensions that they are capable of taking. No more, no less. Don't count on an encompassing, preordained system of retroaction to call them back to order. It is impossible to appeal to the "equilibrium of nature," or to the "wisdom of Gaia," or even to its relatively stable past as a force that was capable of restoring order every time politics divided these scattered peoples excessively. In the epoch of the Anthropocene, all the dreams entertained by the deep ecologists of seeing humans cured of their political quarrels solely through the conversion of their care for Nature have flown away. For better or for worse, we have entered into a *postnatural* period.

Obviously, behind the dreams of global unification there was, there still is, Science. Couldn't we find in Science a unifying principle of last resort that would bring the world into agreement and that could direct a mass of humans toward incontrovertible programs of action? Let's all become scientists – or at least let's spread science everywhere through education – and we'll be able to act in concert. "Facts of all countries, unite!" Unfortunately (I almost said fortunately), this solution is made impossible not only by the pseudo-controversy carried on by the climate skeptics, as we saw in the first lecture,[87] but also by the very singularity of all these disciplines, which depend on a distribution of instruments, models, international agreements, bureaucracies, standardization, and institutions whose "vast machine," to borrow

---

[87] See Edwin Zaccai, François Gemenne, and Jean-Michel Decroly, eds, *Controverses climatiques, sciences et politiques* (2012).

Paul Edwards's title, has never been presented in a positive light to public awareness.[88] The climatologists and the Earth System scientists have been led into a *post-epistemological* situation that is as surprising for them as it is for the public at large: it is as if both groups find themselves thrust "outside of nature."

If there is no unity either in Nature or in Science, this means that the universality we seek has to be in any case woven loop after loop, reflexivity after reflexivity, instrument after instrument. It was to make this effort of composition at least thinkable that I proposed, in the first lecture, to define collective lives through the distribution of agency and through the choice of connections that link these forms of action.[89] This is what I have called a *metaphysics* or a *cosmology*, something that may allow us to escape for good from the Nature/Culture format by leading us toward something like the *world*. These collectives – and this is what makes all the difference – are not *cultures*, as they were for traditional anthropology; they are not unified by being, after all, "children of Nature," as the natural sciences of yesteryear maintained; nor, of course, are they a little bit of both, as the impossible dreams of reconciliation or dialectic would have it.[90] The true beauty of the term Anthropocene is that it brings us very close to *anthropology*, and it makes less implausible the *comparison of collectives* finally freed of the obligation to locate any one collective with respect to the others according to the sole schema of nature (singular) and cultures (plural), where unity would be on one side, multiplicity on the other. Finally, multiplicity is everywhere! Politics can begin again.[91]

Facing the Anthropocene, once the temptation to see it simply as a new avatar of the schema "Man facing Nature" has been set aside, there is probably no better solution than to work at disaggregating the customary characterizations until we arrive at a new distribution of the agents of geohistory – new peoples for whom the term human is not necessarily meaningful and whose scale, form, territory, and cosmology all have to be redrawn. To live in the epoch of the Anthropocene is to force oneself to redefine the political task par excellence: what people are you forming, with what cosmology,

---

[88] Paul N. Edwards, *A Vast Machine: Computer Models, Climate Data, and the Politics of Global Warming* (2010).
[89] See Philippe Descola, *Beyond Nature and Culture* ([2005] 2013).
[90] The word "collective" brings together in a single concept precisely that which *collects* a multitude of agents defined neither by nature nor by society. On all these definitions, see Latour 2005.
[91] See Bruno Latour, *Politics of Nature: How to Bring the Sciences into Democracy* ([1999] 2004b).

and on what territory? One thing is certain: these actors who are making their stage debuts have never before played roles in a plot as dense and as enigmatic as this one! We have to get used to it: we have entered irreversibly into an epoch that is at once post-natural, post-human, and post-epistemological! This makes a lot of "posts"? Yes, but this is exactly what has changed around us. We are no longer exactly modern humans in the old style; we are no longer living in the Holocene!

The redistribution of agency – what used to be called, not so long ago, the "environmental questions"! – is not a way to assemble the concerned parties peacefully. It divides more effectively than all the political passions of the past – it always has. If Gaia could speak, it would say, like Jesus: "Do not suppose that I have come to bring peace to the earth. I did not come to bring peace, but a sword" (Matt. 10: 34). Or, more violently still, as in the apocryphal Gospel of Thomas: "I have cast fire upon the world, and behold, I guard it until it is ablaze."[92]

*

Let me conclude this lecture with another interpretation of the planetary clash at the end of a famous film by Lars von Trier.[93] The plot in part involves a stray planet named Melancholia, which is threatening to crash into the Earth; the threat reveals how the protagonists, each isolated from the rest of the world in their homes, will react to the catastrophe. Without spoiling the suspense for those of you who haven't seen it, I'll just say that it doesn't end well. The fragile tree-branch shelter built by the heroine to protect her sister and her nephew doesn't seem to suffice. Still, it is possible that the lesson of this metaphor is quite different: it might not be the Earth that is destroyed in a final, sublime, apocalyptic flash by a wandering planet; it might be our Globe, the global itself, our ideal notion of the Globe, that has to be destroyed, so that a work of art, an *aesthetic*, can emerge.[94] Provided that you agree to hear in the word "aesthetic" its

---

[92] Apocryphal Gospel attributed to Thomas, Logion 10, www.earlychristianwritings.com/thomas/gospelthomas10.html.
[93] Lars von Trier, *Melancholia* (2011).
[94] "For that reason, Gaia resembles planet Melancholia much more than it does the Earth. Melancholia is an image of the titanic, enigmatic transcendence of Gaia, an entity that suddenly and devastatingly falls on a world, ours, that has suddenly become all too human" (Déborah Danowski and Eduardo Viveiros de Castro, *The Ends of the World*, 2016, p. 41).

old sense of capacity to "perceive" and to be "concerned" – in other words, a capacity to make oneself sensitive that *precedes* all distinctions among the instruments of science, politics, art, and religion.

In one of his many linguistic innovations, Sloterdijk suggested that we need to pass from monotheism and its old obsession with the form of the Globe to monogeism.[95] The monogeists are those who have no spare planet, who have only one Earth, but who do not know its form any better than they know the face of their former God – and who are thus confronted with what could be called an entirely new genre of geopolitical theology. Once the Globe has been destroyed, it has space and time enough so that history may start up again.

---

[95] Not to be confused with monogenism, a theory about the unique origin of humanity! "The proofs of God's existence must inevitably bear the blemish of their failure, while those of the globe's existence have an unstoppable influx of evidence on their side" (Sloterdijk, 2013, p. 6).

FIFTH LECTURE

How to convene the various peoples (of nature)?

> Two Leviathans, two cosmologies • How to avoid war between the gods? • A perilous diplomatic project • The impossible convocation of a "people of nature" • How to give negotiation a chance? • On the conflict between science and religion • Uncertainty about the meaning of the word "end" • Comparing collectives in combat • Doing without any natural religion

When I noticed the new issue of *Nature* on the newsstand, I thought the figure that had been haunting me for four or five years, the colossus whose troubling power I hadn't been able to shake off, was looking at me with his blind eyes and advancing toward me. I thought I was about to merge with this composite body, more colorful than a Harlequin costume.[1] The metamorphic zone in which all the properties we're trying to trace in these lectures are exchanged is this very body, made up of guts intestines mines galleries, arms vegetables fauna, factories wrists and muscles, plexus major discoveries Columbus's sailing ships, cities shoulders missiles, oceans clouds sternum, clavicles atomic explosions, the whole quite strangely framed: above, by the title of the journal, *Nature*; below, by the title

---

[1] *Nature*, March 11, 2015.

How to convene the various peoples (of nature)?

A

Reprinted by permission from Macmillan Publishers Ltd: *Nature*, Volume 519, Issue 7542, 12 March 2015, © 2015.

Figure 5.1a Front cover of *Nature*, March 12, 2015, by Alberto Seveso.

of the cover story, "The Human Epoch," terms that had been opposed for three centuries, before both of them were dissolved by the Anthropocene – the epoch that this issue of the journal is precisely seeking to define and date.

Looking at the cover, I couldn't help being struck by its family resemblance to another monster, the "mortal god," a much more familiar composite image found in the frontispiece of Hobbes's *Leviathan*, a work that in large part determined the religious, political, and scientific history of the Moderns that I shall use throughout the lectures that follow.[2] You surely recall this image, in which, with the sword of civil power in one hand and the cross of spiritual power in the other, this macrocephalic giant – a worthy forerunner of the Giant Marionettes of the Royal de Luxe theater company (an agglomeration of tiny men reflected in a crowned head thanks to

---

[2] The connections among all these realms, which historiography tended to distinguish, became visible starting with Stephen Shapin and Simon Schaffer, *Leviathan and the Air-Pump: Hobbes, Boyle, and the Experimental Life: Including a Translation of Thomas Hobbes, Dialogus physicus de natura aeris*, by Simon Schaffer (1985).

Figure 5.1b Frontispiece of Thomas Hobbes's *Leviathan*, drawn by Abraham Bosse, 1651.

a subtle optical procedure)³ – dominates a vast landscape of cities, countrysides, fortresses, and castles.⁴ As Hobbes explains throughout his book, nothing less will do if people are to stop cutting one another's throats. Only the invention of a State strong enough to obtain incontrovertible assent from all its subjects could put an end to the wars of religion. To re-establish civil peace, the "mortal god" of the State had to take the place of the "immortal God" invoked by

---

³In "Seeing Double: How to Make Up a Phantom Body Politic" (2005), Simon Schaffer showed how the head was enlarged by a simple optical procedure borrowed from Abbé Nicéron.

⁴This frontispiece has fascinated historians of art such as Horst Bredekamp (*Stratégies visuelles de Thomas Hobbes: le Léviathan archétype de l'État moderne*, 2003); Dario Gamboni, ("Composing the Body Politic: Composite Images and Political Representations, 1651–2004," 2005); and also Carl Schmitt (*The Leviathan in the State Theory of Thomas Hobbes: Meaning and Failure of a Political Symbol*, [1938] 1996), to whom we shall return in the seventh lecture. On the cover of their book, Shapin and Schaffer replaced the cross of spiritual power with Boyle's air pump, the scientific instrument that first came to symbolize the new political epistemology.

all the fundamentalists of the period, each in his own way, so as to overthrow the established order.[5]

The frontispiece illustrated Hobbes's new distribution of all the agencies: inert matter, a world governed mechanically by the laws of nature, a society driven solely by the passion of interest, a strictly controlled interpretation of the figurative language of the Bible, and a definition of scientific truth as unquestionable as the propositions of Euclid. This is exactly what the drawing offered by the journal I had in front of me called into question: an animated world, an Earth that vibrates underfoot, no recognizable landscape, no affirmed authority, frightful mixtures, a proliferation of hybrids, scattered members of sciences, industries, and technologies. And, especially, the discouraging impression that this collective headless body is walking blindly with its arms hanging down; the figure stands out against the dark background without knowing where it is going or whom it is going to meet! Facing the Leviathan, you know who you are and before what authority you have to bow; but how are you to behave before this other Cosmocolossus?[6]

Setting up these two idols side by side, I could not help thinking that we were perhaps witnessing the return to war of all against all. Hobbes thought he had settled the question of order by extracting civil society from the state of nature through a solemn contract that made it possible to construct the artificial machinery of the Leviathan out of whole cloth. Is it possible that Hobbes's solution is being called into question today by another monster, the hybrid of geology and anthropology designated naïvely by the journal as the "human epoch," a new amalgam of artifice and nature? Or could it be a matter of inventing, through a new compact, a new contract, a new artifice, something that one could call the State of Nature?[7]

Whereas in the seventeenth century it was necessary, according to Hobbes, that matter be declared inanimate so that order could be re-established, at the beginning of the twenty-first the Earth's

---

[5] "This is the generation of that great LEVIATHAN, or rather (to speak more reverently) of that *Mortal God* to whom we owe under the *Immortal God*, our peace and defence" (Thomas Hobbes, *Leviathan* [1651] 1998, p. 114).
[6] This is the name I first gave to the theatrical project that later became *Gaia Global Circus* (Pierre Daubigny, "Gaia Global Circus," 2013); see the introduction.
[7] The capital letters are going to be important from here on to distinguish the state of nature – a Hobbesian myth required to contrast with the State – and the State of Nature, which is indeed, in fact, the constitution under which the Moderns lived until the emergence of the ecological mutation and the "end" of the notion of "nature." See Bruno Latour, *Politics of Nature: How to Bring the Sciences into Democracy* ([1999] 2004b).

retroaction in response to our behaviors suffices to disrupt order thoroughly. In any case, as in the time of the "Glorious Revolution,"[8] we can no more indulge in the belief that the question of nature has been resolved, that religion is a thing of the past, that science offers an unquestionable certainty, that we can fool ourselves into believing that we know the driving forces that agitate humans or the goals of politics. We may doubt that the Anthropocene marks a geological period, but not that it designates a transition that obliges us to take up all these concerns anew.

It would be more comfortable, as I am quite prepared to recognize, to leave aside the religious question! How we would all like to believe that religion is behind us! Hobbes must have had the same thought. But it is too late. Not only because of what is called the "return of the religious" or the "rise of fundamentalisms," but because Gaia's appearance on the scene obliges us to doubt *all* encompassing religions, including those that have to be called *religions of nature*. The paradox is rather amusing: Gaia is accused of being "a religion taken for a science," when it is the emergence of Gaia, on the contrary, that obliges us to redistribute the features of the preceding epoch, including the strange idea that construed the Nature known to Science as something that had to *oppose* Religion (I am keeping the capital letters here not as a sign of solemnity but as a reminder that we are dealing with figures of speech, not with domains of the world). If we were to try to separate Science and Religion today, from the vantage point of the Anthropocene, it would be a real massacre, given how much Science there is in Religion and how much Religion there is in Science. By trying to separate them, such as they are, before rethinking them both, we would lose any chance of bringing them both back to Earth, *separately at last*.[9] This is one of the strengths of Gaia, this acid powerful enough to corrode the amalgam of any *natural religion*.

In any case, we don't have a choice, since the disaggregation of the old Nature/Culture format forces us to redraw the limits of all the collectives.[10] In the epoch of the Anthropocene, it would be pretty futile to want to do without anthropology. The same question confronts all

---

[8] This is the name the English gave to the end of the civil wars of religion, in 1689, and to the establishment of a new constitutional order.
[9] Here we see all the ambiguity of "natural religion," the term proposed as the topic for the Gifford Lectures. One can see in it either the search for "proofs of the existence of God via science" or the search for a place left for spirituality in an entirely material world (the latter is what a large number of Gifford lecturers have done). But one can also try to uncover the origin of a problem introduced so inauspiciously.
[10] *Collective*, let us recall, is the term that replaces the old asymmetrical concepts of society or culture (see the previous lecture). Society (or culture) is half of a single concept, the other half of which is constituted by nature.

cosmologies: what does it mean, for a people, to *measure*, *represent*, and *compose* the form of the Earth to which they find themselves attached?

In this fifth lecture, I am going to focus, I am afraid, on an operation of science fiction that will be somewhat reminiscent of the television series *Game of Thrones*![11] Of course, I will not be dealing with the realm of Westeros or with the Seven Thrones, or with finding out whether the blond Daenerys Targaryen does or does not regain the iron throne of her ancestors. What I want to draw is a rough map of the territories occupied by peoples struggling against one another. To produce such a drawing, we shall have to learn to see how the collectives badly assembled up to now by the Nature/Culture format could be defined and articulated mutually, using procedures that we might call operations of war or peace – in other words, operations of *risky diplomacy*. We shall try to make collectives comparable by asking them to make explicit, each one for the others, four variables that will define their cosmology, for a while, by answering four questions:

- By what *supreme authority* do they believe they have been convoked?
- What limit do they give their *people*?
- What *territory* do they believe they are inhabiting?
- In what *epoch* are they confident they are living?

To these questions, we shall have to add a fifth:

- What *principle* of organization distributes agency (a principle that I shall call its *cosmogram*)?

Let us agree that we are going to compare different peoples, each one convoked by a different entity that defines, orders, classifies, composes, divides up – in short, *distributes* – different types of agency in different ways, each according to its cosmology.

I recognize that this questionnaire is quite rudimentary with respect to all the variables that anthropology ought to take into account, and that the former list is built around Western concepts, but it is the object of a diplomatic proposition to start somewhere with what the diplomats understand best.[12] I will use the list to stop addressing

---

[11] This HBO series, inspired by George R. R. Martin's fantasy novels, has a cult following.
[12] I follow here the lessons drawn from Richard White to start building what he calls a fragile middle ground: *The Middle Ground: Indians, Empires, and Republics in the Great Lakes Region, 1650–1815* ([1991] 2011).

others by asking them "What is your specific *culture?*," leaving aside their necessarily common nature. This is the only way I have found to shatter the false unanimity that always comes after the appeal to Nature. Thanks to this approach, we're going to be able to begin to trace the new geopolitical – or, better, Gaia-political – situation that will be our concern in the subsequent lectures. We shall find less gore than in *Game of Thrones* (and no sex at all) but only the violence that those who claim to be assembling peoples to defend themselves against those who seek to destroy their land have to learn to confront head on. This will be no surprise, since we are now indeed engaged in a war of the worlds.

*

To begin this delicate task of convocation, it would be useful to have at hand a provisional definition of the term *religion*. I shall turn to Michel Serres for the definition that strikes me as most likely not to irritate contemporary readers at the outset:

> The learned say that the word religion could have two sources or origins. According to the first, it would come from the Latin verb *religare*, to attach.... According to the second origin, which is more probable, though not certain, and related to the first one, it would mean to assemble, gather, lift up, traverse, or reread. But they never say what sublime word our language opposes to the religious, in order to deny it: *negligence*. Whoever has no religion should not be called an atheist or unbeliever, but negligent. The notion of negligence makes it possible to understand our time and our weather [our climate (*notre temps*)].[13]

At this stage, the word "religion" does no more than designate that to which one clings, what one protects carefully, what one thus is careful not to neglect. In this sense, understandably, *there is no such thing as an irreligious collective*. But there are collectives that *neglect* many elements that *other collectives* consider extremely important and that they need to care for constantly. To introduce the religious question again is thus not first of all to embarrass oneself with beliefs in some more or less strange phenomenon, but to become attentive to the shock, the scandal, that the *lack of care* on the part of one collective can represent for another. In other words, to be religious is first of all to become attentive to that to which others cling. It is thus, in part, to learn to behave as a diplomat.[14]

---

[13] Michel Serres, *The Natural Contract* (1995, pp. 47–8).
[14] See Isabelle Stengers, *La vierge et le neutrino* (2005). On the question of diplomacy as a method of inquiry, see Bruno Latour, *An Inquiry into Modes of Existence:*

How to convene the various peoples (of nature)? 153

To address a collective is first to find a way of naming what it respects the most, what it recognizes as its *supreme authority*. If a collective takes care of itself, and sometimes of others, it might be because it invokes a *divinity*, or rather – so as not to shock sensitive readers – a *deity* by which it feels it is being convoked. We have known this as long as anthropology has existed: there is no collective without a ritual during which people discover that the only real way to come together as a group entails being convoked by this authority and invoking it in return.[15] We learned this from Durkheim, who demonstrated that the figure of Society, with a capital S, could play the role of supreme authority for certain modernized peoples[16] – and we have understood, over the course of the last century, that the Market, always with a capital M, could also serve as the authority of last resort over vast territories.[17] In this sense, there is no such thing as a durably *secularized* collective; there are only collectives that have modified the name and the properties of the supreme authority in whose name they gather.

But we also know that the back-and-forth movement that connects a people brought together by *its* divinities with other divinities invoked by the peoples *they* bring together cannot resist the corrosive influence of critique for long. The slightest mark of distance or indifference is enough to reduce divinities to the status of decorative themes. This is what happened to the immortal gods of Antiquity: they disappeared with the people to whom they *belonged* and that they themselves *held in their grip*. They were mortal, after all, and it is only their phantoms that have become a source of amusement or nostalgia. It would be ridiculous, for example, to start invoking the ancient Gaia today with a hymn such as this:

> To Gaia, mother of all, shall I sing:
> The oldest one, firm foundation of all the world.
> All things that move over the face of the earth,
> All things that move through the sea, and all that fly:
> All these are fed and nourished from your store;
> From you all children and all good harvests come forth…

An Anthropology of the Moderns ([2012] 2013b) and the associated site http://modesofexistence.org under the word "diplomacy."

[15] Which does not mean it has a unity or any form of transcendence, simply something that, if it were withdrawn, would mean the collective has ceased to exist.

[16] Émile Durkheim, *The Elementary Forms of the Religious Life* ([1912] 1965), and my analysis of this canonical text, in "Formes élémentaires de la sociologie: formes avancées de la théologie" (2014b).

[17] Michel Callon, ed., *The Laws of the Markets* (1998b).

> Your fertile earth yields up riches to satisfy all their needs;
> ...blessed spirit, may your fruits increase, and with joyful heart
> Look kindly upon us this time and for ever.[18]

Such an invocation would be taken as facile irony or as a futile attempt to resuscitate a cult that disappeared long ago. For such a text to ring true, there has to be a real *people* that feels totally indissociable from this divinity through deeply rooted rituals. Nothing is further from my intention, as you will have understood, than to make you laugh at the evocation of Gaia or to make you believe that Gaia is only a figure of the past – a shade, a phantom. This is why I won't try to invoke this character directly, since we don't share enough of the same culture, don't belong to the same people, don't fall back on the same rituals that would put us in a position to salute the ancient Ge with the name of *justissima Tellus*.[19]

But how are we to go about asking a collective to specify the name, attributes, functions, origins, and figure of a supreme authority of this sort when a given collective announces proudly *that it recognizes no divinity*? On this point, we shall have to take our time and move, as we are now accustomed to doing,[20] from the *name* given to figures to the *behaviors* of these same figures. Divinities, like concepts, like heroes of history, like objects in the "natural world" – rivers, rocks, streams, hormones, yeasts – have competence – and thus substance – only through the performances – the attributes – that give them form *in fine*. To behave diplomatically, when one is manipulating materials as explosive as deities, is to require oneself always to begin with the attributes, so as not to fight right away over the substances.

Jan Assmann, the great Egyptologist and historian of mythic memory, has reminded us that there was a venerable tradition in the various city-states of the Mediterranean and the Middle East, before the advent of Judaism and Christianity, in which *translation tables* were drawn up for the names of the gods that were worshipped.[21] In an epoch that was becoming cosmopolitan,[22] these translations

---

[18] "Hymn to Gaia," I and II, an ancient Greek Homeric hymn.
[19] "The most just Earth," cited from Virgil by Carl Schmitt in *The Nomos of the Earth in the International Law of the Jus Publicum Europaeum* ([1950] 2003), a text that we shall revisit in the seventh lecture.
[20] A glance at the method proposed in the second lecture will help keep readers from getting lost in what follows.
[21] "The gods were international because they were cosmic. The different peoples worshipped different gods, but nobody contested the reality of foreign gods and the legitimacy of foreign forms of worship" (Jan Assmann, *Moses the Egyptian: The Memory of Egypt in Western Monotheism*, 1998, p. 13).
[22] Eric H. Cline, *1177 B.C.: The Year Civilization Collapsed* (2014).

How to convene the various peoples (of nature)? 155

offered a practical solution to the moderate relativism with which every adherent to a local cult recognized its relationship to the local cults of the many strangers that lived among them at that time. "What you, a Roman call Jupiter, I, a Greek, call Zeus," and so forth.

According to Assmann, these translation tables worked by shifting attention from the proper names of the divinities to a series of characteristics that the names summed up in the mind of their worshippers. If, for example, the name "Zeus" was incomprehensible to a listener, the speaker would reel off the list of his *attributes*: "Guide of destinies" (*Moiragétès*), "Protector of the suppliants" (*Ikesios*), or "God of the favorable winds" (*Evanémos*), and, of course, "Bearer of thunder" (*Astrapeios*), until the foreigner found a corresponding divinity in his own language. The precaution that such people took to cohabit without cutting each other's throats was to make sure that, if the list of qualities was similar enough, they could take the proper names to be more or less synonymous – or in any case negotiable: "Your people names him this, my people call him that, but through these invocations we designate the same deity, who carries out the same type of actions in the world." This form of inter-translation thus offered a political solution to ensure civil peace in societies with multiple attachments: as long as people clung to the names, they fought endlessly and in vain. The translation tables of the names of gods in the ancient city-states were at once the result of and the occasion for diplomatic negotiations in the great cosmopolitan cities.

But, as Assmann shows in a provocative and convincing way, the diplomatic situation that allowed inter-translation became impossible after what he calls the "Mosaic division," which he associates with the ancestral figure of the God of Moses – preceded by the still more ancient figure of Akhenaten's god.[23] A completely new relation is then introduced between the question of divinities and the question of truth. Starting from this point of rupture in history, we are going to be able to spot the emergence of religion through the reactions of horror in the face of the moderate relativism that had been authorized by the tables of gods' names and by the multiplication of

---

[23] See Assmann 1998, and especially his subsequent book *The Price of Monotheism* (2010), which catalogues the disputes set off by the first work: "The distinction between true and false religions...was unknown to traditional, historically evolved religions and cultures. Here the key differences were those between the sacred and the profane or the pure and the impure. Neglecting an important deity amounted to a far more serious offense than worshipping false gods, the chief concern of secondary religions. In principle, all religions had the same truth-value and it was generally acknowledged that relations of translatability pertained between foreign gods and one's own" (p. 23).

iconoclastic gestures.[24] Whatever these tables may have allowed in the past, the "one, unique God" could no longer be synonymous with any other deity whatsoever. Translating the name of the one into the name of the Other became not only unfeasible but scandalous and even impious. "True" divinity became untranslatable by any other name; no cult but its own could be tolerated, on pain of idolatry. It is as if the real God had fulminated: "You shall not make my cult *commensurable* with any other, under any circumstances." The old sense of the word "religion" was no longer comprehensible: quite to the contrary, the new injunction required neglecting that to which the others clung! This is why Assmann proposes, for this new association between religion and truth, the apparently counter-intuitive term *counter-religion*, a term that will guide us in this lecture and the next.[25]

But what does this have to do with us today, you will ask? Have we not long since left that "Mosaic division" behind, accustomed as we are to comparing *religions* in the plural without being at all troubled that each claims to be truer than the others? What could possibly prevent comparisons? Have we not really and truly become pluralists? Aren't we in a world that is definitively secular at last? To be sure; but in the previous lecture we began to understand that *believing oneself to be irreligious* did not suffice for one to *be* irreligious. As we saw in the case of Toby Tyrrell, a professor of the sciences of the Earth System, it is not so easy to have a secular vision of the world.[26] One can consider oneself scientific and freed of any particular belief while attributing to Evolution, or to Lovelock's Gaia, properties that make them indistinguishable from the divinities of the Providential Globe. The name assigned to the supreme authority is less important than the qualities attributed to it.

If pluralism is so rare, appearances notwithstanding, it is because there is always a deity waiting in ambush that demands to be made commensurable with no other – its name matters little. Whatever we may think of the Moderns, however non-believing they deem themselves to be, however free of any divinity they may imagine themselves, they are indeed the direct heirs of that "Mosaic division,"

---

[24] "Hate as such did not come into the world with monotheistic truth, but a new kind of hate, the iconoclastic or theoclastic hatred of the monotheists for the old gods, which they declared to be idols, and the anti-monotheistic hatred nursed by those whom the Mosaic distinction excluded and denigrated as pagans" (ibid., p. 67).
[25] Counter-religions, or secondary religions, are thus distinguished from primary religions. See Jan Assmann, *Violence et monothéisme* (2009).
[26] See the fourth lecture.

How to convene the various peoples (of nature)?   157

since they continue to connect supreme authority with truth, with one nuance: the division henceforth passes *between, on the one hand, believing in any religion at all and, on the other, knowing the truth about nature.* We can now understand the odd name *counter*-religion given by Assmann: it is just as applicable to the religions called – for the purpose of simplification – monotheistic as to the new counter-religion that is arising *against* all religions, *including* the monotheisms. To declare oneself without any divinity at all does not suffice to cast into oblivion the voice of the supreme authority that, for its part, fulminates just as violently as the previous one: "You shall under no circumstances make *knowledge* of the laws of nature *commensurable* with any cult." A strange law, requiring neglect of what the others hold dear! Whether we like it or not, we are the descendants of a division that obliges us to associate the supreme authority to which we entrust our fates with the question of truth. Even those who violently reject the monotheistic religions have borrowed from them this quite particular way of violently rejecting idolatry. Iconoclasm is our common good.[27] From the true God fulminating against all idols, we have moved to the true Nature fulminating against all the false gods. The division has remained, as have the lightning, the thunder, and the smell of the storm.

You see where the difficulty lies: it is already hard enough to convoke the religions to make them comparable to one another, even if they are accustoming to bowing down, with more or less good grace, before this now popularized form of pluralism; but how can we hope that the negotiation will not immediately be aborted if one of the collectives indignantly refuses to say what territory it occupies, what supreme authority gathers it together, in what epoch it is situated, and what principle of composition it recognizes?

It is with this problem in mind that I would like to situate the new diplomatic question: is it possible to reinvent this practice of tables translating the names of gods for the purpose of listing other entities, other cults, other peoples, and spotting among these various collectives the relationships that remain invisible as long as we stick to our overly local and overly sectarian viewpoint? If we have to make war – the war of the worlds – we want to assure ourselves that we are cutting each other's throats not over names but over features that differentiate between real friends and real enemies. If these are the territories that are engaged in battle, then we have to be able to trace

---

[27] Bruno Latour and Peter Weibel, eds, *Iconoclash: Beyond the Image Wars in Science, Religion and Art* (2002).

their borders. Even a superficial sketch is preferable to the absence of any map.

*

The very idea of a negotiation among peoples made commensurable by the form of relativism – better still, of relationism – proper to translation tables of gods' names can only arouse, from the start, as I well know, a cry of indignation. "How do you dare make comparable those who believe in more or less bizarre divinities and those who speak of 'Nature,' when these two invocations are totally incommensurable? Even the term 'invocation' is shocking. Invoke Gaia, Allah, Jesus, or Buddha if that amuses you, but it is intolerable that you should speak in the same way of 'invoking' Nature. Between the first four names and the last, there has to remain a gulf that no negotiation can fill!" The intensity of the indignation allows us to recognize the line drawn by this radical division between the false gods and the true one, even if the division now comes between what is said about the gods, on one side, and what is said about "reality," on the other. "You can't compare these entities." "You have to choose your camp." "Nature is not a religion." Or, to parody a famous quip: "When I hear somebody talking about Nature that way, I reach for my revolver!"

But wait! We are here to think, not to fight – at least for now. We want to shift attention from nouns to attributes. Before sending each other to be burned at the stake, let's first set up a list of the characteristics that you bring together under your emblem and that others bring together, perhaps, under a different denomination. "But nature," you'll say, "is neither an 'emblem' nor a 'denomination': it is the matter of which we are made and in which we all live." I know, but I've asked you to wait, to be patient: what you are expressing here is what you require others not to neglect *when they are talking to you*. Very well. Let us agree now to listen to other cries of indignation against other culpable negligences. If you agree to a truce for a moment, I believe that it will not be impossible to propose a suspension of hostilities, since, as we have already seen in the previous lectures, "Nature," despite its reputation of incontrovertibility, is the most obscure concept there is, or in any case the least apt to bring a conflict to a definitive end.

It would not be a bad thing, moreover, to take a bit of distance from this overly fascinating term "Nature," about which we forget too quickly, even when we add a capital letter and quotation marks, that it is not a domain but a concept. I am going to fall back on a stratagem that I promise to leave behind once it has produced its effect: I am going to try to define the *people* associated with this supreme authority whose features we are going to attempt to specify. What name shall we give to the authority? To avoid the word "God," which would be too disrespectful, too provocative, in this context, I propose "Out-of-Which-We-Are-All-Born," "*OWWAAB*." If that sounds very odd, it is just the sort of oddness I need, for it will facilitate inter-translation with other titles and invocations. For a few moments, I need to take on the style of a George R. R. Martin. As in *Game of Thrones*, it can be convenient for strangers to greet one another by saying, for example: "You are the people of OWWAAB, we belong to the people of Zeus; those folks over there guarding the northern border are the people of Odin!"

How are we going to designate the loop that connects the "people of Nature" and that supreme entity? If I fall back on the word "religion," even if I stick to the definition given above, the opposite of negligence, I fear that the negotiation may come to an abrupt end without having shed any light at all either on the ancient cults or on the cult of the "naturalists." The experts will cry out indignantly: "Belonging to the people of Nature is not a religion!" – and they will not be wrong. But if they are not wrong, it is for the simple reason that all the words that have to constitute the vocabulary of titles on the left of the translation table[28] have to be versatile enough to concentrate attention solely on the list of characteristics, on the attributes. This is the only way to allow negotiations to continue. For this reason, let's settle on the word "cosmogram."[29]

In our day, as in Antiquity, it is because we live in cosmopolitan city-states and because we have divergent ways of occupying the Earth that we have to engage in such a risky exercise. If we could stick to our own particularities, to our identities, we would not need to invent some sort of instrument to make the collectives commensurable. We would have no need for this relativism – by which I mean the establishment of relations. But, today, we are globalized through and through, torn between the effort to avoid total war and the requirement of complete harmony, clinging to the hope that we will succeed in spite of everything in forging some *modus vivendi*. In any case, those who are prepared to cross swords have never agreed to sit down at the negotiating table – they have been on the warpath for a long time, armed from head to toe, and we are the ones who

---

[28] See table 5.1, p. 168 below.
[29] A term offered by John Tresch and deployed with great efficacy in *The Romantic Machine* (2012).

are slowly beginning to equip ourselves in the hope of responding to them one day.

*

If we have to start by filling in the portrait of the people of OWWAAB in their absence, *in absentia*, in a sense, it is because they have the strangest way of being and not being of this world. They refuse to be a people and to be limited to a territory. They are at once everywhere and nowhere, absent and present, invasive and stupefyingly negligent. If we set up the table of attributes, we understand right away why they don't constitute a collective. Its adherents depict OWWAAB by six qualifiers: it is *external, unified*, and *inanimate*; its decrees are *indisputable*, its people is *universal*, and the epoch in which it is situated is *of all time*. Except they also assert that OWWAAB is *internal, multiple, animated*, and *controversial*; that its people are reduced to *a few* and that they live in an epoch from which all the others are separated by a *radical revolution*. Between the two columns, there is no discernible link! We can understand why this people divided against itself is so uneasy, so unstable. And we won't be surprised to find that it reacts just as badly to the emergence of Gaia as to the hypothesis of the Anthropocene. Taken together, these two phenomena would oblige it to find an anchorage, to locate itself, to make clear, finally, what it wants, what it is, and to specify, finally, who are its friends and who are its enemies.

Let's start with the expression "external." Apparently, its adherents mean by it something like this: "Which does not depend on the wishes, whims, and fancies of the people who invoke it. OWWAAB is not negotiable!" There is nothing astonishing here. This attribute is common to all entities capable of gathering a people around their supreme authority. It is because they are *beyond* their people that these entities possess the power to convoke them and gather them together. Their apparent transcendence is part of their definition. Which is another way of saying that a supreme authority is an authority that is, in fact, *supreme*.

But if we dig a little deeper, we encounter an apparently contradictory property: OWWAAB is at once outside and beyond, to be sure, but it is also *within* refined networks of practices that seem indispensable and that are called "scientific disciplines." Every time we indicate a characteristic of the "natural world" that corresponds to certain properties of OWWAAB, we are obliged also to follow the complicated path by which objective knowledge is produced. Our sights are focused simultaneously on infinity and on the foreground

– unsuccessfully, of course, as we saw in the previous lecture. The tension between the outside and the inside of this entity is extreme: insofar as it is a set of *results*, OWWAAB is *outside*. We might even say that its decrees are like the icons called *acheiropoietes* – that is, "not made by human hands."[30] Insofar as this entity is a *process of production*, its decrees are found *inside* conduits where numerous human hands helped by numerous instruments are bent on making it an *external* reality.

It is as though the public could not accommodate – in the optical sense of the word – these two levels at the same time: the first always remains blurred when the second is in sharp focus. We have already come across many examples of this bifocalism, but I can't keep from thinking in particular about the false controversy over what has been called "Climategate," a debate that arose just before the big 2009 climate meeting in Copenhagen, COP 15.[31] The climate skeptics thought they could weaken these scientific truths by "revealing" that they had been produced by men and women! As if such a revelation ought to provoke a scandal! As if it were impossible to accept the idea that global warming was actually real, "outside," in nature, without any manipulation of the data, and that such a certainty came *nevertheless* from within the networks of scientists exchanging millions of emails and sharing interpretations of data concerning computer models, satellite views, and fragments of sedimentary cores obtained at great cost from dozens of expensive explorations! As if it were still impossible to solve this problem of bifocal vision and to follow the way facts are at once carefully *fabricated* and made *factual owing to the care* taken in such fabrication. There should be no more contradictions here than in the so-called automated technologies, whose engineers know perfectly well that they are only auto-matic provided that a crowd of assistants accompanies them to *make* them work automatically – in the final analysis, nothing is more *heteromatic* than a robot.

Whereas many other cultures are bent on exploring this contradiction, the people of Nature have not given it a thought. It is as though

---

[30] Recognizing the human hand at work in the production of the sciences and ignoring it in the production of beliefs underlies the ambiguity of all constructivism. This was precisely the object of my essay *On the Modern Cult of the Factish Gods* ([1996] 2010a).

[31] The artificially constructed controversy over the existence of a link between human activity and global warming depended solely on the "revelation" of researchers' everyday work. See https://en.wikipedia.org/wiki/Climatic_Research_Unit_email_controversy.

these people had to make their cosmology revolve around *two focal points* at the same time: one in which everything is external, where nothing is made by humans; the other in which everything is internal and made by humans. Like an unstable Copernican revolution with two suns at the same time, around which the Earth is zigzagging erratically without ever coming to rest.[32] Here is a clear indication, for the other peoples who are trying to translate this entity into their own language, that the behavior of this collective is bizarre and even dangerous. They could ask its members: "On what Earth do you live, then?"

That this people might belong to *no Earth at all* becomes probable when the second attribute is taken into consideration. "OWWAAB is unified and all agents obey its universal laws." And yet it is just as hard to reconcile this universality with the prodigious diversity of the scientific disciplines, the specialties, subspecialties, thematic networks and domains in which these "unified" and "universal" laws are applied in practice. Naturally, practice could be left out of the description, but we are involved in moving from ideas to practice, from names to characteristics, from concepts to agencies.

Viewed this way, the jungle of scientific descriptions looks more like the legal institution, with its complex casuistry of diverse codes and intermingled jurisprudences, than the unification implied in the traditional expression "laws of nature." To be sure, at the local level there are some processes of unification in which a phenomenon is explained, justified, digested, absorbed, and understood by another, more encompassing solution – and this is fortunate. But the process of totalization and inclusion is itself always local and costly, and it has to be accomplished by the immense efforts of multiple organizations, multiple theories, multiple paradigms.[33] The process resembles the way legal precedents gradually take on importance through the multiplication of cases, trials, appeals, and counter-appeals, until the precedents invoked by the various courts of justice acquire the status of warranted, relatively universal principles – at least for as long as they are cited, archived, and interpreted.[34]

If, in the course of negotiation, those who frequent this strange people have been surprised by the first two attributes of OWWAAB

---

[32] An instability well identified by Peter Sloterdijk; see the fourth lecture.
[33] See Nancy Cartwright, *The Dappled World: A Study of the Boundaries of Science* (1999).
[34] For examples of gradual unification of universal laws, see Peter Galison, *Einstein's Clocks and Poincaré's Maps: Empires of Time* (2003).

How to convene the various peoples (of nature)? 163

– exteriority and universality – what are they going to think of the third: that OWWAAB deals only with *inanimate* agents? All the other peoples will find something even more enigmatic here. As we have seen, starting with the first lecture, the contradiction lies in the words themselves: an agent, an actor, an actant, by definition, is *that which acts*, that which has – is endowed with – agency.[35] How can one render the entire world "inanimate"? It turns out that this is not a mystification but a *mystique*, a very interesting and respectable mystique in many respects, as well as a very spiritual form of contradiction – let's say, an unexpected form of piety. Once again, every discipline, every specialty, every laboratory, every expedition *multiplies* the surprising agents of which the world is made – agents that can easily be followed through the proliferation of technical vocabulary that pervades scientific articles. Such proliferation might surprise us if we accepted the stunning vision implied by the term "reductionism." Normally, if one really achieved the *reduction* implied by that term, one ought to be prepared to read *fewer and fewer* articles, shorter and shorter articles written by fewer and fewer scientists, each one explaining more and more phenomena better and better, until someone reached a minuscule equation from which all the rest would be deduced, a prodigiously powerful flash of information that could be written on a bus ticket, a real Big Bang on the basis of which all the rest could be generated![36]

Now practice, once again, does exactly the opposite. Scientific literature constantly *multiplies* the number of agents that have to be taken into account in order to follow a course of action to its endpoint. If we replace the technical name of each of these agents by what they *do*, as the most elementary semiotic method requires, we do not find ourselves facing the oxymoron "inanimate agents"; on the contrary, we face a prodigious *multiplication* of potential agents. The clear result of the scientific disciplines is an immense *increase* in what moves, acts, heats up, boils over, and becomes complicated – in sum, in what actually *animates* the agents that constitute the world and in the continuous refinement of the *metamorphic zone* that we encountered in the earlier lectures. Even if you want to explain,

---

[35] *Trans*.: See the second lecture, note 20.
[36] This is the contradiction in all causalist discourse: if the cause really enjoyed the *textual* role attributed to it by the discourse, one would not really need what follows – the consequences would be superfluous, as it were. Hence the disconnect between what the text does and what the epistemology says. To put it differently, epistemology would be maintained only through indifference to textuality. Every causal account is thus also a narration: this is what brings it closest to the world.

account for, or simplify, it always requires an *addition* and not a *subtraction* of agents.[37]

"Why are these three contradictory characteristics not better instituted themselves, more effectively recognized or even better ritualized?" This question might well be asked by the other parties to the negotiations who seek to inter-translate "people of OWWAAB" into their own language. "Faced with such contradictions, here is certainly what *we* would have sought," they might say. The answer lies in the fourth property attributed to this entity: the indisputable quality of its decrees. In itself, this attribute is unremarkable. The "brute facts" – what the English language, which invented the idea, calls "matters of fact" – are only the final results of very complex assemblages that allow reliable witnesses to validate the testimony of laboratory tests. These assemblages are in no way contained in the word "fact" – unless one remembers its etymology. Isolated, left to its own devices, cut off from its network of practices, a "matter of fact" is a weak injunction, too readily ignored. It *maintains* its indisputability only if support teams accompany it throughout its career.[38]

But what makes the attribution of indisputability to OWWAAB even stranger is the unexpected expansion of the discussions well beyond the narrow limits of specialists and experts. The controversies have developed to such an extent that laboratory scientists have been forced to increase drastically the number of those who contribute to the fabrication of the facts. They have had to engage many other members from the public at large, members who earlier would have been solicited only to learn, study, repeat, use, or simplify the established facts, never to debate them or participate in their production, evaluation, or revision.[39] *Matters of fact*, to use my wording, have become so many *matters of concern*.

We can understand the reaction of the other peoples in the face of this series of contradictory injunctions: "Who are they, really, those people who are capable of alternating, without even being aware that they are doing so, between such radically opposed requirements?" And things don't get any better with the fifth attribute that

---

[37] Let me recall a passage from Alfred North Whitehead: "We are instinctively willing to believe that by due attention, more can be found in nature than that which is observed at first sight. But we will not be content with less" (*The Concept of Nature*, 1920, p. 29).
[38] This is one of the best-documented tenets of science studies; see E. Hackett, O. Amsterdamska, M. Lynch, and J. Wacjman, eds, *The Handbook of Science and Technology Studies* (2007).
[39] See Tommaso Venturini, "Diving in Magma: How to Explore Controversies with Actor-Network Theory" (2010).

How to convene the various peoples (of nature)? 165

the adherents to OWWAAB ascribe to their deity. At first glance, *everyone* can invoke the deity as supreme authority, since the people who invoke it define it as "Out-of-Which-*We*-Are-*All*-Born." "We" and "all": the ambition to gather together is not a modest one! But, from another standpoint, we quickly notice that this gathering does not involve everybody, but only those who are sometimes called the "rational people" or the "educated public," or even, in a still more restrictive way, "those who have studied these questions," the specialists, the experts. Nevertheless, this restriction does not yet delimit the form of the actual people, since these "proof workers"[40] need to be well equipped; they need to have the right materials and the right financing, they need to have agreed to long years of training, and they need to subscribe to a system of evaluation, certification, standardization, and verification of data that reduces their number, on each somewhat delicate question, to a few dozen. The human race is shrinking down to a happy few!

This people is decidedly indefinable, all the more so in that it is as impossible to situate in time as in space. To what epoch does it belong? To none, since it is indifferent to history and it has access to universal truths that exist for all eternity. But at the same time, of course, this people has a history, and it sees itself as the heir to a radical break, a recent break, which has allowed it to escape from an archaic, obscure, confused past in order to enter into a more luminous epoch that makes radical distinctions possible between past, present, and a glorious future: something like a scientific revolution. But, from another standpoint, there is nothing less easy to simplify than the history of each science, each concept, each instrument, each researcher, each as contingent, as multiform, as full of steps backward, zigzags, losses, forgettings, rediscoveries, as the rest of the history with which these scientific adventures find themselves, in any case, completely mixed.[41] This people without a history does indeed have a history that it is unable to reckon with and that it views as something as shameful as being limited to a specific time and space, or as being sure of nothing as long as it's not based on data obtained at great expense.

*

---

[40] In Gaston Bachelard's excellent expression; see *Le rationalisme appliqué* (1998, chapter 3).
[41] This is what is visible in the massive enterprise led by Dominique Pestre in particular: see Christophe Bonneuil and Dominique Pestre, eds, *Histoire des sciences et des savoirs*, vol. 3: *Le siècle des technosciences* (2015).

If the people of Nature cannot be convened, this is because it is precisely not a collective, since no process of composition makes it possible to *collect* the scattered members. How can we be surprised that it feels incapable of occupying the Earth, of knowing where it is and what it can do there, even as it claims to be grasping the Earth "in its globality." Torn between these two lists of features, it never sees how to reconcile them: its status of extraterritoriality prevents it from defining its territory; its universality prohibits it from understanding the relations that it must establish; its quest for objectivity paralyzes it in the face of controversies from which it no longer knows how to escape; its claim to encompass everyone leaves it disconcerted before the small number of those who truly belong to it; as for its history, it never knows whether it is supposed to escape from the present time through a new revolution or escape from the very idea of radical revolution. The strangest thing of all, what has most surprised all the other peoples, is that *it believes that it is alone in finally inhabiting this material world, the true inanimate world here below*, whereas it comes from elsewhere and still resides in the lovely global space of nowhere! Here is the proof that it contains in itself something ferocious, dangerous, unstable, and – why not say it – profoundly unhappy. Yes, the people of Nature are wandering souls who never stop complaining about the irrationality of the rest of the world.

It is not surprising that this people never agrees to present itself as a collective, and especially not as one collective *among* the others, by spelling out its mode of collection, its cosmogram. And yet we have to try to bring it back to the negotiating table, to imagine a peace process. And thus we have to try to address it with some chance of being understood by its adherents. Let us take care not to hurt the feelings of persons who seem very sensitive to these contradictions but also seem to lack any resources for *overcoming* them. It is moreover because their researchers cannot overcome the contradictions that they appear so susceptible, so sensitive, in a constant state of anxiety, and that their sensitivity is so easily upset by any suspicion of "relativism."[42] But by the same token, if we wish to move on in our diplomatic parley, we cannot allow ourselves to say: "Oh! You are the ones who agree to live under the auspices of an external, unified, inanimate, indisputable, and thus indestructible external entity." We

---

[42] This sensitivity was tested during what has been called, with considerable exaggeration, the "war of the sciences"; see especially Isabelle Stengers, "La guerre des sciences: et la paix?" (1998).

How to convene the various peoples (of nature)? 167

cannot do this, since the attributes on which those adherents insist also reveal that Nature is inside, that it is multiple, that it agrees to come to grips with animate and highly controversial beings, that it has a confused history, and that its compass is as limited as it is variable.

To pacify them and offer them a bit of reassurance, we have to be in a position to address the people of Nature respectfully, in all its authority, *as an entity that is strong enough to resist any profanation*. (You will understand that I am not indulging here, despite appearances, in a bit of irony, but that I am embarking on a very delicate task of *composition*. Even if these people respect no one, we must try to speak to them with respect; this is the only way to struggle against any form of fundamentalism. We must especially avoid imitating their bad manners.)

What is certain is that it is impossible to address them with enough respect when one invokes their divinity in a tone that might be called *epistemological*, since, in that case, only the six attributes – externality, unity, inanimate agents, indisputability, universality, and atemporality – would be taken into account. We would only be indulging their illusion of extraterritoriality. But this people would not be invoked with enough respect, either, if we stressed only the six attributes in a tone that could be called critical or, better, *anthropological*.[43] We would not have resolved the break between the two columns. To succeed in calming the members of this people, to pacify them and bring them back down to Earth, we would need to manage to speak to them in a tone that could be called *secular* – or, better, *terrestrial* – which would make it possible to gather together the sixteen characteristics at the same time. If that remains impossible, it is because of the radical rupture that has been introduced between the two columns. As long as we have not understood the origin of that rupture, it will be impossible for us to pacify the relation of the people of Nature to the Earth and, incidentally, to offer scientists a version that does not oblige them to believe in the portrait the epistemologists have drawn of them.

---

[43] The anthropology of the sciences is a better term to designate the domain of "science studies," especially since the diplomatic turn allows for numerous connections with anthropology; see Julie Cruikshank, *Do Glaciers Listen? Local Knowledge, Colonial Encounters, and Social Imagination* (2010), and Anna L. Tsing, *The Mushroom at the End of the World: On the Possibility of Life in Capitalist Ruins* (2015). Learning to live in the ruins "on the edge of extinction" is also the experience to which we are invited by Thom Van Dooren's astonishing book *Flight Ways: Life and Loss at the Edge of Extinction* (2014).

You don't need to tell me that no known repertory exists for pacifying this people impossible to convoke: I know this all too well! Scientists – column one – and researchers – column two – are two different species. This is why I am seizing the occasion of the Anthropocene to go in search of the origin of that impossibility, right where it is to be found, namely, in the *counter*-religion that the people of Nature have inherited without wanting to sort out its components. Yes, Nature is indeed *against* religion, but in two distinct senses, of which its people is conscious of only one. The matter is too important to rush through it. If we are truly seeking a *modus vivendi*, then we have to invent new ways to tolerate each other or to decide who are really our enemies. Who ever said that geopolitics would be simple, especially when the prefix "geo-" hides less and less well the formidable inclusiveness of Gaia? To speak of the people of Nature, in one of these three tonalities – epistemological, anthropological, or *terrestrial* – is to prepare to *redistribute* from top to bottom our capacities for mobilization as well as the definition of the front lines and the forces in presence.

Table 5.1 Table comparing the main features of two versions of the concept of nature, the epistemological and the anthropological, showing how strongly they differ.

	People of Nature	
	Nature one (epistemological)	Nature two (anthropological)
Deity	Laws of nature	Multiverse
Cosmogram	Exterior	Interior
	Unified	Multiple
	Deanimated	Animated
	Indisputable	Controversial
People	Everyone	Scientists
Ground	Off the ground	Attached to networks
Epoch	Radical break	Multiple temporality

*

What makes the people of Nature so incapable of situating themselves is that this people has constructed itself in reaction to another one, which, for its part, announces itself clearly *as a particular people*; however, as we continue to set up our translation tables, we're going

How to convene the various peoples (of nature)? 169

to notice that this other people does not necessarily know any better than the first where it resides. To continue in the same vein as the *Game of Thrones*, let's call this one the people that calls itself *Children of the Grand Design* or the *People of Creation*. This will allow us to understand that the "conflict between Science and Religion" rather resembles the famous war between the Little-endians and the Big-endians in *Gulliver's Travels*, even as it hides another much more important conflict, which is for its part directly political, over the occupation of the Earth. When one speaks of a "religious vision of the world" in "radical opposition" to a "strictly scientific vision" of that same world, one is appealing to another supreme authority that is not so different from the first column in table 5.1: it has the same characteristics, in fact, except that the first stubbornly overanimates what the second stubbornly deanimates.

We no longer have to be tripped up by the fact that the one is determined to call "God" what the other insists on calling "Nature," since it is their attributes and those alone that must allow us to make these two supreme authorities comparable. Now the God that orders the religious view of the world bears a very close resemblance to the Nature that orders the scientific view of the world. Three of their features are in fact exactly the same: truth is external, universal, and as indisputable as it is indestructible. Even the question of the delimitation of the people is not very different, since the Children of the Grand Design are recruited by an explicit procedure – a form of conversion – that gives their people the more precise name of Church, just as diplomas, examinations, and the continual reduction of the number of the elect operate a selective triage for the people of Nature. In each case, "everyone," at least in principle, is called to belong to the people in question, but in practice there are few thurifers.

Nor does the question of epoch allow us to differentiate them radically, for these two peoples share the idea that a radical break has occurred in a more or less recent past – a rupture that has propelled them into a totally new history called *Light* by one group, *Enlightenment* by the other. What matters is that both peoples locate themselves in *the time that follows* a radical break – Revelation or Revolution (we'll come back to this crucial point in the next lecture). As for their relation to the ground, it is lacking in both cases: in the first because the people is in any case removed from the ground, in the second because the people belongs to a different world, the world, apparently, of meanings and goals, of a Grand Design, a Providence toward which this population aspires to beam itself.

The only real difference, the one that for both peoples justifies going to war, total war, is whether the agents that populate the world are totally deanimated – mere concatenations of causes and consequences – or whether they obey a design that makes it possible to add to them, if not a soul, in any case a goal, a program, a plan. The opposition appears to be radical, unless we recall the argument that I have continually sought to spell out in these lectures: both deanimating and overanimating still mean not respecting the *animation* proper to the discoveries of the world made by the sciences. Deanimation, let us recall, is not a primary process but, rather, a secondary treatment, polemical and apologetic, which attributes to the sciences and the world they describe the behavior of inert and obtuse things that are as unlike them as the overanimation proposed by their adversaries.

If, for example, the People of Creation draft a moving elegy on the structure of the eye, "so obviously conceived by a benevolent Creator, since no accumulation of random encounters could have produced it," they are preparing for a magnificent battle against the people of Nature, who are just as eager to cross swords, and who are ready to demonstrate without the slightest doubt that the structure of the eye is "*nothing more than* the unanticipated result of small changes accumulated over generations of purely contingent chance events."[44] The problem is that the appearance of a radical conflict rests entirely on that little "nothing more than," that mystique of reductionism – and we have learned to doubt that its kingdom is of this world.

The harmony between the protagonists can be spotted as soon as we seek to identify *what quantity of action*, animation, activity, has been developed by each argument. We immediately notice that both narratives have managed to *lose* what was original about the evolution of the eye. We rediscover here quite precisely, as we did in the third lecture, the loss of agency, of narration, of geohistory that transforms Gaia into a self-regulated System. We won't be surprised to learn that the "admirable structure of the eye," in the Creation argument, does strictly nothing more than serve as a redundant example to celebrate the Creator's benevolence. It may be pleasant and exalting to know that "the flowers of the field sing the glory of God," unless the song never varies from one creature to another! The insistence on creatures that were "designed" rather than produced "by chance" generally has no result except to demonstrate one more time the same creation by the same mysterious hand of the same

---

[44] This claim of pure contingency was revised and stabilized again in the twentieth century by Jacques Monod, in his famous book *Chance and Necessity: An Essay on the Natural Philosophy of Modern Biology* ([1970] 1972).

Creator. The *Creator* acts: not the eye, not the flower of the field. To fall back on my own jargon, the Creator is a mediator; the flowers of the field are mere intermediaries. In terms of actantial roles (an ugly term for such a beautiful thing),[45] the net result is *zero*, since the quantity of animation hasn't increased by one iota. A Creator, yes, but no creation for all that.[46] Everything lies in the cause, nothing in the effect. In other words, literally, *nothing happens*. The passage of time does nothing to the world. There is no history.

But what is particularly disconcerting for those who respect, as I do, people who sing the glory of God as much as people who celebrate the objectivity of the sciences is that the second narrative, by eliminating all the surprises that we find proliferating as soon as we follow the history of the structure of the eye, strives to be *as impoverished* as the other one. By claiming to do nothing but align concatenations of "purely objective agents that are only material," it loses the creative capacity of the agents that are strewn all along its path.[47] When a Richard Dawkins compares the design of his blind Watchmaker to the design of the seeing Watchmaker of his religious enemies, he fills his First Cause with all the creative capacities of which he wants to deprive the Creator.[48] In the "nothing more than" of reductionism, the blind Watchmaker introduces a great number of steps that gradually annihilate the difference between his activity and the providential act of Creation that Dawkins set out to oppose.

And yet how many words have been expended on the distinction between "spiritualists" and "materialists"! After a while, we no longer see where the dispute lies: a design and an Engineer versus a design and a Creator, what a fine combat indeed, worthy of spilling one's guts over! It is no easier to grasp the origin of this dispute than the origin of the one that set Catholics and Protestants at each other's

---

[45] Algirdas Greimas and Joseph Courtés, eds., *Semiotics and Language: An Analytic Dictionary* ([1979] 1982, p. 6). "Actantiality" is uglier still, although it could be a synonym for "agency" without being immediately linked to the limited repertoire of the human.
[46] Creation – which is the inverse of creationism – presupposes that the cause-consequence relation is modified in such a way that the consequence slightly exceeds the cause. This amounts to saying that time flows from the future toward the present, and not from the past toward the present. Or, to put it still differently, that the consequences, in a way, always "choose" what their causes will be.
[47] Unless we read Stephen Jay Gould, *Wonderful Life: The Burgess Shale and the Nature of History* (1989), or the astonishing Jan Zalasiewicz, *The Planet in a Pebble: A Journey into Earth's Deep History* (2010).
[48] Richard Dawkins, *The Blind Watchmaker* (1986).

throats, or the exact doctrinal point in the name of which today's Shiites and Sunnis have chosen to kill one another.

As soon as we avoid deanimation, the little "it is nothing but" fills up with a multiplicity of events, all contingent, to be sure, but all surprising, which oblige each of the followers to take them into account in their own way. Of course, these are not the lessons that would have been drawn from the flowers of the field, but neither are they those that would have been drawn from the First Cause, the famous intelligence of the blind Watchmaker capable of "steering" all this evolution. Who best follows the process of creation? The one who reaches the same conclusion about every course of action or the one who multiplies the agencies of which the worlds might be composed? Obviously, the second.

Except that, unfortunately, at the end of the demonstration, when he is challenged by his "religious" adversary, the naturalist, too, will make an effort to draw the same repetitive lesson from the structure of the eye, according to which evolution "demonstrates yet again without the shadow of a doubt" that there is neither a grand design nor a designer. This is where we end up – but belatedly and without any relation to the real practice of the sciences – with Whitehead's desolate summary: "so that the course [of human history] is conceived as being merely the fortunes of matter in its adventure through space."[49] A sad triumph on the part of our clever naturalist, who has done everything possible to make himself as stupid as his adversary, his left hand trying to take away from the world the agents that his right hand had so intelligently multiplied there. The scientific vision of the world has managed quite an exploit: nothing more happens in this world than in that of the Creator God!

We can understand that it is not by adding the word "soul" to an agent that you are going to make it something *more*, nor by calling it inanimate are you going to make it something *less*, depriving it of its action or its animation. Agents act! One can try to "overanimate" them or, on the contrary, to "deanimate" them: they will stubbornly remain agents. In any case, the difference between *overanimated* and *deanimated* elements is not a cause for which we have to live, pray, die, fight, or build temples, altars, or globes. If we have to fight, let's at least fight for goals that are worth it.

When we look at table 5.2, we thus note that the term "natural religion" has hardly any meaning. We are dealing with *two forms of counter-religion*, with two peoples that are basically very close to

---

[49] Whitehead, 1920, p. 20.

Table 5.2 Table comparing the main features of two versions of the concept of natural religion – one from science, the other from religion – showing how little they differ, except on the question of animation.

	Natural religions	
	Nature no. 1 (People of Nature)	Religion no. 1 (People of Creation)
Deity	Laws of nature	Ordering God
Cosmogram	External	External
	Unified	Unified
	Deanimated	Overanimated
	Indisputable	Indisputable
People	Everyone	Everyone
Ground	Off the ground	From another world
Epoch	Radical break	Radical break

one another, in which the ones believe they are celebrating their God with dignity while depriving themselves of access to the sciences and to the diversity of the world, whereas the others multiply in practice things having to do with the world but deprive themselves of this multiplicity in believing that they honor their deity by the "nothing but" of reductionism. "Nothing but," really? Why embrace this form of nihilism?

We understand why it is useless to accuse Science of being a substitute for religion or to seek in a natural religion what might convince unbelievers of the existence of Providence. One can neither oppose nor reconcile the scientific and religious versions of the world. They are not different enough to be opposed, not similar enough to be fused together. It would be useless to ask Science to be kind enough to leave a little room for another "dimension," the "religious," understood either through its spiritual localization in the soul or through its cosmic extension into what is called "Creation." It is better to try to do just the opposite and dissolve the amalgam between the two that is created by the ambiguity of the term "counter-religion." The people of Nature believe they are fighting against the people of Religion, whom they resemble, and they cannot reconcile themselves with their own anthropological version, which is nevertheless their strength. But, as we are now going to see, the People of Creation believes that it is struggling against the people of Nature, whom it resembles, while the people of Nature too has forgotten the very meaning of its quite specific vocation. By fighting Religion, Science has lost its connection with itself; by fighting Science, Religion has lost track of its most valuable asset.

*

Why this insistence on affirming or negating a Design that seems so essential to the relationships maintained between the "scientific vision" and the "religious vision" of the world? These are two ways, as we now understand, of *seeing* nothing about the world, either by depriving it of all action, and thus deanimating it, or by adding to it a soul for which it has no use, and thus overanimating it. Since I am convinced that this is what prevents us from having access to the world, from coming back to Earth, from giving a terrestrial vision of science and an even more secular definition of nature, I have to ask you to agree to take one further step and explore the meaning of this counter-religion whose emergence has confounded the fortunes of those who were to be its heirs.

If the idea of Design is so important, it is because it captures one of the features of the counter-religion that has to do with the question of *ends*. The intuition of the counter-religion, such as it can be reconstituted through multiple metamorphoses, is that, despite the passage of time, *the world has an end*, not in the sense that it is going to end – even though the idea of the *end of the world*, as we shall see in the next lecture, can translate this intuition in part – but in the much more radical sense that *the goals it pursues would be definitively achieved*. That the world has an end does not mean that it has a goal in the sense of having being "created with a goal," but that it is possible to experience it as having achieved the end – which can be translated by a whole host of formulas, strange ones for many of our contemporaries, all of which have the same meaning: to be "saved," to be "children of a God who cares for us," to be "God's chosen people," to "find ourselves in the Presence," "to have been created," and so on. These are all provisional, awkward formulas that are immediately attacked as insufficient, deceitful, or impious by other versions of these same counter-religions.[50]

---

[50] The instability of these forms of expression and the impossibility of speaking "well" about them or of gathering them into "beliefs" are at the heart of their definition. See Bruno Latour, *Rejoicing: The Torments of Religious Speech* ([2002] 2013a). What makes this section difficult is that the specific mode of existence it tries to register has become difficult to detect, much like politics today. For a way to recover those differences, Latour 2013b may offer some help.

The problem with such an intuition is that it is fundamentally *unstable*, for the excellent reason that the end times have come, *but that time is lasting*! There is no way to escape from this tension.[51] The end has been reached, and it is unreachable. We are saved, and we are not. Enough to drive us mad. The counter-religions are powers whose radioactivity no one has yet been able to control. Millennia have gone by, and the power of the counter-religions has not diminished. We know this well, we Moderns, since we are their more or less direct heirs, and since we are stunned witnesses of the return of the wars of religion that we thought we had left behind several centuries ago, along with wars over the occupation of the Earth, wars whose planetary scope reduces the world wars of the twentieth century to the dimension of local conflicts.

Despite the multiplicity of their self-described "Revelations," these counter-religions have no content other than the stupefying realization of the endlessly explored truth that the end has been reached, the goals achieved, the times judged – and judged *definitively*. Assmann is right to say that, with such an intuition, the question of truth is introduced into the traditional religions, which had never had to deal with it before. But this truth did not have the vocation of entering into head-on competition either with the truth of knowledge or with that of the divinities belonging to the so-called traditional religions.[52] This new form of truth, this new mode of existence, explored a quite different relation to the mundane, to the ordinary, to the passage of time, by distributing differently the relations of goals and means. If the ends can be achieved *in time, even though* the times *go on*, and *thanks* to time, then everything in the meaning of history and the manner of occupying the Earth changes radically.

But without anything changing: here is the whole mystery of this form of truth, the source of enthusiasm and of fright and fury at once. Because of this instability, the introduction of truth into the counter-religions introduces at once a powerful opening – what Freud calls "progress in the life of the mind"[53] – but also unleashes a cascade of

---

[51] In the next lecture, we shall encounter Eric Voegelin's decisive argument in *The New Science of Politics: An Introduction* ([1952] 2000a); we can also find the argument in a number of other texts, such as Hans Jonas's "Immortality and the Modern Temper: The Ingersoll Lecture, 1961" (1962, p. 15).
[52] Assmann explores this anew in *Violence et monothéisme* (2009); as I see it, this is what explains iconoclasm as well as the extreme difficulty involved in stabilizing the meaning of concepts of construction and creation (see Latour 2010a).
[53] See Bruno Karsenti's commentary, *Moïse et l'idée de peuple: la vérité historique selon Freud* (2012b).

more or less violent battles, as if truth did not know how to coexist with any other value. From this cascade, we have not escaped. As of now, every counter-religion has only added its virulence to the earlier virulence, for want of having achieved the cohabitation of truths.[54]

It would take more than one lecture to list the features of this counter-religion, but let's say that it corresponds no more closely to what the people of the Grand Design celebrate than the anthropological vision of Nature corresponds to its epistemological version. One may call it "God," but it is also the *end* of all the gods and divinities, and even in a sense *the end of God*, in the well-known sense of the death of God.[55] In this sense, the counter-religion is indeed "counter" – against – itself, engaged in a continual struggle over the figure that it is to give its supreme authority. When one begins with iconoclasm, one never ends. In any case, the reassuring figure of an ordering God who protected the earlier people makes no sense, since order precisely does not pre-exist in relation to its own history. No Providence precedes it – not any more than a world made of deanimated matter, indisputable, universal, and external laws would make sense.

But the counter-religion has no use, either, for an overanimated matter that would shift attention toward another world while imposing neglect of the radical alterity that it is a question, on the contrary, of sensing.[56] Unlike the other two, this counter-religion is profoundly embodied, since it constantly renews its participation in a present world, definitively judged, achieved, saved, celebrated, and situated, but from which it is not a matter of extracting oneself for another world, since everything goes on as before. No world detached from the ground, no ultra-world, and thus no lower world either.

It is especially in the conception of time that the originality of this other counter-religion stands out: there is indeed the feeling of a radical break, but with the crucial nuance that the break must constantly be *taken up again*. One cannot escape from this fundamental instability, from this indecision: "The end time has come," yes, but it goes on. And this prolongation gives decision the same lacunary,

---

[54]The impossible pluralism of modes of veridiction is the object of my *Inquiry into Modes of Existence: An Anthropology of the Moderns* ([2012] 2013b).
[55]Among the most significant expressions of the prefix "counter" in "counter-religion," we find the theme of the putting to death of a crucified God as well as the theme, taken up without many modifications, of the "death of God." This is the sense in which secularization continues the movement that explores the ferocious enigma of the *counter*-religion.
[56]We shall return to this theme in the sixth lecture: this is what Eric Voegelin calls "immanentization," a quite specific way of failing to achieve either immanence or transcendence.

incomplete, fragile, mortal character it had before the end time came. This contradiction must not be overcome.[57] We shall see in the following lectures why *not overcoming this contradiction* is essential in order to avoid the poisons of science, politics, and religion – or, rather, why the distinct virtues of science, politics, and relation become poisons when one starts to confuse them.

You find this very strange, very contradictory, and very unstable? Yes, but there's nothing I can do about it, it is this end of history – in every sense of the word "end" – that has been introduced into history and that continues to act as much in every conception of religion as in every conception of going beyond religion.[58] If the Moderns – who have never been modern! – are so unsure of themselves, it is because they have inherited this ferocious contradiction.

*

The little game of drawing up lists of peoples in order to compare them with one another, so that they will stop facing off against one another, is obviously overly simplistic, even childish. But it is the only way I have found to combat two entrenched prejudices: the first involves the connection between *nature*, in the singular, and *cultures*, in the plural; the second involves the curious conception of a temporal break that lulls us with the illusion that the question of religions has already been resolved. The two prejudices are closely linked: this is because nature, through a sort of *translatio imperii*, has inherited almost all of the features of (the counter-)religion, because it has appeared as a universal against whose background only cultures that are certainly multiple but without any intimate link with the unified nature of things could stand out. True nature against multiple cultures: there is our counter-religion. And it is because it has inherited not the old religions of the past but a particularly ardent, triumphant, indecisive, sometimes fiercely iconoclastic form of counter-religion that the struggle of nature against religion can be mistaken for the definitive annihilation of all religious questions.

---

[57] This is the sense of the quite particular theology explored tirelessly by Péguy through the detour of style; see Bruno Latour, "Nous sommes des vaincus" (2014c); Marie Gil, *Péguy au pied de la lettre: la question du littéralisme dans l'œuvre de Péguy* (2011); and also Camille Riquier's chapter "Charles Péguy: métaphysiques de l'événement" (2011).
[58] The attitude toward iconoclasm is a much better guide for diagnosing the immense question of "secularization" than the attitude taken toward the gods: "Tell me with which hammer you are going to strike which idol, and I'll know which divinity you serve."

Table 5.3 Table summarizing the contrasting features of the concepts of science and of religion, showing that the contradictions are not between science and religion but between two different versions of each of those domains.

	Natural religions			
	Science		Religion	
	Nature one (epistemological)	Nature two (anthropological)	Counter-religion one	Counter-religion two
Deity	Laws of nature	Multiverse	Ordering God	God of ends / ends of God
Cosmogram	External	Internal	External	Local
	Unified	Multiple	Unified	Multiple
	Deanimated	Animated	Overanimated	Animated
	Indisputable	Controversial	Indisputable	Interpreted
People	Everyone	Scientists	Everyone	Church
Ground	Off the ground	Attached to networks	From another world	Embodied
Epoch	Radical rupture	Multiple temporality	Radical rupture	Reprise

The map is sketchy, I know, but it at least allows us to escape from the unanimism that is always associated with the idea that the religious question has been definitively settled by the emergence into history of "the Nature known to Science." If we now consider the more complete table in table 5.3, we see that the term *"nature" does not define what is assembled in practice, any more than the term "religion" qualifies the type of people, rites, and attachments proper to these practices.* This is the point, even if it is purely negative for the time being, that I wanted to reach. There is no natural religion, and one cannot continue to invoke Nature in the hope of resolving conflicts between peoples whose interests are so clearly divergent.

By embracing Nature as the ultimate truth, its people have done no more than prolong ever so slightly the movement of the counter-religions themselves, along with their particularly toxic conceptions of truth. The solution proposed by Hobbes in the seventeenth century in order to put an end to the state of nature by shifting toward the State, as a way of getting out of the wars of religion, appears to us now as a stopgap solution, a simple *armistice*, but not at all as a *peace treaty* that would have let us reach the end of the demands of these counter-religions whose violence and fruits we are harvesting simultaneously, but without managing to distinguish between them. How can we achieve a peace treaty if the peoples involved cannot invite one another to the negotiating table? The two figures of Cosmocolossus with which I began this lecture have indeed come to blows.

I have never spoken about Gaia without someone objecting immediately that I risked "confusing religious questions with ecological or scientific questions." Yet just the opposite is true. It is because I have an ear for the religious questions that I very quickly detect those who put religion where it has no business being, in particular in science or in politics. What has always alerted me is the extent to which the order of nature, its distinction from culture and politics, its obsession with deanimating agents, stems from a particularly troubling form of religion. It is the ecological mutation that obliges us to secularize – perhaps even to profane – all the (counter-)religions, including that of nature.

In any case, ecology obliges those who are gathered together by "Nature" to consider the sixteen features in the table at one and the same time. It is totally unrealistic to confuse the peoples assembled in the epistemological mode with those who are assembled in the anthropological mode, even if both can invoke the *same entity* called "Nature" and declare themselves "naturalists" by insisting on their radical separation from all the other peoples assembled by other entities, thanks to the qualities of their sacrosanct "reductionism."[59] Were we really to follow the injunctions of this supreme authority, we would have to attend not just to the left-hand column but also to the one on the right. We would need to dig down into the scientific networks, absorb the dizzying multiplicity of the forms of agency of that supreme authority, note the long concatenations of its agents, so surprising every time, and assimilate ever more numerous controversies over multiple "matters of concern."

The real surprise is not that the distribution of agency under the auspices of "Nature" is so complex, but that the people that situates itself under the auspices of "religion" grasps *so few* of the characteristics of what is of vital importance for the people that this entity is supposed to convoke. If you find it disorienting that the invocation

---

[59] All the more so in that, from now on, we have to defend the sciences as victims of a generalized pollution, on the same basis as water, air, land, and food. See Isabelle Stengers and Thierry Drumm, *Une autre science est possible! Manifesto pour un ralentissement des sciences* (2013). See a paper in English by Stengers: "Another Science Is Possible! A Plea for Slow Science" (2011b).

180   *Fifth Lecture*

of "Nature" does not include any of the real attributes to which its practitioners are so passionately attached, I find it much more disorienting that the very ones who are said to be gathered together by the entity that they often call "God" grasp nothing more through this invocation than the externality, the unity, and the indisputability of Creation – that is, quite precisely the epistemology of those whom they consider their enemies (more or less the question, basically a superficial one, of the presence or absence of a factitious *Design*). This is the problem with amalgams: once they are mixed together, it is impossible to recognize the original values.

*

To extract in a lasting way the values that are blended in this amalgam, we would have to undertake a new operation of engendering peoples, a demo-genesis, in an even more mystifying fiction than the previous one. And yet I can't resist the temptation, in concluding this lecture, to have a go at concocting this ultimate chimera.[60] Let's now suppose – the supposition is extravagant, I know, but the times we live in are no less so – that we subject our table to a little operation of *reordering*! In table 5.4, I have done nothing but *invert* two columns. I took the one that sums up science as it is done (the anthropological and no longer the epistemological version) and moved it further to the right, next to the one that summed up the original, active version of religion. And I took the liberty of moving the epistemological version of religion to the left, right next to the epistemological version of science! Don't you find that this reorganization makes things much more logical – yes, more logical?

When we juxtapose them, it becomes clear that, as in table 5.2, the two left-hand columns belong to the same *natural religion*. They share the same fundamental postulate, in effect: they proceed as if the task of unifying the world had been accomplished, as if there were no difficulty in speaking of the universe as a unified whole. For these two peoples, the universe – Nature or Creation – has already been entirely assembled by the same regime of causality, except that blind Cause reigns over deanimated things and Providence over overanimated things.[61] The people of Nature, like that of Creation, embrace

---

[60] Remember that the initial remit of the Gifford Lecture series at the origin of this book was to try to "reconcile" Science and Religion in the sense bequeathed to those two terms by the nineteenth century, especially by the intrusion of Darwin...It's about time to move the discussion to another century.

[61] It is obviously this complicity that creates the whole dynamism of David Hume's dialogues concerning natural religion, in his *Principal Writings on Religion*

How to convene the various peoples (of nature)? 181

Table 5.4 Table shifting the columns shown in table 5.3 so that the contrast is no longer between science and religion but between "natural religions," on the one hand, and what could be called "terrestrialization," on the other, each of the two domains including versions of science and religion.

	Natural religions		Terrestrialization	
	Nature one (epistemological)	Counter-Religion one	Nature two (critical)	Counter-Religion two
Deity	Laws of nature	Ordering God	Multiverse	God of ends / ends of God
Cosmogram	External Unified Deanimated Indisputable	External Unified Overanimated Indisputable	Internal Multiple Animated Controversial	Local Multiple Animated Interpreted
People	Everyone	Everyone	Scientists	Church
Ground	Off the ground	From another world	Attached to networks	Embodied
Epoch	Radical break	Radical break	Multiple temporality	Reprise

the world *in toto*, as if the "point of view from nowhere" were a real place offering a comfortable seat and a good viewing angle. Both peoples are full-fledged members of what Peter Sloterdijk calls the "age of Spheres" – that is, an epoch in which it wasn't at all difficult to hold the Earth in one's fingers.[62] They are equally off the ground, and both are located in the epoch that follows a radical break, making any backward movement impossible.

The chimera that interests me involves imagining groups of people who would not remain insensitive to the features of the *two right-hand columns*. It would no longer be a question of natural religions, since the shared feature would be that of no longer having an ordering principle. There would certainly be a supreme authority, but this would lie no longer in unity – capable of designing a universe – but in connection or composition. More precisely, every time any entity whatsoever has to extend itself, it has to pay the full price of its

---

including *Dialogues concerning Natural Religion; and, The Natural History of Religion* ([1779] 1993).
[62] See Peter Sloterdijk, *Globes: Macrospherology* ([1999] 2014), discussed in the fourth lecture.

extension. Which is another way of saying that it has a history. In other words, the members of these peoples would no longer feel that they are living under a Globe, but in the middle of relations that they have to compose one by one without any means of escaping historicity. To accentuate the contrast, I propose to say that such population groups would share the same feeling of *earthboundedness*. If there's no such word, it's precisely because we have yet to bring into existence the thing that it designates! Such groups would share the need to protect each other against the temptation of unifying too quickly the world that they are exploring step by step. Both groups, indeed, find themselves on a ground whose materiality and fragility they are discovering more and more every day. Neither of the two believes itself to be located outside of the time that is passing.[63]

The reason it was so important for us to get rid of the amalgam of "natural religion" is that here, in the cosmopolitan situation that I have taken as our point of departure, we are dealing not with only *two* "distributions of agents," as was still the case when David Hume was writing his *Dialogues*,[64] but rather with as many distributions as there are entities convoking peoples today. When the naturalists proclaim themselves the children of *That*-of-*Which*-We-Are-All-Born, and Christians proclaim themselves to be the children of *The-One*-of-*Whom*-We-Are-All-Born, there can be virulent disputes between the "which" and the "whom," but I would like us to remain sensitive to the request of those who say: "So what is this 'we'? What about this 'all'? Don't count 'us' in! We belong to neither of these peoples. Your entities do not convoke us at all. We live under conditions that distribute agents entirely differently. Don't unify the situation so prematurely! Please don't implicate us in your planetary wars; we don't want to play any role in your intrigues." We haven't finished absorbing the diversity of ways of occupying the Earth. The Anthropocene is first of all the opportunity to listen seriously at last to what anthropology teaches us about other ways of composing worlds – without depriving us, nevertheless, of the sciences, which are radically different only in the epistemological version.[65]

---

[63] That would amount to capturing the historicity common to the world, the sciences, and the religions.

[64] In the initial version of the 2013 Gifford series, I devoted one lecture to imagining a role for poor Pamphilus, a non-speaking character in this well-known, magnificent dialogue. I regret having had to abandon the re-enactment of David Hume's famous text.

[65] This is one of the cries that reverberate throughout the work of Eduardo Viveiros de Castro, especially in *Cannibal Metaphysics: For a Post-Structuralist Anthropology* ([2009] 2014) – and we are truly dealing with matters of metaphysics here.

Going beyond the number two, putting in place a sufficiently ample comparison among the mechanisms that make it possible to distribute agency, avoiding the quarrel between "nature" and "religion," all these could constitute vital resources for discovering the exact form of the Earth when the time comes to find a way of participating in the institution, or rather in the founding, of Gaia. There is no doubt about it: we have become nations divided, often divided internally, because we are convoked by many different entities to live under very different models of the Earth.[66]

As a first approximation, it is obvious that the people assembled under Gaia will resemble neither those who invoke Nature nor those who say that they worship a deity with all the appurtenances of religion. None of the eight attributes we have recognized up to now seems to be an attribute of Gaia. As we saw in the third lecture, Gaia is not only external but also internal; it is not universal, but local; it is neither overanimated nor deanimated; and, beyond that, unquestionably, it remains totally controversial. Gaia is probably other Earths, other Globes, invoked by another people, as foreign to what used to be called "nature" and "naturalists" as to what was called religion. How can it be invoked respectfully?

This is what we now have to discover, by returning to the big question of the "end time," which is at the origin of the very idea of counter-religion. For it happens that those who accuse ecology of being too often "catastrophist" and of indulging in "apocalyptic" discourse are those who, not content with having triggered catastrophes, have obfuscated the very notion of apocalypse.

---

[66] It's worth pointing out that Clive Hamilton is making exactly the opposite argument in *Defiant Earth: The Fate of Humans in the Anthropocene* (2017) because of what he sees as the return of a necessary unified anthropocentrism, precisely because of the advent of the Anthropocene.

SIXTH LECTURE

How (not) to put an end to the end of times?

> The fateful date of 1610 • Stephen Toulmin and the scientific counter-revolution • In search of the religious origin of "disinhibition" • The strange project of achieving Paradise on Earth • Eric Voegelin and the avatars of Gnosticism • On an apocalyptic origin of climate skepticism • From the religious to the terrestrial by way of the secular • A "people of Gaia"? • How to respond when accused of producing "apocalyptic discourse"

How could I not have been stunned to read, in the issue of *Nature* curiously titled "The Human Epoch" with which I began the previous lecture, that 1610 was one of the possible dates to use as a marker for the beginning of the Anthropocene?[1] Why 1610? Because the reforestation of the American continent had, by that date, led to the stocking of so much atmospheric CO_2 that climatologists could use it as a minimum quantity on the basis of which they could measure

---

[1] *Nature*, March 11, 2015. Let us recall that stratigraphers seek to determine from a transition in sediments where to put the "golden spike" that distinguishes one geological period from another. In the still disputed case of the Anthropocene, the question is whether to identify it as a very long period (essentially the entire Holocene), a very short period (since 1945), or something in between. See Simon L. Lewis and Mark A. Maslin, "Defining the Anthropocene" (2015). Needless to say, this new date is being fiercely disputed (see the third lecture).

its regular increase. But why this massive reforestation? Very simply, according to the authors of the article, because of the extermination by the sword, but also by contagion and disease, of nearly fifty-four million Native Americans, in the wake of Columbus's "discovery of America." The "great discoveries," colonization, the fight to occupy territories, forests, carbon dioxide – it's all here, defining the Anthropocene: anthropology plus climatology in a violent land grab.[2]

But 1610, as you surely recall, was also the year Galileo published his *Sidereus Nuncius*, the "Messenger from the Stars" that is said to have brought universal history out of its "closed world" to propel it into the "infinite universe."[3] Remember what Brecht said: "Today is 10 January 1610. Today mankind can write in its diary: Got rid of heaven."[4] We have to acknowledge that these two references to 1610 resonate quite well together, since the first brings us to the limits of the Earth from which the second had initially pulled us away; whereas we believed we were in a nature finally indifferent to human action, we find ourselves plunged back onto an earth that has never stopped reacting in response to the unforeseen consequences of our acts of domination.

But I had entirely forgotten that 1610 – more precisely May 14, 1610 – was also the date on which Henry IV was assassinated by François Ravaillac; the latter was condemned for regicide a few days later (most French schoolchildren have undoubtedly shuddered as they contemplated the classic image of the assassin drawn and quartered, pulled apart by four horses). What's the connection, you'll ask, between this event and the two previous ones? I didn't see any, I confess, until I reread *Cosmopolis: The Hidden Agenda of Modernity*,[5] by Stephen Toulmin (1922–2009), a historian of science and a specialist in casuistry.[6] The coincidence of certain dates in history is so striking that one is inclined to see a fateful sign.

In this lecture, which will probably be more difficult than the others, I'm going to try to continue exploring the religious – or, more precisely the (counter-)religious – origin of our contemporaries' remarkable indifference to the ecological mutation. What makes this

---

[2] On what Charles Mann calls the "Columbian exchange" and the transformation that followed, see Charles Mann, *1493: Uncovering the New World Columbus Created* (2011), a sequel to his very useful *1491: New Revelations of the Americas before Columbus* (2005).
[3] An allusion to the title of Alexandre Koyré's book *From the Closed World to the Infinite Universe* (1957; see chapter 3).
[4] Bertolt Brecht, *The Life of Galileo* ([1945] 2001, p. 24).
[5] Stephen Toulmin, *Cosmopolis: The Hidden Agenda of Modernity* (1990).
[6] Stephen Toulmin, *The Uses of Argument* ([1958] 2003).

exploration difficult is that it requires us to mix the history of the sciences, the Christian religion, and politics, beginning with the great crisis of the religious wars and then moving back in time – this will strike you as even stranger – to the history of Gnosticism. Something is happening around the seemingly bizarre theme of "the end of times" that it would be useless to try to avoid. It is in a certain relation to the notion of immanence that we are going to find the key to the prevailing indifference to the terrestrial. This indifference is indeed of religious origin, but not at all for the reason usually invoked, which aims to make Christianity responsible for the forgetting of the material world.[7]

*

Let us begin with the chapter Toulmin devotes to the assassination of the good King Henry, in which the author thinks he can spot the end of one epoch and the beginning of another, as surely as geologists think they can place a golden spike between two layers of sediment to distinguish the Holocene from the Anthropocene. "In practical terms, Henry's murder carried to people in France and Europe the simple message: 'A policy of religious toleration was tried, and failed.' For the next forty years, in all the major powers of Europe, the tide flowed the other way."[8]

Let's do away with tolerance! This was the beginning of a terrible century, the seventeenth, foolishly designated, according to Toulmin, as the "century of reason," the century of the scientific revolution, while in fact it was the century of the dreadful Thirty Years' War, which ravaged Europe in the way that wars of religion are ravaging Syria, Iraq, and Libya today – and which ended with the Treaty of Westphalia and the contested invention of sovereign states. If the death of France's Henry IV can serve as a marker, in Toulmin's view, it is because it separates two periods: one that had been characterized by pluralism and skepticism[9] and one that was characterized by a new form of absolute certainty. People confronted by the horrors

---

[7] I am taking as virtually providential the unexpected support brought to this chapter by the appearance of Pope Francis's encyclical *Laudato Sí: On Care for Our Common Home* (2015) at the very moment when I despaired of making my text understandable to my readers!
[8] Toulmin 1990, p. 53.
[9] In the old positive sense (the one reclaimed for example by Frédéric Brahami, *Le travail du scepticisme: Montaigne, Bayle, Hume*, 2001), and not at all in the sense of those who strut around brandishing the expression "climate skeptics."

of war don't want to hear any more talk of open minds, relativism, experimentation, or tolerance:

> By 1620, people in positions of political power and theological authority in Europe no longer saw Montaigne's pluralism as a viable intellectual option, any more than Henry's tolerance was for them a practical option. The humanists' readiness to live with uncertainty, ambiguity, and differences of opinion had done nothing (in their view) to prevent religious conflict from getting out of hand: *ergo* (they inferred) it had helped *cause* the worsening state of affairs. If skepticism let one down, certainty was more urgent. It might not be obvious what one was supposed to be certain about, but *un*certainty had become *un*acceptable.[10]

You were expecting Montaigne, or Erasmus? You are going to find yourselves, in science, with Descartes;[11] in religion, with Reformation and Counter-Reformation; in politics, with Hobbes's invention of that sort of sovereign state that has been called "Westphalian" ever since.[12] You were hoping to be done with religious wars, through accommodation, tolerance, negotiation, diplomacy, and the exploration of shaky forms of composition? You are going to be asked to choose your side among several types of absolute certainties. It matters little what you will be certain about: a political order, an interpretation of the Bible, mathematics, law, experimental narrative, obedience to the Pope or the Sun King; what counts from this point on is being certain. It is hard not to read this passage without relating it to the present time. For what new Thirty Years' War must we prepare ourselves if, four centuries later, the "political and theological authorities" start considering pluralism as "totally unacceptable" in order to struggle against the aggravation of wars of religion? Today, as was the case yesterday, the reaction to the various forms of fundamentalism can blind us.

Toulmin is so persuaded of the importance of the year 1610 that he uses it to relocate by a century what is usually called the scientific revolution – henceforth firmly defined as a Counter-Renaissance.[13]

---

[10] Toulmin, 1990, p. 55.
[11] Toulmin makes the intriguing suggestion that a sonnet in praise of the "good king" Henry written in the Collège de la Flèche might have been the work of a brilliant young student named René Descartes (ibid., p. 60).
[12] I know that the adjective "Westphalian" simplifies a huge question about the history of the State, but it is convenient for emphasizing all the difficulties that those who aspire to "govern the climate" will have to address (see Stefan Aykut and Amy Dahan, *Gouverner le climat? Vingt ans de négociations internationales*, 2014) while preserving the model of the Old Climate Regime. We shall come back to this problem in the final lecture.
[13] Toulmin addresses this topic in *Cosmopolis* (1990), chapter 2.

Sixth Lecture

It was in the sixteenth century, according to him, that all sorts of innovations were tried out in a truly experimental spirit, in the joyous havoc of an Erasmus, a Rabelais, or a Palissy:

> The received view of Modernity thus tried, anachronistically, to credit 17th-century philosophers with the toleration, and the concern for human welfare and diversity, that belonged rather to 16th-century humanists: positions that were linked with a skeptical philosophy that rationalist philosophers like Descartes were bound, in public at least, to reject and abhor.[14]

We won't be astonished to learn that, in that period as in our own, everything hinges on the animation or deanimation of matter, in science and politics alike. For subscribers to absolute certainty, it has to be possible to link public order to the definitive silence both of the masses and of matter. The key term here is *autonomy* of movement. What is going to be invented is the inertia of matter, the matter that will serve to form *matters of fact*. After the disorder of the Republic, after Cromwell, after the beheading of King Charles, order will reign only if both things and the people are deprived of any autonomous capacity for action:

> Commonwealth sectarians [the radical challengers of the period] read any proposal [by the naturalists] to deprive *physical* mass (i.e. Matter) of a spontaneous capacity for action or motion, as going hand in hand with proposals to deprive the human mass (i.e. the "lower orders") of the population of an autonomous capacity for action, and so for social independence. What strikes us as a matter of basic physics was, in their eyes, all of a piece with attempts to reimpose the inequitable order of society from which they had escaped in the 1540s. After 1660, conversely, English intellectuals stopped questioning the inertness of matter, *for fear of being tarred with the same brush as the Commonwealth regicides*.[15]

Doesn't that sound familiar? That the Earth may react to our actions bothers today's intellectual elites as the autonomy of matter once bothered the supporters of the established order! With the New Climate Regime, the same question arises: how to distribute agency by parceling out powers, aptitudes, and capacities, among things, gods, humans, and classes, in order to impose one cosmology over another. Everything is reshuffled: the order of nature as well as the political order, and, as always, what one must think about religion and who has the right to interpret God's word – which has since become the word of the Market. The defense of the autonomy of things, like the defense of the autonomy of peoples – the refusal to

---

[14] Ibid., p. 80.
[15] Ibid., p. 121, emphasis added in the final sentence.

let others, whoever they may be, impose their laws on you – remains the big question, scientific as well as political.

Toulmin goes so far in his revision of the usual periodization that he does not hesitate to describe the seventeenth century as the century of the *scientific counter*-revolution.[16] Attention to the particular becomes an obsession with the universal; rootedness in time is replaced by an atemporal vision, skepticism by dogmatism, subtle casuistry by an obsession with general principles; the body is set aside in favor of the mind, facetiousness in favor of seriousness, collages by coherency, the disputable by the indisputable. And yet how much this Renaissance has been roundly mocked! What the humanists had conceived has been aborted by the rationalists.[17] In Toulmin's hands, the very term "epistemological break" changes meaning: it is no longer that which purports to found reason through a radical move that would clear the slate of the past but, rather, that which, out of despair in the face of violence, has cut all the threads that would allow thinking. The epistemological break is still there, but it no longer marks, as it did for Michel Foucault, the start of a "classical age" of reason built on the ruins of the "prose of the world"; instead, it marks the beginning of a counter-revolution – let's say a Counter-Reformation of thought – that is going to make science, religion, politics, and the arts mutually incomprehensible.[18] Rationality becomes a prohibition against applying reason.[19]

Toulmin errs out of optimism. In his book published in 1990, he thinks he can rejoice in the fact that the modernist parenthesis is finally ending, owing to the surge in ecological questions.[20] According to him, we have left behind the epoch of absolute certainty and returned to a modest pluralism, attentive to the Earth as well as to people, open to religion as well as to the arts, to casuistry, to subtle relativism, to skepticism, to the reasonable more than to the rational; this pluralism characterized the sixteenth century, in his view, and it

---

[16] Ibid., p. 80.
[17] Lorraine Daston and Fernando Vidal, *The Moral Authority of Nature* (2004).
[18] This is the object of Horst Bredekamp's revision, throughout his work but especially in *The Lure of Antiquity and the Cult of the Machine: The Kunstkammer and the Evolution of Nature, Art, and Technology* ([1992] 1995), of the theme of the "prose of the world" in total rupture with the classical age as it is described by Michel Foucault in *The Archaeology of Knowledge* ([1966] 1972).
[19] This seeming opposition to rationalism, which is actually an extension of the paths of reason, is the object of my *Inquiry into Modes of Existence: An Anthropology of the Moderns* ([2012] 2013b).
[20] His book and my own, *We Have Never Been Modern* ([1991] 1993), came out shortly after the events of 1989, which lent themselves to a new periodization of history.

also characterizes the destruction of the Old Climate Regime. After this long parenthesis, the movement of the true scientific revolution, continually delayed,[21] could finally start up again. In particular, still according to Toulmin, because the ecological questions and the rise of a worldwide civil society make state borders – monstrosities invented to put an end to the wars of religion – obsolete. The Westphalian states have finally been encompassed within the countless networks of other territories acting in the name of other legitimacies that are gradually erasing the borders.[22] We have passed from Leviathans at war with one another to Lilliputians at war with states: "If the political image of Modernity was Leviathan, the moral standing of 'national' powers and superpowers will, for the future, be captured in the picture of Lemuel Gulliver, waking from an unthinking sleep, to find himself tethered by innumerable tiny bonds."[23]

A quarter of a century later, we can hardly share Toulmin's optimism. He had not foreseen the extent to which people could simultaneously ignore the rapidity of the ecological mutations and plunge back into a new cycle of wars. But what he did see he saw well: if the scientific counter-revolution had the effect of interrupting for a time the course of religious wars (and this was a good thing), it was at the price of a paralysis of thought, which was frozen for several centuries in an unfortunate distribution of functions among politics, science, and religion, under the protective authority of the State. And it is because of this paralysis that the ecological questions drive us mad.

But what Toulmin felt, before and better than anyone else, was our current closeness to the sixteenth century, a period made so unstable and so inventive by the shock of the discovery of *new lands* – and so tragic for those who were "discovered." For our part, it is the shock of discovering *new ways* of being on Earth that destabilizes us, perhaps, but that could make us just as inventive – all the more so in that, this time, we too, are finding ourselves "*un*covered," exposed.

*

And yet, facing the ecological mutation, instead of getting all excited, as our ancestors did facing the discovery of new lands, we remain frozen, indifferent, disillusioned, as if, at bottom, nothing could happen to us. This is what we have to understand.

---

[21] See Toulmin, 1990, chapter 4, "The Far Side of Modernity."
[22] The course of history, since the publication of Toulmin's book, has not followed this line – at least not yet; this will be the subject of the eighth lecture.
[23] Toulmin, 1990, p. 198.

One can of course blame the inertia of habits, the fear of novelty, the heady benefits of consumerism, the iron cage of capitalism; one can point to the influence of the lobbies that work actively on disinformation; or one can take into account the work of psychosociologists on the fear that paralyzes instead of provoking a reaction.[24] Those arguments may well hold up. But ultimately, if someone tells you your house is on fire, whatever your indolence, your psychology, or your ancestry, you are going to rush outdoors, and the last thing you'll be inclined to do as you dash down the stairs is to stop on the landing to quibble about whether the firefighters who are setting up their big ladder are really firefighters and if they are 90 percent or 95 percent likely to get you out safely. If we were in a normal situation, the smallest warning about the state of the Earth and its feedback loops would have already mobilized us, just as any question of identity, security, or property would surely have done.

Here is the question, then: Why do ecological questions not seem of *direct* concern to our identity, our security, and our property? Why are we not in a normal, banal, everyday, ordinary situation? Don't tell me that it's the scope of the threat or the distance from our daily preoccupations that makes the difference. We react as one to the slightest terrorist attack, but the notion that we are the agent of the sixth extinction of terrestrial species evokes only a jaded yawn. No, reactivity and sensitivity are what have to be considered. Collectively, we *choose* what we are sensitive to, what we need to react to quickly. Moreover, in other periods, we have been capable of sharing the suffering of perfect strangers very far removed from us, whether through "proletarian solidarity," in the name of the "communion of saints," or quite simply out of humanism. In this case, it is as though we had *decided to remain insensitive* to the reactions of *beings of a certain type* – those who are connected, broadly speaking, to the strange figure of matter. In other words, what we have to understand is why we are not true materialists.

This insensitivity is ancient in origin. Jean-Baptiste Fressoz has proposed to call "disinhibition" the attitude through which, since the eighteenth century, every time a warning has been sounded about the dangers of some industrial action (manufacturing lye, lighting with gas), some scientific development (vaccination, inoculation), some colonial appropriation of land (deforestation, plantation), the decision will be made, in a more or less subterranean but always explicit way, to go ahead anyway. After a terrible railroad accident (the first of its

---

[24] Mike Hulme, *Why We Disagree about Climate Change: Understanding Controversy, Inaction and Opportunity* (2009).

kind), Lamartine, the great French Romantic poet, exclaimed: "We must pay with tears the price that Providence puts on its gifts and its favors... Gentlemen, we know that civilization is a battlefield where many succumb in the cause of the advancement of all. Pity them, pity them... and let us go forward."[25] This "let us go forward," "let us go on," is admirable, and how valiant it is to accept bravely the consequences of a risky action – especially when they fall, generation after generation, on the heads of other people's children!

So it is not as though people haven't been warned, not as though the alarm systems have been angrily unplugged; no, the sirens have been blaring full blast, but a virile decision has nevertheless been made not to let oneself be *inhibited* by the dangers. If there is inhibition, in contrast, it concerns the speed of reaction to catastrophes generated later on. The two attitudes clearly go hand in hand: disinhibition for action where the future is concerned; inhibition when reckoning with retroactive consequences.[26] Virility on one side, impotence on the other. Time has so little influence on this attitude that we find it intact, two centuries later, in the "hopes" of geo-engineering or post-humanism: the disastrous consequences are indeed identified, but the experts, accusing their opponents of excessive spinelessness, are prepared to forge ahead nonetheless, even faster if possible, so as to make the factual situation irreversible – always in the name of "necessary modernization."[27] Where does this strange way of leaping headlong into an adventure with one's eyes closed come from?

In this lecture, I want to explore the religious – or, more accurately, the *counter-religious* – origin of this choice, this decision in favor of

---

[25] Cited in Jean-Baptiste Fressoz, *L'apocalypse joyeuse: une histoire du risque technologique* (2012, p. 273). At the same moment, the schema of innovation versus resistance has come into play, making it possible to condemn all resistance in the name of the heedless fears of innovation – fears that turn out, every time Fressoz studies them, never to have existed! What existed was the opposition to enterprises of domination to which it was entirely sane to want to resist. From then on, "the train of progress" one should not miss for fear of being left behind has started its journey forward. And still does!

[26] Ulrich Beck undertook to analyze these contradictions, starting with *The Risk Society: Towards a New Modernity* ([1986] 1992).

[27] See Clive Hamilton, *Earthmasters: The Dawn of Climate Engineering* (2013). In the "ecomodernist manifesto" we find the same idea of accelerating instead of reversing the movement of the modernization front; see www.thebreakthrough.org. The parodic version (or the critical version, depending on the reader) is offered by Alex Williams and Nick Srnicek in "#Accelerate Manifesto for an Accelerationist Politics" (2013) (not to be confused with the "great acceleration," a term proposed by Will Steffen et al. in "The Trajectory of the Anthropocene: The Great Acceleration," 2015a). The question is whether to continue the competition to determine who will be the most "resolutely modern."

disinhibition. To do this, we have to go even further back in time, before the tangle of science, religion, and politics became inextricable. If you recall the previous lecture, the term "scientific counter-revolution" used by Stephen Toulmin must have reminded you of the term "counter-religion" proposed by Jan Assman to emphasize the contrast between the so-called traditional religions, which are relatively indifferent to questions of truth or falsity, and those for which the question of truth becomes essential.[28] The "true" God cannot be made commensurable with any other; in contrast, however, one can call many other supreme authorities "God" – for example, the protective State, or Nature as known by Science.[29] This is what happened when it became necessary, in order to bring the wars of religion to a close, to shift the source of absolute certainty from one agent to another.

So that people would stop cutting one another's throats in the name of absolute certainties, all mutually contradictory, the collective was to be stabilized around a call for certainty, although, as Toulmin puts it so amusingly, without being sure *about what* we must be certain![30] Is it the political ideal? Scientific progress? Established religion? Economic progress? For fear of violence, we take refuge in certainty, but at the same time we don't allow ourselves to distribute levels of confidence on the basis of what each domain really requires – and especially on the basis of the type of assurance it can provide. How could religion, politics, science, nature, and the arts tell the truth in the same way, with the same degree of certainty? To discover the origin of the disinhibition in question, we need to go even further back in time, to a period well before the State offered its solution. That solution froze the battle lines but did not bring real peace; it paralyzed the Moderns, particularly in the way they registered reactions to the materiality of their innovations.

Why am I so sure that we have to look to religion to find the origin of this curious form of indifference to warnings about the current state of nature? Because of the resurgence, or even the omnipresence, of the term *apocalypse*. As soon as you speak with some degree of seriousness about ecological mutations, without even raising your

---

[28] Jan Assmann takes up a more radical version of this theme more concisely in *Violence et monothéisme* (2009). See the previous lecture.
[29] Let us recall that the name given to an agent is less important than the functions with which the agency is endowed. This is what allows for translation between seemingly distinct forms of supreme agents and thus for an outline of a geopolitics (see the fifth lecture).
[30] "It might not be obvious what one was supposed to be certain about, but *un*certainty had become *un*acceptable" (Toulmin, 1990, p. 55).

voice, you are immediately accused of "apocalyptic discourse" or, in a somewhat attenuated version, "catastrophist discourse." You may as well face the question directly and respond: "Well, yes, of course, what do you want us to be talking about?! Modernity is living entirely within the Apocalypse or, more precisely, as we shall soon see, *after* the Apocalypse. This is why Modernity has condemned itself to understanding nothing about what history is bringing it that is really new. So we have to agree finally to engage for real in an apocalyptic discourse *in the present time*."

*

If it's hard to talk about religion, this is not only because of the widespread belief that the religion question is definitively behind us, but also because it has become almost impossible for us to go back to what religion could have meant before the armistice of the seventeenth century – that is, before its mutation into forms of absolute certainty for which it is no better suited, at bottom, than science or politics. As *belief in* something, religion is of little interest, and we are right to pay it very little attention. The forms that have translated it over time, if we separate them from the movements that gave rise to them, can only leave us feeling that they are an accumulation of old keepsakes, whose only value is ethical, aesthetic, or patrimonial.

And yet, if religion – as counter-religion – remains active, remains fruitful, it is because of the discovery that one can live, that one must live in the "end times," in the sense – at once very specific and very unstable – that the ends have been definitively achieved, within time, and can only be realized *thanks to time*. As we have identified it in the previous lecture, the truth expressed by such a discovery does not come from a particularly strong degree of certainty, quite the contrary, but rather from the unfolding, the reprise, the embodying of the term "definitive." If something is definitive, then, in effect, it can be translated by "absolute," "certain," "assured," "present," except that, as we are talking about an end of time *within time*, to experience this truth is to make oneself aware of the fact that it is equally uncertain, ill-assured, relative, fragile, absent, and always to be recommenced!

As long as we live in this tension, we understand what may be signified by the emergence of counter-religion and the new form of historicity that has imposed itself in the course of history.[31] It is

---

[31] This has been a classical theme since Karl Löwith's *Meaning in History: The Theological Implications of the Philosophy of History* (1949).

How (not) to put an end to the end of times? 195

paradoxical, in fact, to experience the time that is passing at once as what is radically distinct from the end times and nevertheless as what is achieving these same ends. As soon as we lose this ever-so-bizarre sense of history, even for a moment, we lose the sense of religious truth. Until we understand it again, a moment later. Counter-religion, as its name indicates, ceaselessly *struggles against itself*. This is what makes it so difficult to grasp, and it is also the source of its power, which is at once liberating – the ends are achieved – and toxic – there is always a risk of being mistaken about the ends!

That this end time has been expressed in countless continuously amended beliefs, and that these beliefs, starting in the seventeenth century, have become certainties to be defended against the competition of science and politics, need not concern us here; these observations would only be distractions. Anyway, I know of nothing more discouraging than the task of tracing the gradual degradation of religious innovations into simple beliefs to be defended – or, worse, enforced by some form of morality police.[32] What counts for our analysis is the fact that, at the moment when this paradoxical historicity stopped being understood, it was as though the enigma posited by the counter-religion had been split in two. The *end time* was retained, and so was the idea of *definitive truth*, but the two notions were brought together from then on in the most improbable form: *a certain number of peoples tell themselves henceforth that they are absolutely certain that they have reached the end of time*, have arrived in another world, and are separated from the old times by an absolute break. To these peoples, obviously, nothing serious can happen any longer, since they believe they have always been within the "end of history."[33] It is thus completely useless to speak to them in apocalyptic terms announcing to them the end of their world! They will reply condescendingly that they *have already crossed over to the other side*, that they are already *no longer of this world*, that nothing more can happen to them, that they are resolutely, definitively, completely, and

---

[32] This is what makes one perk up every time one hears, within the ecclesiastical institution itself, a different music that recalls the radicality of the movement from which it arose – as is the case with the encyclical *Laudato Sí* (Francis 2015), whose originality can be measured by the efforts that have been made to stifle its impact.
[33] Quite unintentionally, in *The End of History and the Last Man* (1992), Francis Fukuyama provided a very accurate diagnosis of the post-apocalyptic situation of America and the impossibility of re-engaging with historicity in which the nation had found itself for the previous thirty years. How could those who were done with history take an interest in – or even comprehend – the new geopolitics of a multiple Earth? I take this to be the deep reason for the attachment of Americans to climate skepticism: "something like that cannot happen to us anymore."

forever modernized! That their only movement is to keep on going forward, never backward. Their motto is that of the Spanish Empire: *Plus ultra*.[34]

For here is what is most extraordinary: these peoples who call themselves non-religious and nonbelievers, lay and secular, have extracted from the counter-religion that preceded them its deepest meaning – it is true that one can live in the end time – by reversing the sense of that discovery, turning it into its exact contrary: *there is no longer any doubt* that the end time has actually come about! What has disappeared along the way? Doubt, uncertainty, fear and trembling before the radical *impossibility* that time can end and that this achievement can get along without the temporal flow. Everything depends on a minuscule misinterpretation of the term "definitive." The Moderns are the ones who have managed to shield themselves from passing time, by appropriating for themselves the most dangerous, the most unstable of all forms of counter-religion. How could they not be disinhibited? Believing they are fighting religion, they have become irreligious in the sense recalled in the previous lecture: they have made *negligence* their supreme value.[35] Nothing more can happen to them. They are already and forever in another world! There is no direction except straight ahead; it is as though the option of turning back had been cut off.

It was Eric Voegelin (1901–1985) who put his finger on this operation of inversion, in *The New Science of Politics*, a brilliant, underappreciated book.[36] The end times, in the Jewish and Christian traditions alike, had already been subjected to numerous transpositions in the form of an *end of times*, a possible, foreseeable, and, of course, hoped-for end. It was not the time of the end *within* the time that passes; it was the end, the final interruption *of* the time that passes. But this slippage led to ongoing doubt about the veracity of such a translation. The apocalypse, in the sense of the revelation of a certain regime of historicity, gradually became, in particular thanks to the numerous glosses on the Apocalypse of John, a discourse about the expectation of the end of the world.

Now if you have followed me this far, you can see that nothing authorizes anyone to foresee – to predict – the end of the world; one can only preach it or pray for it. "End" means first of all achievement,

---

[34] How could Moderns, whose whole pride and sole ideal consisted in *passing between* the Pillars of Hercules, find the taste for, the pride in, the ideal and the politics of "*giving* themselves limits"?
[35] See Michel Serres, *The Natural Contract* (1995), p. 81.
[36] Eric Voegelin, *The New Science of Politics: An Introduction* ([1952] 2000a).

then finitude, finally revelation, but always in and with time, and especially with the passage of time as its necessary medium. This is actually what gives an entirely new value to time that passes: it bears, *and bears alone*, the final achievement, which is never final! What lasts forever lasts *only through* what does not last. To remain in the spirit of this upsetting situation, the last thing from which one would have to escape is time. If one begins to do this, then one is going to *oppose* the time that passes with the time that has to end in order to reach what lasts. This is the case of the millenarians; or else, in an even stranger reversal, you start declaring that the waiting time is over, that history has ended, that it is about to end! As soon as one translates "the time of the end" by "the end of the times," one finds oneself on the brink of a dizzying metamorphosis – and an irresistible temptation to *shift to eternity* while *abandoning* the time of finitude and mortality.

Voegelin credits Joachim de Flore (1130–1202) with a central role in this gradual misunderstanding of the apocalyptic message – I ought to say, in this gradual modernization that is simply going to wipe out, little by little, the Jewish and Christian origin of the message.[37] Joachim in fact adds to the traditional Christian division (already quite debatable) between the epoch of the Father and that of the Son – and thus between the Old and the New Testament – a new epoch, which he called the Kingdom of the Spirit. It is with this Kingdom that things, if I dare put it this way, are going to go wrong!

We must tread carefully here, because the point of divergence is, at the outset, minuscule, so minuscule moreover that the popes find nothing to reproach in Joachim's slightly borderline orthodoxy: waiting for the Kingdom of the Spirit seems to be a perfect interpretation of the dogma of the Incarnation, which is after all defined by eternity *in* time. With this nuance: Joachim makes the waiting period, by definition impossible to control, the realization *within* history of the *end* of history. But that's exactly what Incarnation means! No, listen carefully here: it is exactly the opposite. The relations between the end of times and the finitude of time have been reversed.[38] History begins to bear, in its very movement, the transcendence that puts an end to it! This means, then, that we are going to be able to transform immanence into what is able to bear eternity for good – to the point

---

[37] See the overview by Henri de Lubac, *La postérité spirituelle de Joachim de Flore* ([1981] 2014), and Thierry Gontier, *Politique, religion et histoire chez Eric Voegelin* (2011).
[38] "In the course of history, for our salvation 'we have nothing more to wait for' (which does not mean: to exploit, to explore, to put to work); nothing, and certainly not a 'Spirit' that would 'surpass' Christ, destroying along with his Church the means by which *his* Spirit could continue to live" (Lubac, 2014, pp. 159, 194).

of pushing Joachim to establish not only correspondences among the *figures* of the Old and New Testaments, as had always been done,[39] but to formulate veritable historical *forecasts* that he purports to verify by a dizzying exercise in numerology. From now on, the course of history charged with eternity becomes *controllable* by those who know how to predict its path with certainty.

In the hands of Joachim's commentators, the minuscule nuance grows into a radical transformation of the message: the continuing expectation of the return of the Son – of which "you know neither the day nor the hour" (Matt. 25: 13) – becomes the certainty that the Kingdom of the Spirit will be realized here below. But to realize here below the promise of the beyond inevitably means passing from a definition that could be called spiritual to a form of politics. One then abandons St Augustine's wise and precarious solution, which consisted in expecting nothing from the earthly City but everything from the heavenly City. The monks of subsequent generations, enthusiastic readers of Joachim, dreamed for their part of actually *realizing* the heavenly City right here, by radically transforming the earthly one. And who was to manage this kingdom – which thus became politico-religious? These same monks, leading ascetic lives inspired by Scripture! As imperceptible as it was radical, the transition began to pervert both religion and politics. From that moment on, poor politics, so impotent, so modest, so concrete, always so disappointing, was charged with the crushing weight of making the kingdom of the Spirit realistic! Religion, so fragile, so unsure of itself, was going to have to take it upon itself to direct the course of the world! What unleashed all the furies of Western history was that, clearly, neither politics nor religion could bear such burdens. One must never allow politics to degenerate into mysticism, for fear that mysticism will degenerate into politics.

Does this remind you of something? You will be perfectly right, Voegelin tells us, if you recognize in this figure of counter-religion what it has continued to keep on becoming among the Moderns. Pull off the monks' robes; forget the archaic terms "Son," "Spirit," and "Kingdom"; forget the mention of a New Testament; you have before your eyes the terrifying prospect of entrusting to militants, inspired by the certainty of truths from on high, the achievement of Paradise on Earth. Yes, exactly: the exercise of terror. No longer the Earth vibrating under the presence of a Paradise that it alone can achieve provided that the two are not confused, but an Earth that has become the reality (always virtual) of Paradise itself. The promises of the beyond have been turned into *utopias*. This would not be too serious

---

[39] See Erich Auerbach, *Figura* (1959).

if no one had come up with the idea of transcribing them into reality! A realization led by militants – not to be confused with activists[40] – definitively immunizes against doubt, since they will have passed *to the other side of uncertainty* concerning time and its direction. The ends are no longer what you expect but what you have – and what, of course, will inevitably betray you.

According to Voegelin, one cannot play with the kingdom of the Spirit with impunity. Joachim de Flore, good monk that he was, believed that adding a new epoch to universal history to complement that of the Son was a very pious thing to do. Yet he succeeded only in *bringing an end* to that of the Son, thus introducing into Christianity itself the programmed disappearance of Christianity.[41] Modernization retains all the apocalyptic features but deprives itself of the uncertainty that was required to keep science, politics, and religion *from getting mixed together*. The Moderns, according to Voegelin, began to believe that one could finally move from trembling before the incompleteness of the world – the political theology proposed by St Augustine – to a new possibility that would be the completion, the achievement, of the world here below by the intrusion of the Spirit – and its successors. *Living in the expectation* of the Apocalypse is one thing; living *after* its realization is something else again. Such was the momentum given to the counter-religion before the Reformation. And the Reformation and Counter-Reformation became more and more violent, since they could lead only to reactions, ultimately inevitable, to the prior politicization of the religious mind by Joachim's interpreters. Once the wars of religion began, there was no solution other than the one so well analyzed by Toulmin: the State was quickly shored up by Science, and both were soon gobbled up whole by the Market.

*

You may well be asking what connection there can be between this detour through the history of political theology[42] and ecological

---

[40] John Dewey's whole political philosophy, especially in *The Public and its Problems* (1927), consisted in managing to distinguish experimentation, linked to the practice of investigation, from the application of a truth. This is what makes it possible to distinguish activists from militants. See also, on the relation between politics and truth, Walter Lippmann, *The Phantom Public* (1925), and my introduction to the French edition, *Le public fantôme* (2008b).
[41] See Eric Voegelin, "Ersatz Religion: The Gnostic Mass Movements of Our Time" (2000b), for a brief summary of his argument.
[42] The term "political theology" was introduced by Carl Schmitt to designate the archaeology of the principal political concepts that the modern period considers

questions. In fact, the link is as direct as it is dazzling, and it rests entirely on the word *immanentization*, which Voegelin uses to sum up the reversal of meaning of the word *definitive*. This is what led Westerners to lose the Earth by cutting off access to immanence. For the history related by Voegelin does not move from transcendence to immanence but, rather, from an epoch in which the link between the two remained unstable to another epoch which no longer saw anything in the immanent but the definitive insertion of the transcendant – and its failure. It was as though immanence as well as materiality were going to disappear, crushed under the weight of this ersatz immanence.

If the history of the Moderns had consisted in moving from the abandonment of illusions about the beyond to the solid resources of the here below, it would have become wholly *attentive to the terrestrial*. But *for those who have immanentized Heaven, there is no longer any accessible Earth*. The whole paradox of modernization is that it has lost sight, more and more, of any contact with the down-to-earth, with materiality: it no longer sees anything in this world below but the other world simply *immanentized*. This is what explains why the Moderns feel so lost – to the point that they never know whether they have been Modern or not![43] In other words, if they miss out on the world, these Moderns, their failure results not from excessive materialism but, rather, from an overdose of ill-placed transcendence.

Let us look at the way Voegelin proceeds. He tries, first of all, to understand where the instability of the counter-religion comes from (counter-religion is Assman's term; Voegelin obviously doesn't use it, but it clarifies the movement he is describing quite well): "What specific uncertainty was so *disturbing* that it had to be overcome by the dubious means of *fallacious immanentization*?"[44]

In order to grasp the solution Voegelin offers, we have to shed the entrenched prejudice according to which religion – Christianity in particular – is only a tissue of fables swallowed whole. This prejudice may be valid, but only after the armistice that, by mixing all the

---

secularized but that always lead back to still active theological schemas (Carl Schmitt, *Political Theology: Four Chapters on the Concept of Sovereignty*, [1922] 2005). I am using it here to underline one of the constitutive features of the counter-religion, which has to do with uncertainty about what is secular and what is religious.

[43] I am always accused of not specifying the precise limits of the Modern people, or what country they live in, and in what period. I hope it will now be clear why these questions cannot be answered. The Moderns actually don't know *where and when* they are. This is exactly what is at stake in the attempt to re-root them by way of Gaia.

[44] Voegelin, 2000a, p. 187, emphasis added.

distinct sources of truth together in a competition lost in advance in order to reach uncontestable certainty, pushed religion into dogmatism. Voegelin starts from the principle – this is his huge contribution – that one has to be able to go back to the source of the vibration that is proper to counter-religion and to the time of the end. A rare mind, he is capable moreover of accepting ontological pluralism where religion is concerned. He invites us, in fact, to recognize *three different types* of supreme authority:

> Terminologically, it will be necessary to distinguish between three types of truth. The first of these types is the truth represented by the early empires; it shall be designated as "cosmological truth." The second type of truth appears in the political culture of Athens and specifically in tragedy; it shall be called "anthropological truth."...The third type of truth that appears with Christianity shall be called "soteriological truth."[45]

In his book, Voegelin maintains that Western history has never managed to keep these three forms of religion together. Augustine understood nothing about the Roman gods. Hobbes had no feeling for Augustine's God.[46] What interests Voegelin is the history of that *loss of feeling* and the means of recovering a "maximal differentiation" that would make it possible not to neglect any of the forms of religion invented over the course of history.[47] Thus he takes very seriously the type of truth-telling, the mode of existence, proper to this particular form of counter-religion associated with Christianity. But he also stresses that this mode depends on an uncertainty so great that it will not resist the temptation to get rid of it: "One does not have to look far afield for an answer. *Uncertainty is the very essence of Christianity*. The feeling of security in a 'world full of gods' is lost with the gods themselves;[48] when the world is de-divinized, communication with the world-transcendent God is reduced to the *tenuous bond* of faith."[49]

---

[45] Ibid., pp. 149–50. One would speak today of civic religions, moral or humanistic religions, and religions of salvation. The terms are unimportant here; what counts is the pluralism of the types of supreme agents that allow people to orient themselves. Voegelin's argument is that the West has never succeeded in maintaining all three at once.
[46] Ibid., p. 159. But it is especially the decisive passage on Hobbes that is most pertinent here (pp. 217–18).
[47] Ibid., p. 152.
[48] This is the well-known argument of internal secularization within the Christian tradition itself, which will then be turned against that tradition. Although Voegelin does not use the term, it is another way to define the "counter" in "counter-religion."
[49] Ibid., p. 187, emphasis added.

The former divinities, those of religions capable of being compared to one another through the translation tables I discussed in the previous lecture, the ones Voegelin calls "cosmological," have been consumed by the biting fire of the counter-religion. The religions of salvation – this is the meaning of the word "soterological" – start by destroying the divinities – this is what "de-divinized" means here – before being carried away later on by the same movement of religion rising up against itself.[50] In the intermediate period, between the vanished cosmological religions and the new (counter-)religion of irreligion, Voegelin draws the picture of a Christian making a great effort to hold onto his vocation:

> The bond is *tenuous*, indeed, *and it may snap easily*. The life of the soul in openness toward God,...*trembling on the verge of a certainty that if gained is loss* – the very lightness of this fabric *may prove too heavy a burden for men who lust for massively possessive experience*.[51]

If it is true that being a Christian requires one to live in fear and trembling, then you can easily understand that there will be a strong temptation to jump on any opportunity to *stop* fearing and trembling![52]

If you are having trouble with this passage, it is probably because you have transformed the situation of fear and trembling before the presence of the time of the end into the assured *belief* that there are *two worlds*, well separated: that of the here below and that of the beyond, toward which, according to the critics of religion, believers can only aspire to be beamed up. But this solution, in which transcendence becomes Heaven and immanence becomes Earth, is an easy way out, a fallback solution of indolence and loss. The bond between immanence – the time that passes – and transcendence – the achievement of the ends – was invented by the counter-religion and then lost by its modernized version; it requires a *vertical* relation between the two, and not at all the superposition, sandwich-style, of a layer of materiality over a layer of spirituality. This is the eternal misunderstanding between the "spiritualists" and the "materialists": they believe that they oppose one another but they speak of exactly the same thing, all sides unaware that spreading the supernatural on top of the natural is already to have lost both. But we have to recognize

---

[50] Ariadne's thread is always the attitude toward iconoclasm, and not the variable nature of the icons that are offered to the idol-smasher's hammer.
[51] Voegelin, 2000a, pp. 187–8, emphasis added.
[52] This cessation was what unleashed Kierkegaard's furor as well as the irony he directed against the religious attitudes of his time (*Fear and Trembling; and, The Sickness unto Death*, [1843] 2013).

that the tendency is irresistible: "The more people are drawn or pressured into the Christian orbit, the greater will be the number among them who do not have the spiritual stamina for the *heroic adventure* of the soul that is Christianity; and the likeliness of a fall from faith will increase [with the progress of civilization]."[53]

Voegelin's hypothesis is a radical one: peoples who have unquestionably become Christianized but who see their wealth growing and their cities expanding, and who, starting in the fifteenth century, discover an abundance of new lands and new horizons while still remaining under the sway of Christianity, are going to transfer the weight of this crushing burden to something else. What can that be? A much older tendency, still more or less present in the Jewish and Christian traditions, that of *Gnosticism*.[54] The very term recalls the slippage that strikes Voegelin as both inevitable and calamitous: whereas faith is uncertainty (a vibration of presence and absence proper to counter-religion), Gnosticism, as its etymology indicates, is *assured knowledge*. Faith is what grasps you; knowledge is what you grasp.

We can readily understand that the Gnostic temptation became irresistible during the period Toulmin defined as one of indisputable certainty. And there was even more pressure in that direction, starting in the seventeenth century, owing to the seeming resemblance between that form of certain truth and the new form of incontrovertibility offered by the sciences.[55] From this moment on, religion presents itself as nothing but an effort – obviously futile – to resemble assured and indisputable knowledge.

> The attempt at *immanentizing* the meaning of existence is fundamentally an attempt at bringing our *knowledge of transcendence* into *a firmer grip* than the *cognitio fidei*, the cognition of faith, will afford; *and gnostic experiences offer this*

---

[53] Voegelin, 2000a, p. 188, emphasis added.
[54] Adolf Harnack's *Marcion: The Gospel of the Alien God* (1990) seems to have been the work that sparked the interest of German philosophers in the analysis of Gnosticism, especially Hans Jonas, *The Gnostic Religion: The Message of the Alien God and the Beginnings of Christianity* ([1958] 2001), where the connection with ecology is obviously crucial (see Clara Soudan, "Théologie politique de la nature: l'ontologie théologique de Hans Jonas au fondement de son éthique environnementale de la responsabilité" (2015).
[55] "And, finally, with the prodigious advancement of science since the seventeenth century, the new instrument of cognition would become, one is inclined to say inevitably, the symbolic vehicle of gnostic truth.... Scientism has remained to this day one of the strongest gnostic movements in Western society;...the special sciences have each left a distinguishable sediment in the variants of salvation through physics, economics, sociology, biology, and psychology" (Voegelin, 2000a, pp. 191–2).

firmer grip in so far as they are an expansion of the soul to the point where God is *drawn into* the existence of man.[56]

The interpretation of the Moderns depends on the meaning of the term "immanentization," which makes it possible to explain both "secularism" and "materialization." Voegelin does not say, as the usual grand narrative does, that we have passed from Obscurantism to Enlightenment, from the expectation of the illusory goods of Heaven to the grasp of earthly realities – in short, from a life inspired by religion to a secular life. No, he tells us that we have passed from a situation in which immanence and transcendence, the passage of time and the time of the end, the terrestrial City and the celestial City, were in a relation of mutual *revelation* – this is the literal meaning of the word apocalypse – to an entirely different situation, in which we believe we can grasp, realized here on earth, the promised presence of the world beyond. According to him, the Moderns have not been secularized – and this is the object of a vast dispute[57] – but, conversely, immanentized. The inevitable result: *they have no sort of possible contact with the terrestrial*, since they can see in it only the transcendent, which would be trying awkwardly to fold itself into the immanent. And necessarily failing! Fundamentalism was born, and has never stopped metastasizing.

A recent example may make my borrowing from the too littleknown history of Gnosticism more comprehensible. The recent emergence of Islamic fundamentalism, which is pushing to maximum intensity both the counter-religion of Islam and that of modernization, allows us to grasp the movement pinpointed by Voegelin. In the film *Timbuktu*, an old imam is trying to explain the meaning of the word "jihad" to the militants who have come wearing Kalishnikovs to "modernize" the ancestral city of Timbuktu by fire and the sword.[58] "You want to prevent us from waging jihad, and you're an imam!" the militant exclaims indignantly. To which the imam responds with humility that he could not allow himself such arrogance, for he has been waging jihad *against himself* for sixty years, and he is still not exactly sure what God is telling him to do... Here is the whole

---

[56] Voegelin 2000a, p. 189, emphasis added.
[57] On the controversy as presented by Hans Blumenberg in *The Legitimacy of the Modern Age* ([1976] 1983), along with Voegelin, see the excellent article by Willem Styfhals, "Gnosis, Modernity and Divine Incarnation: The Voegelin–Blumenberg Debate" (2012). What interests me in this dispute, in any case, are its consequences for the contempt for matter that turns out to be linked to a fascination with the material.
[58] Abderrahmane Sissako, *Timbuktu* (2014).

difference: a soul trembling *under the hand* of God is not at all the same thing as the spiritual certainty of someone who believes that *his hand is God's hand*! The old imam is living in the old Islam, which does not yet see itself as completely inseparable from politics; the new militant combines religion and politics in a single, radical certainty, in which the roles of preacher, judge, researcher, police chief, and enforcer are merged.[59] The "expansion of the soul to the point where God is drawn into the existence of man" has resulted in certain men taking themselves to be God, no longer measuring the distance that separates them.

Although Voegelin did not talk about the "Islamic revolution," he would have had no trouble extending the line of analysis that goes from the first still Christianized Puritans to the various forms of utopian militantism that are violently anti-Christian but fiercely modernizing. From the aspersorium to the Kalishnikov, then from the Kalishnikov to the suicide belt, the logic is unmistakable. Nihilism has more than one weapon in its arsenal.

> A line of gradual transformation *connects medieval with contemporary Gnosticism*. And the transformation is so gradual, indeed, that it would be difficult to decide whether contemporary phenomena should be classified as Christian because they are intelligibly an outgrowth of Christian heresies of the Middle Ages or whether medieval phenomena should be classified as anti-Christian because they are intelligibly the origin of modern anti-Christianism.

And he concludes: "The best course will be to drop such questions and to recognize the essence of modernity as the growth of Gnosticism."[60] Unfortunately, we haven't finished measuring the scope of this "growth." Whereas the theme of the apocalypse arose from the feeling of Presence from which one should not separate oneself, it has now become the Absence that the Moderns have imposed on the rest of the world – and now, in an unexpected reversal, on themselves.

> However fatuous the surface arguments may be, the widespread belief that modern civilization is Civilization in a pre-eminent sense is experientially justified; the endowment with the meaning of salvation has made the rise of the West, indeed, *an apocalypse of civilization.*[61]

There is no doubt about this point: the West has landed on all other civilizations like an Apocalypse that has put an end to their existence.

---

[59] See Gilles Kepel and Jean-Pierre Milelli, eds, *Al Qaeda in its Own Words* (2008).
[60] Voegelin, 2000a, p. 190, emphasis added.
[61] Ibid., p. 194, emphasis added.

By believing oneself to be a bearer of salvation, one becomes the apocalypse for others. Do you understand why we have to be suspicious of those who accuse ecological discourse of being too often apocalyptical? They are the ones who, on the contrary, by refusing to continue to live in the time of the end, have imposed a violent end on all the other civilizations. Joseph Conrad and Francis Ford Coppola were right: one must not say "Apocalypse yesterday," but always "Apocalypse now."

*

If you were to ask why the so-called ecological questions don't interest very many people, in spite of their scale, their urgency, and their insistence, the answer might not be too hard to find if you were to take their (counter-)religious origin into account. Telling Westerners – or those who have recently become Westernized, more or less violently – that the time has come, that their world has ended, that they have to change their way of life, can only produce a feeling of total incomprehension, because, for them, the Apocalypse *has already taken place*. They have already gone over to the other side. The world of the beyond has been achieved – in any case for those who have become wealthy. They have already crossed the threshold that puts an end to historicity.[62]

They know, they hear, but deep down *they do not believe it*. Here is where we have to seek the fundamental source of climate skepticism, I believe. It is not skepticism bearing on the solidity of one's knowledge, but skepticism about the skeptics' own position in existence. If they doubt or deny, it is because they take those who are crying out, in a timely and counter-timely fashion, that we have to change our way of life *totally* and *radically* as nutcases who are no more worthy of attention than Philippulus the Prophet who scares Tintin, in *The Shooting Star*, with his gong and his white sheet. A "total and radical change of life style?" They have already accomplished this, precisely, by *becoming resolutely modern*! If modernity were not so deeply religious, the call to adjust oneself to the Earth would be easily heard. But because modernity has inherited the Apocalypse, simply shifted

---

[62] This is why it is more or less futile to try to do without an analysis of the theme of the end of times; in fact, it occupies the entire history of Western inspiration, right up to the young woman who exclaimed, on June 22, 2015, at the colloquium that introduced the "ecomodernist manifesto": "It's time to go beyond *that doomsday mood!*," picking up the line from Lamartine cited earlier in this lecture: "Let us…go forward."

a bit into the future, the call elicits only a shrug of the shoulders or an indignant reply: "How can you come and preach the Apocalypse to us *yet again*? Where is it written in the Books that there will be another Apocalypse *after* the first one? Modernity is what we have been promised, what we have reached, what we have conquered, sometimes by violence, and you think you can take it away from us? Tell us that we were wrong about the meaning of the promise? That the Promised Land of Modernity should remain promised? This is nonsense!"

And nowhere, indeed, is it written that the Apocalypse may be followed by another. Hence the entrenched certainty, the total calm, the icy coldness of those who nevertheless read announcements of various catastrophes every day. It seems that they have a *right* to the Earth that has in fact been promised them, they feel entitled – but there is nothing terrestrial about this Earth, since what is denied, precisely, is that it has a history, a historicity, a retroaction, capacities – in short, agency. Everything trembles, but they don't, nor does the ground on which they stand. The framework in which their history unfolds is necessarily stable. The end of the world is only an idea.[63] How do they manage to believe in this stability while everything is vibrating underneath them? Because this apparent stability is imposed on materiality by an idea of matter borrowed from the world beyond, which they have confused with the world here below.[64] And this is where we come back to the astonishing amalgam between the counter-religious idea of modernity and the just as counter-religious idea that Science has inherited. *Matter is materiality plus* (I mean to say *minus*!) *immanentization.*

What doesn't manage to get through to people bombarded by bad news about the ecological mutation is the activity, the autonomy, the sensitivity to our actions, of the materials that make up the critical zones in which we all reside. These people seem incapable of responding to the agency of these materials. You remember how we have often been astonished, since the beginning of these lectures, by the deanimation of the world imposed by the epistemological view

---

[63] The clever solution consists, in Kantian fashion, in making the end of the world a constant of the mind but without any relation to the state of the world, as we can see with Michaël Foessel's book *Après la fin du monde: critique de la raison apocalyptique* (2012). The title is quite revealing: to place oneself "after" is to ensure oneself against the danger of being *at the time of the Apocalypse*.

[64] Matter is an idealism completely opposed to materiality. On the genealogy of the extension of the *res extensa*, see Alfred North Whitehead, *The Concept of Nature* (1920), and especially the commentary by Didier Debaise, *L'appât des possibles: reprise de Whitehead* (2015, p. 33).

of scientific activity.[65] Now we can grasp the religious – and, more precisely, apocalyptic – origin of such deanimation. It results from the narratives of causality that attribute all action to the cause – going back step by step to the First Cause – and all passivity to the consequences. A strange competition between Nature and Creation, between the blind Watchmaker and the all-seeing God, in order to try to empty the world, as much as possible, of any activity. Hence the extreme resistance to taking into account the Earth's activity on the part of those who look at materiality as something inert and passive, and who believe that the world they live in is made up only of objects, of simple matters of fact caused by other equally inert matters of fact.

The most serious consequence, however, is that these Moderns superimpose on materiality the *contempt for matter* that is one of the ancient features of Gnosticism. You have surely noticed that the same individuals who remain insensitive to the ecological crises are very touchy about any question concerning morality and identity, and they're prepared to go out and demonstrate as soon as their interests are threatened. If they have chosen to be negligent, it is only toward beings that belong to the realm of "nature." Why this choice, so contrary to what is so obvious? It is as though Gnosticism had rendered matter at once desirable and contemptible – *desirable* because it has to embody the ideal, *contemptible* because in the long run it proves unsuited for that task!

The only thing that the world here below cannot do, in fact, is fulfill the promises of the world beyond, immediately and completely! If what is not happening can only be realized through the intermediary of what is happening, this is possible only under the conditions imposed by the passage of time. And thus slowly, with difficulty, with loss, aging, care, and concern. However, in the Gnostic tradition there is a Manichean feature that has persisted throughout the epochs: mistrust, disgust, hatred even, toward matter, the aborted result of a failed project conceived by some perverse demiurge.[66] This tradition is reactivated every time matter disappoints the utopians. Every time, that is to say always! By seeking to achieve Paradise on

---

[65] See the first lecture.
[66] The strangeness of the Gnostic schema is that it so distanced the good God that, in order to account for Creation and explain why everything works so badly in this world below, it was necessary to imagine an exceedingly clumsy and perverse demiurge. See Harnack, 1990, and Voegelin, 2000a.

Earth, one succeeds only in realizing Hell on Earth – not always for oneself, but certainly for others. The failure of these projects – religious, scientific, technological, revolutionary, economic, governmental, the adjective hardly matters – leads those disappointed in Gnosticism to scorn matter even more, for its inability to rise to the level anticipated by the Ideal.[67] Hence the strange position of objects, conceived at once as the sole reality and as the target of the deepest scorn.

This is the most dangerous consequence of a counter-religion that, after turning against the divinities, then against the idea of God, turns once more *against nature*. What is called the demiurgic spirit of the Moderns would be of little import if that demiurge were not the one of the Gnostic tradition, brimming over with the malignity that has transformed this earthly world into a cesspool from which one has to try to escape by all possible means. The Gnostics can no longer enter into contact with the terrestrial. They may aspire to escape toward the transcendent by way of a utopia, they may try to create their utopia for real, they may despise the world and violently reject matter as unfit to be transformed by Ideas: whatever they do, every solution they invent is more calamitous than the last!

You rightly suspect that it would be totally useless to talk to these Gnostics about ecology, about the terrestrial world, about uncertainty or fear and trembling before the ongoing distribution of agency. Don't expect to interest them in the metamorphic zone that has occupied us from the beginning of these lectures! They have ended up in the implausible but, alas, very real situation of being assured of their own salvation, even as they inhabit a material world which, at bottom, they hold in contempt! By losing the vertical axis, they have also lost the horizontal. Hence the astonishing claim on the part of these peoples, already identified in the previous lecture, of being the only ones who live in the real inanimate world here below, which is for them at once the only desirable one and the only one totally deprived of meaning! Here we find the origin of the abject object, rejected with horror by most philosophies, which hasten to turn away so as to go back to the illusory grandeur of liberty and subjectivity. On the

---

[67] What is most interesting in Bernard Yack's book *The Longing for Total Revolution: Philosophic Sources of Social Discontent from Rousseau to Marx and Nietzsche* (1992) is the way it traces the political consequences of this despair on the part of revolutionaries in the face of the inability of matter to realize the ideal. For immanentization blinds people to the possibilities of immanence.

root of a tree metamorphosed into matter, Roquentin, as we readily understand, can only vomit.[68]

*

The religious origin of the ecological crisis is indisputable; I hope you understand this, but not at all for the reason given in Lynn White's overly famous article that accused Christianity of having reified matter and given man absolute mastery over living beings.[69] Something has indeed happened that has made a very large number of pious minds indifferent to the fate of one specific type of being, the type usually associated with materiality interpreted as matter. But if there is a historical origin of the ecological crisis, it is not because the Christian religion has made the created world contemptible[70] but, rather, because that religion, sometime between the thirteenth century and the eighteenth, lost its initial vocation by becoming Gnostic, before passing the torch to the superficially irreligious forms of counter-religion.

If White is not wrong, though, it is because the Christians, having lost the race for the most indisputable type of certitude, have gradually abandoned all concern with the cosmos in order to devote themselves to the salvation of humans alone, and then among the humans to the salvation of the soul alone, before abandoning the soul itself to the exclusive benefit of morality. A slow degradation that has led them to lose the world, not only in the trivial sense that fewer and fewer inventive minds have been interested in their message, but in the much more serious sense that they have become increasingly indifferent to the fate of the cosmos.[71] Believing themselves to be attached to the Spirit, they have lost the Earth. Believing that they are defending religion, they have driven everyone to assault the Earth through negligence. Led astray by the supernatural, itself a delayed reaction to the

---

[68] *Trans.*: Roquentin is the protagonist of Jean-Paul Sartre's novel *Nausea* (trans. Lloyd Alexander, introduction by Richard Howard [New York: New Direction, 2007]).
[69] Lynn White, Jr, "The Historical Roots of Our Ecological Crisis" (1967).
[70] See Hélène Bastaire and Jean Bastaire, *La terre de gloire: essai d'écologie parousiaque* (2010); Christophe Boureux, *Dieu est aussi jardinier* (2014); and Michael S. Northcott, *A Political Theology of Climate Change* (2013).
[71] This is the strange misunderstanding of the (counter-)religions engaged in a struggle that they deem necessary against the cosmological religions. This is why it is so astonishing that after a long association between ecology and "paganism" – a phantasm – Pope Francis addresses the Earth as "sister" and "mother." See the opening paragraph of *Laudato Sí*.

invasion of "nature," they are no longer in a position to do their duty by *defending materiality, unjustly accused, against matter, unduly spiritualized*. They need to be reminded of the celebrated evangelical injunction, inverted: "What use is it if you save your soul, if it means losing the world?"[72]

The fate of Christianity is nevertheless of little importance compared to the loss of meaning imposed on materiality by the move to force it to become matter. It is here, really, where the major injustice lies, and this is what ultimately explains the Moderns' insensitivity to what they do. There is something frightening in contemplating the accumulation of sedimentary layers that are gradually going to cover over the proliferating agencies to the point of making them inaccessible. Materiality – active, historical, multiple, complex, open – becomes first of all, through the process of immanentization, an ersatz Paradise. Then, grasped by science in its struggle against religion, it undergoes a layer of idealization and becomes that which is "nothing but" the concatenation of causes and consequences in strict obedience to the "laws of nature." Deprived of any autonomous agency, after having served as a playground for human ingenuity, materiality is ultimately accused of being unfit to accommodate the ideal. The Moderns are irreligious only in this: they neglect materiality.

And this has been going on during the three or four centuries when the sciences, the real ones, have done nothing but multiply forms of agency! For forty years now I have been measuring the gap that separates Science from the sciences, matter from materiality, and I keep on being stunned by it. Only religious passions are powerful enough to make those who are in the process of discovering the world lose it. Is there any chance at all of restoring it to those by whom and for whom it has been discovered? We would have to go back to 1610, seeking a way to stop confusing the contrasting virtues of science, religion, and politics. Which would mean, if we follow Toulmin, that we have to agree to plunge back into the maelstrom of the Renaissance – "great discoveries" and wars of religion included. This isn't very appealing? No, of course not, but it is the only hope we have of recapturing what was lost at such a time by the demand for undifferentiated certainty; the only way, after 1610, to prevent wars of religion.

To move forward, we would have to be able to establish a new contrast between, on the one hand, the terms *religious* and *secular* and,

---

[72]Cf. Matt. 16: 26. See Bruno Latour, "Si tu viens à perdre la Terre, à quoi te sert d'avoir sauvé ton âme?" (2010b), and the collective volume edited by Pasquale Gagliardi, Anne-Marie Reijnen, and Philipp Valentini, *Protecting Nature, Saving Creation: Ecological Conflicts, Religious Passions, and Political Quandaries* (2013).

on the other, the term *terrestrial*. The terrestrial is *immanence freed of immanentization*. If we could manage this, we could finally dispense with the religious, but not in the sense of secularizing existence. On the contrary, it would be a matter of reactivating the potentially active and fruitful aspects of the old theme of counter-religion: *uncertainty as to the ends*. The terrestrial is neither profane nor archaic nor pagan nor material nor secular; it is just what is still out ahead of us, like an Earth that is in effect new. But not in the sense that it would be a geographical space to discover and measure – in the sense, rather, of a renewal of the same old Earth, once again unknown, to be composed. This is indeed one of the possible injunctions of Gaia. It would be the only way to achieve what Voegelin called a "maximal differentiation" – in short, a civilization. It would amount to shedding intoxication with the notion of matter, rediscovering materiality and thus restoring autonomy, temporality, and history to all forms of agency and their distribution.

But, to rediscover history, we have to be able to break away from the strange theme according to which history has already ended, that there has been a total and radical rupture, as if we have definitively burned our bridges behind us. This is the well-known cliché of the irresistible "headlong flight forward."[73] What makes the ecological mutation incomprehensible to those who have been modernized is that there is no possible turning back, since the Moderns believe that they are in a post-apocalyptic epoch – it hardly matters whether it is the Enlightenment of Revelation, the Enlightenment of Science, or the glare of Revolution. In the most profound sense of the term, history for them is always over. Without any way to regain the present, there will be no exit, since they will hear every call to come back to Earth as a return to the archaic or the barbarous.[74]

It may seem paradoxical, but to shatter the Apocalypse – and thus to keep it from falling on us as we have fallen, we Westerners, as an apocalyptic rain on other cultures – we have to return to apocalyptic language, we have to become present again to the situation of terrestrial rootedness, and this no longer has anything to do, as you will have understood, with a return to (or respect for) "nature." To

---

[73] It is amazing that the exact same theme may be detected, barely modified, in a book with the perfectly gnostic title *Homo Deus*, in which Yuval Harari tries to restart the same irresistible "train": "Those who ride the train of progress will acquire divine abilities of creation and destruction, while those left behind will face extinction" (*Homo Deus: A Brief History of Tomorrow*, 2016, p. 273).

[74] This is what makes the theme of "degrowth" inaudible (see Nicholas Georgescu-Roegen, *La décroissance: entropie, écologie, économie*, 2011).

become sensitive – that is, to feel our responsibility, and thus to turn back on our own action – we have to position ourselves, through a set of totally artificial steps, *as though we were at the End of Time*, and thus give Paul's warning its meaning: "Those who mourn [should live] as if they did not; those who are happy, as if they were not; those who buy something, as if it were not theirs to keep; those who use the things of the world, as if not engrossed in them. For this world in its present form is passing away" (1 Cor. 7: 30–31).[75]

*

To end this lecture, I should like to introduce one more people onto the map of the philosophical "Game of Thrones" that I started in the previous lecture, a people that would describe itself not as "of Nature" or "of Creation" but "of Gaia." Others may be shocked by the introduction of a "goddess" into what ought to be a "strictly naturalistic description," but this can no longer make us uncomfortable. It is not hard to attribute a proper name to the entity by which this people is delighted to be convoked. Gaia, as we now understand, is *much less* a religious figure than Nature is. Consequently, there is no need to hide the personification: let us give It the capital letter and the gender It deserves while reserving for "Nature" the personal pronoun "She." For Gaia *puts an end to the hypocrisy* of invoking a Nature about which the fact that She was the name of a divinity was being hidden; that She failed to mention by what right She convoked peoples; and especially the particularly deanimated way She had of distributing her series of causes and consequences.

"Nature" held the strange ability to be at once "internal" and "external." She had the fascinating capacity to be mute and at the same time to express Herself through facts – with the advantage that, when the naturalists spoke, one never knew who was speaking. More surprisingly, She was organized in successive levels, starting with atoms, molecules, and living organisms, all the way to the ecosystems and social systems, in a well-ordered progression that allowed those who evoked Her always to know where they were and who guaranteed the best foundation for what was to follow. This architectonic quality allowed Her (or them) to exclude (or to "explain," as they say) any particular level in the name of the level immediately below,

---

[75] The drama of the Gnostics is that, by forgetting all the links with the (counter-)religious tradition against which they are fighting, they also lose all the benefits that could be drawn from that tradition. They have absorbed the poison, but have given up all hope of taking the antidote.

according to a "reductionism" that seems highly implausible today. More surprisingly still, She allowed them to decree that the things in the world *must be*, even as they claimed never to confuse what *must be* with what is. A touching but quite hypocritical modesty, as if it were riskier to say that something *must be* than to define its "*essence*."

In the great repertory of the history of religions, it is hard to find a divinity whose authority has been less contested than Nature and the laws through which She could compel all things to *obey* Her. It is not surprising that politicians, moralists, preachers, legal experts, and economists still aspire to such an indisputable source of authority. Ah! If only we could profit from the models offered by natural laws! One more source of authority that the drought attributable to the warming of the climate seems to have dried up.

Thus, if we now faithfully compare the attributes with which Nature and Gaia are endowed, I find it much more secular, more terrestrial (I was about to say "more natural"!) to assert that "I belong to Gaia" than "I belong to Nature." At least you know that the individuals who greet you with such an invocation stem from a specific people visibly assembled under the auspices of a personified entity whose properties they can list, as with the old names of Zeus or Isis. If you meet someone who comes from Gaia, you can be sure that he or she is not going to sell you either a totally implausible mechanism for speaking or a pre-constructed architecture so well ordered that it is going to tell you what you *must* do under the veil of what is. Freed from the fact/value divide and from the brutalizing architecture that goes from A, as in atom, to Z, as in Zeitgeist, you can clearly state your goals, describe your cosmos, and finally distinguish between your friends and your enemies.

What other virtues can we attribute to the people of Gaia? This people might escape from the *bifocal vision* from which the people of Nature suffered so badly.[76] What made the situation of Nature's people so implausible was that it seemed to glide in space without having a body, or even a mouth – sometimes completely conflated with objectively known things, sometimes a detached spectator contemplating Nature from nowhere. But scientists cannot survive in such a void, no more than astronauts can survive in the interstellar vacuum without space suits. The two conceptions are about as irreconcilable as the claim of access providers to store our data in the cold, ethereal "Cloud" even as they carefully conceal the numerous power plants that have to be built on Earth to cool the countless

---

[76] See table 5.4, in the fifth lecture; on bifocal vision, see the fourth lecture.

banks of servers that are always threatening to overheat. Clearly, it is this divergence that has made Science, since the seventeenth century at least, so difficult to assimilate into the general culture, and that has made so many scientists as morally naïve as they are politically impotent. If, for the people of "Nature," the two conceptions seem irreconcilable, for the "people of Gaia" *this is not the case*.

Here, too, the sciences of the Earth System could introduce a decisive change, by offering us a particularly sharp and specific reference point. When, for example, the same Charles D. Keeling we met earlier has to defend his long-term series of data about the daily, monthly, and annual rhythm of carbon dioxide in the atmosphere, it would be nonsensical for him not to foreground the instrumentation with which he spent forty years working on the Mauna Loa volcano in Hawaii.[77] If he had to fight for so long against the governmental agencies, against the National Science Foundation itself, against the oil lobbies, it was to save his instruments and the data they supplied. Without them, it would have been impossible for the rest of his community to detect the rapid rhythm with which carbon dioxide was accumulating.

Speaking about the climate objectively and deploying the "vast machine" of the climatologists are one and the same thing, or, to use Paul Edwards's terms, the same movement creates an "epistemic culture" and the "knowledge structure" that accompanies it.[78] The more climate skeptics maintain the old idea of a Science spread more or less everywhere without costs, the more climatologists are compelled in turn to keep foregrounding the scientific institutions on which they depend, and the more they consider themselves as a people endowed with specific interests trapped in a conflict with another people over the production of a series of pertinent data.

Am I mistaken in thinking that, for the first time in the history of science, it is the very visibility of their network that could make scientists more credible? Precisely because they are being more violently attacked by the climate skeptics in the name of epistemology, for the first time they have to count on the institutions of science as their own way of attaining objective truth. Perhaps they will finally agree to acknowledge that, the more precisely their knowledge is situated, the more solid it is? Instead of alternating abruptly between an impossible universality and the narrow limits of their own "point of view," it is because they extend their set of data from instrument

[77] Charles David Keeling, "Rewards and Penalties of Recording the Earth" (1998).
[78] Paul N. Edwards, *A Vast Machine: Computer Models, Climate Data, and the Politics of Global Warming* (2010).

to instrument, from pixel to pixel, from reference point to reference point, that they may have a chance to *compose* universality – and to pay the full price for this extension. The geologists, geochemists, and other geographers would be less schizophrenic if they agreed to call themselves Gaia-ologists, Gaia-chemists and Gaia-graphers! If the problem of composition is so crucial, it is because we can find in climate science not the "*gaya scienza*" evoked by Nietzsche but a *science of Gaia* that would finally be compatible with anthropology and with the politics for which we have to struggle.

Why is it so important to define peoples, when we were talking about a Nature known by science or a Creation preached by religions? To be able to make room for other peoples, other occupations of the ground, other ways of being in the world. We don't emphasize enough that the New Climate Regime is astonishing in that it imposes a terrible and totally unforeseen solidarity between the victims and the responsible parties. Henceforth it is at the heart of the Beauce, in France, as well as in New Guinea, in California as well as in Bangladesh, in the middle of Beijing as well as in the vast territories of the Inuits that land grabs are happening most violently and that the retroactions of the aforementioned Earth are the most dizzying.[79] The New Climate Regime is refreshing, if I dare say so, in that it is beginning to bring together peoples that are similarly impacted. As Davi Kopenawa declared: "Unlike us, white people are not afraid of being crushed by the falling sky. But one day they may fear that as much as we do!"[80] The fact that all the collectives from now on, like "our ancestors the Gauls," share the certainty that they fear nothing but "being crushed by the falling sky" gives a totally different idea of universal solidarity than that of the erstwhile humans occupying the erstwhile "nature."

It has taken anthropologists a long time to realize that "nature" was not a universal category; that most people have never lived "in harmony with nature";[81] and, something still more puzzling, that even the so-called naturalists had never lived in nature, since they have not succeeded in reconciling the epistemological version of their sciences with its practice. In other words, the "naturalists" have never succeeded in living in the idealized materiality that justifies, for some

---

[79] See Nastassja Martin's stunning book *Les âmes sauvages: face à l'Occident, la résistance d'un peuple d'Alaska* (2016).
[80] Davi Kopenawa and Bruce Albert, *The Falling Sky: Words of a Yanomami Shaman* (2013), cited in Déborah Danowski and Eduardo Viveiros de Castro, *The Ends of the World* (2016), p. 74.
[81] Philippe Descola, *The Spears of Twilight: Life and Death in the Amazon Jungle* ([1994] 1996).

of them, their "materialism" and their "reductionism." As for the religious folks, they have not yet noticed how vain their battle was against the so-called pagans, who had long since preceded them into the terrestrial world in which it was going to be necessary, in any case, to continue to live.

Don't flatter yourselves in the hope that you'll be able to drag the Moderns away from the effects of the counter-religion. It has been agitating them for too long, and, like Ptolemy's eagle on his burning rock, it keeps on gnawing away at their stomachs! You think perhaps it would be better to do without tackling religious questions altogether? But that would amount to continuing the very movement of the counter-religion, adding yet another iconoclastic gesture to those that have come before. The best we can do is to remain acutely aware of the link between theology, science, and politics – what I have called the distribution of agency – and to try to figure out how to rediscover the thread of history, that of things as well as of people.

*

If you have followed me, you will understand that the response that has to be prepared to counter those who accuse the ecologists of "engaging in apocalyptic discourse" must take the form of a question: "And you, do you situate yourselves before, during, or after the Apocalypse?" This is the shibboleth that might allow you to sort out the forms of attention to the world. If you situate yourselves *before*, you are living in sweet innocence or in crass ignorance – unless, by an incredible stroke of luck, you have still avoided all forms of modernization and are thus ignorant of the bite of the counter-religion. If you situate yourselves *after*, no trumpet of the Apocalypse will ever be able to arouse you from your slumber, and you will head down like sleepwalkers toward more or less comfortable forms of annihilation. You are interesting to me only if you situate yourselves *during* the end time, for then you know that you will not escape from the time that is passing. Remaining in the end time: this is all that matters.

> We have the opportunity to play the role of apocalypticians of a new type, namely, "*prophylactic* apocalypticians." If we distinguish ourselves from the classic Judeo-Christian apocalypticians, it is not only because we *fear* the end (which they *hoped for*) but especially because our apocalyptic passion has no goal but to *prevent* the apocalypse. We are apocalypticians only in order to *be wrong*. Only to enjoy every day anew the opportunity to be here, *ridiculous* but still standing.[82]

---

[82] Günther Anders, *Le temps de la fin* (2007), pp. 29–30, emphasis added.

This passage is by Günther Anders, an unjustly neglected writer who is defined most often only as the first husband of his celebrated wife Hannah Arendt. In a book published in 1960, aptly titled, in the French translation, *Le temps de la fin*, Anders proposed a deeply arresting analysis of the future of political theology in the epoch of the mushroom cloud.[83] People of my generation in fact passed – this is too often forgotten – from what was called the threat of the "nuclear holocaust" (the well-named MAD, for mutually assured destruction) to the ecological mutation,[84] just as climatologists passed, for the same reasons, from the earliest models for exploring the (fortunately virtual) planetary effect of nuclear winter to the (quite real) effects of global warming.[85]

Without making the threat *artificially* visible, there is no way to get us to move into action. This is what Günther Anders calls a "prophylactic" use of the Apocalypse; it has the same content as Clive Hamilton's argument that we must first of all give up hope – which projects us *from* the present *toward* the future – in order to be able to turn ourselves around – being reoriented by some powerful representation *of* the future *in order to* transform the present.[86] All are trying to shift the eschatology of a too-remote future toward the present, without always being aware that those they are addressing believe themselves to be immune to any eschatology since they have moved to the other side. The last ends? Not really; they don't see what that means.

Be that as it may, the fusion of eschatology and ecology is not a plunge into irrationality, a loss of composure, or some sort of mystical adherence to an outdated religious myth; it is a necessity if we want to face up to the threat and stop playing at conciliating the adherents to pacification who keep on deferring, yet again, the imperative to prepare for war on time. *The apocalypse is a call to be rational at last, to have one's feet on the ground.* Cassandra's warnings will be heard only if she addresses people who have their ears attuned to the racket made by the eschatological trumpets.

---

[83] Developed more fully in Gunther Anders, *La menace nucléaire: considérations radicales sur l'âge atomique* ([1981] 2006).

[84] This is why, no matter how apt the date of 1610 could be for the beginning of the Anthropocene in addition to all the scientific arguments in favor of the 1945 date (Colin N. Waters, Jan Zalaciewicz, et al., "The Anthropocene Is Functionally and Stratigraphically Distinct from the Holocene," 2016, pp. 137–48), there is a generational reason: it frames exactly the existence of the baby boomers.

[85] Spencer Weart, *The Discovery of Global Warming* (2003).

[86] Clive Hamilton, *Requiem for a Species: Why We Resist the Truth about Climate Change* (2010).

How (not) to put an end to the end of times? 219

This is why it is so important, in my view, to try to face up to Gaia, which is no more a religious figure than a secular one. Gaia is *an injunction to rematerialize our belonging to the world*, by obliging us to re-examine the parasitic relations of Gnosticism to the counter-religions. Or, to put it still another way, Gaia is a *power of historicization*. Still more simply, as its name indicates, Gaia is the signal telling us to come back to Earth. If one wanted to sum up its effect, one could say that, by requiring the Moderns to start taking the *present* seriously at last, Gaia offers the only way to make them tremble once again with uncertainty about what they are, as well as about the epoch in which they live and the ground on which they stand.

SEVENTH LECTURE

The States (of Nature) between war and peace

The "Great Enclosure" of Caspar David Friedrich • The end of the State of Nature • On the proper dosage of Carl Schmitt • "We seek to understand the normative order of the Earth" • On the difference between war and police work • How to turn around and face Gaia? • Human versus Earthbound • Learning to identify the struggling territories

Even though I had a reproduction right under my nose, it took my friend the art historian Joseph Koerner to point out again, on the painting by Caspar David Friedrich, the shape of a bend in the Elba before I finally realized all of a sudden, as in a Rorschach test, that what I had first taken to be a swampy area in the foreground, with puddles and mud reflecting the sun's rays, was actually the globe itself, as if it were buried in the Earth. Not the mapmakers' globe, the only one that Friedrich could have spun around with his fingertips, at the start of the nineteenth century, but the meteorological globe, the one that astonished the first astronauts, so different from maps, with its raking lights, its mountains in relief, its iridescent oceans, and the enigmatic presence of unrecognizable continents, as if they belonged to another planet. And of course one has to live on another planet in order to occupy the vantage point of someone seeing the unrecognizable globe gradually sinking down – unless it is rising

The States (of Nature) between war and peace 221

Figure 7.1 Painting by Caspar David Friedrich, *Das grosse Gehege bei Dresden*, Galerie Neue Meister, Staatliche Kunstsammlungen Dresden. © Photo Jürgen Karpinski.

up – in the guise of an Earth inserted into the confines of an ordinary landscape in the vicinity of Dresden. A landscape that the same spectator is supposed to be contemplating head on, but in which he can no more reside than he can penetrate the gold-tinged sky. The symmetrical curve in the clouds creates the impression that the sky is a great orb, but one whose immensity is amplified on one side while it is tugged back, reduced, on the other, inverted by the hodgepodge of pools and puddles in the foreground.

A barge with wind in its sails goes slowly down – or perhaps up – the river, retracing in reverse the bend drawn by Joseph's finger, the limit of *Das grosse Gehege*, the *Great Enclosure* – this is the name of the painting – without giving the viewer any clear sense of what may be delineated here. Is it the terrestrial globe with its edge plunging into the river? The Elba that demarcates pastures, fields, and forest – but where we see no people or animals? Or is the border rather the tiny, paler line above the trees, on the horizon, signaling a second time, at the point where the entire landscape is fleeing the sun, the tipping of the whole painting into the definitive enclosure of night?

But what is most extraordinary is that it seems impossible to fix one's gaze calmly, peacefully, on the bank, under the groves of trees, since this idyllic spot, this Arcadia, is as visually inaccessible as the view of the foreground; it corresponds, as Koerner points out, to the vanishing point – the infinity – of the visual rays.[1] Moreover, it is useless to hope for some bucolic return to a local habitat, since the bank of the bend in the river appears compressed, as if laminated by two immense rollers: the globe in the foreground, which seems to be going under, and the other in the background, the sky where the sun has set – or is perhaps rising – and which seems to revolve around the first like the two screws of a press. No, this is not a landscape that someone might contemplate. It offers no possible stability, except perhaps on the barge, but then one would be in motion.

If I am so fascinated by this painting, it is because a brief moment of inattention is all it takes to miss what Joseph Koerner is convinced he has spotted in it. As proof, he notes that an engraver, Johann Philipp Veith, thought he was doing a good thing by rectifying the impossible vantage point of the virtual spectator of this painting to make it more reasonable and more coherent: by slightly diminishing the curve of the foreground in his copy, making the terrestrial globe a simple bank of the Elba, with mud, puddles, and streams, he succeeded only in spoiling the whole effect.[2] Let's not imitate the engraver: as you look at this painting, you should not try to simplify the place where you should position yourself to contemplate it. Instead, you need to plunge into yourself and ultimately call yourself into question. *In* "nature," no one has a place. Two centuries later, but for reasons entirely different from those of the Romantic era, we too have come to understand this.

Of course I have no idea what Caspar David Friedrich meant to enclose by this painting and by its title, *Das große Gehege*. I chose it as my starting point because it seems to me to sum up one of the arguments of the previous lectures better than any other work of art: one can grasp nothing about the intrusion of Gaia – or perhaps, here, its extrusion – if one confuses it with the contemplation of a globe. The person who believes he sees the terrestrial globe from on high takes himself for God – and since God himself, of course, does not see the world this way, the global vision is at once deceitful and impious. Woe as well to those who think they can escape the vast expanses of heaven and Earth by taking refuge in a grove of trees, their feet

---

[1] Joseph Leo Koerner, *Caspar David Friedrich and the Subject of Landscape* (2009).
[2] Philipp Veith (1768–1837), copy of *Das große Ostra-Gehege an der Elbe* (1832), Dresden Museum.

in the water, on the bank of a river, to contemplate the world as a spectator: they will be crushed!

What is brilliant about this painting is the way it marks the instability of every point of view, whether it's a matter of seeing the world from above, from below, or from the middle. With the Great Enclosure, the great impossibility is not being imprisoned on Earth, it is believing that Earth can be grasped as a reasonable and coherent Whole, by piling up scales one on top of another, from the most local to the most global – and vice versa – or thinking that one could be content with one's own little plot in which to cultivate a garden. In other words, those who claim to be ordering neatly the various dimensions of the Earth don't deserve to be called Earthbound.

*

In these lectures, we are trying to respond to the intrusion of Gaia by learning to shed, one after another, the habits of thought associated with what I have been calling the *Old* Climate Regime. We shall try to rematerialize our existence, which means first of all reterritorializing it or, better, though the word does not exist, *reterrestrializing* it. Obviously a surprising thing for people who complain about being too "down to earth" but who, in the final analysis, have hardly been "down to earth" at all! What we are doing amounts to *repoliticizing* our concept of ecology. This is the task we have to tackle now.

I have prepared this return to politics, in the two previous lectures, by insisting on our diplomatic obligation to introduce ourselves to one another in the form of newly defined peoples. These peoples should make explicit, as clearly as possible, what supreme authorities convoke them, on what lands they believe they are localized, in what time periods they situate themselves, and according to what cosmograms – or cosmologies – they have distributed their agencies.[3] This is the importance of the metamorphic zone for which I tried to have you develop a feel in the first two lectures, by exploring more deeply the notion of *agency*.[4]

As we shall now see, the Old Regime did not really make it possible to do politics, since it never encountered real opponents; it was enough to struggle against *irrational* people or *infidels*, who needed to be educated or *converted*, but never fought. In any case, they didn't need to be combated in the radical sense that they might, in turn, put

---

[3] In the fifth lecture, p. 151, see the list of features that I used to imagine such a convocation of peoples.
[4] See the second lecture.

us in danger of losing our own values. These values remained protected, in Nature, in inevitable Progress, in the Meaning of History, in indisputable Science. To us, really, no, nothing could happen. We could undergo reversals but no real crises. No second thoughts. The last judgment had already taken place. Fundamentally, we were as lacking in history as in politics. Hence our stupefaction, our unpreparedness, our skepticism before the emergence of the strange couple introduced in the third and fourth lectures: Gaia first, then its most recent complication, the Anthropocene.

To understand the repoliticization of the ecology that follows, I am going to ask you to undertake a little soul-searching by asking you the following question: "Have you ever had enemies?" If you agree to reach deep down into yourselves and reflect on the meaning of the battles you fight, I'm almost sure you'll notice that you have never had any. *Adversaries*, yes, of course, but *enemies*, no. You're doubtless fighting the climate skeptics or the capitalists whose ascendancy is destroying the planet, perhaps the banks, or else the politicians incapable of looking beyond their next electoral campaign; or perhaps you're struggling rather against the ecologists, those kill-joys "who want to forbid all innovation," the advocates of de-growth, or even the scientists who have become "a lobby of model-makers without a grip on reality." Yes, adversaries – we all have plenty of these.

And yet, whatever camp you have joined, you are obliged to acknowledge that you have no enemy, if the supreme authority in the name of which you are fighting – the one that has sent you on a mission and whose ministers, militants, and armed forces you have become – already knows with certain knowledge what the story is with history and its assured judgment. You're only proceeding with a clean-up. You are only the avant-garde of an inevitable movement. Time has no hold on the cause we serve since it cannot modify the content. History may advance more slowly than you expected; it cannot radically change its direction. In the literal sense, the cause you serve transcends history.[5]

Have you had enough time to take this little test and to verify which of your adversaries have the capacity to make you tremble with uncertainty about the solidity of your own values? Don't worry, I'm not asking you to make the results public! I'm only asking us all to make ourselves sensitive to the lowering of political intensity that we hope for every time "nature" comes on stage, as if we thought we

---

[5] This unsatisfactory transcendence was at issue in the previous lecture in relation to Voegelin's proposition.

were throwing water on a fire to put it out – when in fact we were pouring on oil.

If the appeal to "nature" has such a power of *depoliticization*, it is precisely because those who fight for it – it hardly matters in what camp – can only carry out, over time, a plan that does not depend on the vagaries of the time that is passing. "Nature" immunizes against the risks of politics. It was conceived for the purpose. This is why, in the proper sense, there has never really been a political ecology.[6] What we call most often by that name is the *application* to reality of principles whose self-evidence comes from a different source, most often from Science, against the obstinate resistance of those who do not obey these principles because they don't really understand them. Nothing in their resistance obliges you to pull up stakes and start over: these people are simply archaic, backward, uncultivated, perhaps corrupt, surely duplicitous. None of them is going to force you to redesign what you call your ecology from top to bottom, nor to decide in what it ultimately consists. Even if you claim to be "at war" against such adversaries, this war won't be a real one, since it will remain *pedagogical*. You will remain basically convinced that, if only you could have explained things clearly to them, they would have been convinced of the rightness of your struggle. When one calls upon "nature" this way, it is almost always because one wants to explain yet again to dunces, within the virtual walls of a classroom, what they are going to end being forced to understand.

If there are no politics, in the sense that we never encounter enemies but only people who are wrong and whom we are going to have to punish or rehabilitate, this signifies that we are not only within the confines of a school but also *within the boundaries of a quasi-State*. The citizens of such a State squabble over the details, to be sure, but they are in agreement about the essentials. Nation-states may well be in conflict with one another – there is no shortage of examples! – but this does not keep them from finding themselves all under the aegis of an authority that has the power to bring them back to reason and that has to be called *sovereign*. Everything happens as if rational people had agreed to live under the aegis of a State whose precise form is never spelled out but which nevertheless fulfills an essential function by serving as arbiter of last resort in all disputes. It was under this strange regime, as we saw in the first lecture, that "nature" had ended up playing the role of the Supreme Court for every moral

---

[6]Bruno Latour, *Politics of Nature: How to Bring the Sciences into Democracy* ([1999] 2004b).

decision.[7] This is what explains the sluggishness of every discussion about ecology: the stupefying idea that, if we turn toward "nature" and its laws, we are necessarily going to reach agreement, *as if* we were citizens of a *single body politic*. Every rationalist, in this sense, *is a citizen of the State of Nature*. Who would dare challenge the Spirit of its Laws? To live under the Old Regime is to pretend that, if Science had actually demonstrated something about Nature, then, obviously, the nation-states, all together in unison, could only comply with its laws! (If you doubt that this is the case for physics, medicine, or biochemistry, think about the sovereign power of the Economy; what empire has ever enjoyed such absolute authority?)

What has put an end to this Old Regime are the fierce disputes about the climate that made all of us realize retrospectively that, no matter how strong the science, it was not enough for nation-states to change their ways. This is when we realized that we had left the State of Nature, a State whose universal laws could be invoked by any rational individual to bring disputes to an end and bring their adversaries around. Before the Anthropocene, we were not as clearly aware of the existence of this virtual Dome, for we limited the existence of states to human assemblages alone. If these assemblages had an ecology, it was on the outside, in the environment, and it served only to situate them somewhere on a map. This fiction has vanished with the plunge into geohistory, with the proliferation of controversies (of which the generalization of climate skepticism is only a symptom) – in short, with the intrusion of Gaia. For the first time, it has become clear that the universality of laws, the robustness of facts, the solidity of results, the quality of models, could no longer be used, even in dreams, to ensure the agreement of minds and bind nation-states under a single yoke. It is because Gaia is not "nature," or any of nature's substitutes, that it obliges us to go back to the question of politics and look for a different principle of sovereignty. If Gaia has such a powerful effect as a political lever, it is because it raises anew the now familiar question: in the name of what supreme authority have we agreed to give our lives – or, more often, to take those of others?

This is why I allowed myself, in the two previous lectures, to engage in the odd exercise of replacing the false universality of the State of Nature – inoperative in any event – with the convocation of distinct peoples, collectives capable of entering into diplomatic relations. What we lose on one side – the indisputable appeal to the

---

[7]See especially the passage on the impossibility of distinguishing between description and prescription.

The States (of Nature) between war and peace

Science of Nature – we shall perhaps regain on the other, provided that we *agree to pass from a regime of apparent peace to a regime of possible peace*. Between the two, it's true, there's no point pretending otherwise, we have to agree to talk about war. We shall never be able to repoliticize ecology without first agreeing to recognize that there is indeed a state of war – a war between worlds – and that the Old Climate Regime was nothing but an armistice in the expectation of a peace treaty that has never been concluded, for it would have obliged us to distinguish precisely between the contrasting truths of religion, politics, and science. I hesitate to insist on the point, but it is in this sense that the "resumption of hostilities" might strike us as a good sign. Finally, thanks to the disputes over the climate and how to govern it, we are asking the political question again in terms of life and death. What am I ready to defend? Whom am I ready to sacrifice?

By an unexpected twist on Hobbes's famous concept, we have entered into the *state of nature* that he located in a mythical past that *preceded* the social contract; he found the model for this past in the (poorly understood) mores of Native Americans: "Hereby it is manifest, that during the time men live without a common power to keep them all in awe, they are in that condition which is called war; and such a war, as is of every man, against every man."[8]

Today, what is strange is that this state of nature is not situated, as it was for Hobbes, in the past; *it is coming toward us*, it is *our present*. Worse still: if we are not inventive enough, it could well become our *future*, too. Now that there is no longer the "common power" of the State of Nature and its laws to keep all the nations "in awe," we have a war of all against each, in which the protagonists henceforth may not be just the wolf and the lamb, but also tuna and CO_2, plant nodules or algae, in addition to the numerous human factions that disagree about almost everything: "It may seem strange to some man, that has not well weighed these things; that Nature should thus dissociate, and render men apt to invade, and destroy one another."[9]

Contrary to what Hobbes said, it no longer surprises us at all today that "nature" can in no case pacify the "political animal"! "Nature," as we now know, divides – and divides radically. It is thus not at all surprising that we are terrified at the idea of having lost the security of the great Leviathan and find ourselves facing this other Cosmocolossus whose adventures we have been following from the beginning of these lectures: the Anthropocene.[10]

[8] Thomas Hobbes, *Leviathan* ([1651] 1998), p. 84.
[9] Ibid.
[10] See the fifth lecture.

If we must not abandon the project of seeking security and protection, peace and certainty, under another Leviathan yet to be invented, it is because the security brought by the State of Nature has never been achieved in reality. The desire to build the Republic, the veritable *res publica*, is always *before* us. Thanks to the emergence of Gaia, we are becoming aware that we had not even *begun* to outline a realistic contract, at least a contract that might hold up on this sublunary Earth of ours. This is why we feel so strongly that we are Hobbes's contemporaries, confronted by the same old question of how to bring an end to civil and religious wars – with the difference that he sought to reconstruct civil society after the guarantee of a truly *catholic* Religion (catholic in the etymological sense of universal) had disappeared, while we have to do the same thing now that the authority of a truly catholic Nature has also collapsed. In the new Leviathan, violent disputes over the exegesis of scientific literature are replacing the disputes at knife-point over the exegesis of biblical literature. Let us recall, in the play called *Gaia Global Circus*, the way Virginia, the climatologist, responded to Ted, the mouthpiece of the climate skeptics: "Go tell your masters that the scientists are on the warpath."[11]

*

To move ahead with these delicate and risky questions, I am going to turn to the author the least apt to reassure you, the toxic and nevertheless indispensable Carl Schmitt (1888–1985). The Nazi legal scholar can be likened to a poison kept in a laboratory for the moment when one needs an active principle powerful enough to counterbalance other even more dangerous poisons: it is all a matter of dosage! In the case in point, the drugs we have to counter are so strong that I invite you to desensitize yourselves with small doses of Schmitt, taken advisedly. In any case, how can we get along without someone who wrote, in the middle of the twentieth century, a sentence so perfectly adjusted to the crisis we are experiencing now? "In mythical language, the *earth* became known as the mother of law... This is what the poet means when he speaks of the infinitely just earth: *justissima tellus*."[12] "The Very Just Earth!" For those of us who are trying to face up to Gaia, and trying to understand what law it might generate, let's admit that we have to look more closely at this text.

---

[11] Pierre Daubigny, "Gaia Global Circus" (2013), p. 30.
[12] Carl Schmitt, *The Nomos of the Earth in the International Law of the Jus Publicum Europaeum* ([1950] 2003), p. 42.

The States (of Nature) between war and peace 229

I am not interested in Schmitt so much as the inventor of the overly celebrated *principle of exception*, however.[13] In seeking to react to the gradual disappearance of politics, squeezed out by management, organization, and the economy (what would today be called "governance"), Schmitt proceeded as though the political exception were a rare moment, reserved for a Leader, who would be above the law. The idea was obviously correct; politics has nothing to do with the simple application of a pre-established rule. But Schmitt truncated this idea by emphasizing just one segment of the quite particular trajectory of political speech – the moment when the Leader decides and cuts the Gordian knot. Now, the political mode of existence is exceptional in *all its segments*, since it traces a curve that never, of course, goes in a straight line.[14] So there is no longer anything exceptional about the principle of exception as soon as we agree to follow the very specific way in which the political distinguishes what is true from what is false, at every moment.

Unfortunately, instead of accepting the originality of this mode by bringing out the way it contrasts with the modes of scientific information or organization, Schmitt singled out just one of its moments – associating it, moreover, with the role of a Führer; he thus dissimulated its paradoxical banality. In other words, Schmitt confused the state of exception with the specificity of this mode. To avoid being contaminated by this limited version of the principle of exception, his readers, purporting to be horrified, began to replace the sinuous discourse proper to the political with the application of rules of good governance.[15] By trying to save the strangeness of the political that was being squeezed out, Schmitt offered such an exotic, Teutonic version of it that he succeeded, in the end, only in hastening its disappearance!

What ought to interest us, rather, is the oddly titled book *The Nomos of the Earth in the International Law of the Jus Publicum Europaeum*, a book written during the war and published shortly afterward.[16] What connection can there be between political ecology

---

[13] Carl Schmitt, *The Concept of the Political* ([1932] 1976).
[14] See Bruno Latour, "What If We Talked Politics a Little?" (2003), and a large part of *An Inquiry into Modes of Existence: An Anthropology of the Moderns* ([2012] 2013b).
[15] On this category error between organizations and politics, see Latour, 2013b, and the corresponding entries on the AIME website, http://modesofexistence.org.
[16] On the way the book was written, see the introduction by Peter Haggenmacher in *Le nomos de la terre dans le droit des gens du Jus Publicum Europaeum* (2001).

and that old reactionary thinker, you ask? None at all![17] It is precisely because Schmitt does not think for even a second about what will become of the ecological question that his way of talking about the Earth and its law, its *nomos*, as he says, can appear so useful to those who are trying to shed *the weight that the concept of* "nature" has imposed on the issues of the Earth, law, sovereignty, war, and peace that have become our questions with the intrusion of Gaia. It is because Schmitt doesn't give a single thought to the Globe that *The Nomos of the Earth* can be used to conceptualize the *successor* to the political, scientific, and theological notion of "nature."[18] When Schmitt looks at the Earth, he sees the matrix of a possible regime of law. Someone who is ignorant of nature to this extent is exactly what we need!

So if Schmitt can help us, even as we limit what we draw from him, it is because as a good legal scholar he understood that one cannot make any distinction between facts and values if one does not go back to a stage prior to the modern form that produced the bifurcation between natural law and positive law, between *phusis* and *nomos*.[19] But it is also because Schmitt too understood – although without Voegelin's generous luminosity – the importance of the Apocalypse in any philosophy of history, and because, unlike the Moderns, he did not believe we had definitively gotten rid of religion. Behind the clutter of his mythology, he understood perfectly that one cannot conceptualize politics if one is trying to avoid the *end time*.[20]

What is more astonishing, for someone of his day, is that Schmitt took the sciences, particularly cartography, not as what describes the world objectively from the outside but as what, *from within the world*, formats – surveys, calculates, draws – the world, represents it in a particular way. In other words, Schmitt has not let himself be taken in by the compelling figure of the Globe: when he talked about

---

[17] I benefited from a seminar organized at Sciences Po in May 2015 on the use of *Nomos of the Earth* in political ecology with Pierre-Yves Condé, Dorothea Heinz, Bruno Karsenti, Michael Northcott, Claudio Minca, Kenneth Olwig, and Rory Rowan, whom I thank for their stimulating remarks.
[18] See Dorothea Heinz, "La terre comme l'impensé du Léviathan: une lecture de Carl Schmitt en juriste de l'écologie politique" (2015).
[19] "Despite the change in the way *nomos* was conceived and expressed, which was already evident in the classical age, originally the word did not signify a mere act whereby *is* and *ought* could be separated, and the spatial structure of a concrete order could be disregarded" (Schmitt, 2003, p. 69).
[20] See Heinrich Meier, *The Lesson of Carl Schmitt: Four Chapters on the Distinction between Political Theology and Political Philosophy* (1998). For a directly religious and even spiritual use of Schmitt in Christian ecology, see the audacious book by Michael S. Northcott, *A Political Theology of Climate Change* (2013).

The States (of Nature) between war and peace   231

the global, it was always because he saw in it the hand of a scientific, economic, or institutional hegemony in the process of expansion – or, as he put it, "land-appropriations."[21] As in Friedrich's painting, for Schmitt the globe was inserted into the world. Through all these features, Schmitt resisted the scientism of his time.

This would be enough, as we can see, to make him useful for our quest, but it is the consequence that he drew from it for the understanding of space that interests me most. Schmitt is probably the only political thinker who did not let himself be *taken in* by the spatial framework. Space, for him, was the provisional result of a phenomenon of expansion, of spacing, of gaining ground, which depends on other political and technical variables. For him, as for more recent historians of the sciences, the *res extensa* is not a space *in which* politics is situated – the background of the map of every geopolitics – but, rather, something that is generated by political action itself aided by its technological instrumentation. In other words, for him, too, space is the offspring of history. Schmitt thus resolutely ignores the canonical distinction between "physical" geography and "human" geography. Precisely because he was both a legal scholar and a political theologian, he tried to reach back to a stage *before* the invention of territory conceived as a transparent space that a sovereign would contemplate from the window of his palace.[22] Note that I am saying "before" and not "after." In fact, in contrast to so many critiques of space, Schmitt is trying not to add the sense of space "experienced" to "objective" space – which would amount to extending the bifurcation between human and physical geography[23] – but, rather, to generate as many other spaces, in the plural, as there are political situations and concrete technologies. To territory conceived as a *space*, an *undifferentiated container*, he contrasts the territories conceived as *places*, *differentiating contents*.

Consequently, when Schmitt speaks of the Earth, he is speaking not of the Globe on which one would then deposit the warring

---

[21] Schmitt, 2003, p. 87.
[22] Two books also published in 2015 make similar connections between the conception of space and political ecology on the basis of Schmitt's *nomos*: Claudio Minca and Rory Rowan, *On Schmitt and Space*, and Federico Luisetti and Wilson Kaiser, eds, *The Anomie of the Earth: Philosophy, Politics and Autonomy in Europe and the Americas*; see also the earlier book by Stephen Legg, ed., *Spatiality, Sovereignty and Carl Schmitt: Geographies of the Nomos* (2011). Unfortunately, the bifurcation between physical space and society remains taken for granted and unsurpassable.
[23] Kenneth Olwig, "Has 'Geography' Always Been Modern? Choros, (Non) Representation, Performance, and the Landscape" (2008); Stuart Elden, *The Birth of Territory* (2014).

nation-states like pieces on a chessboard but of multiple instances of territorialization, some of which would provisionally entail particular relations of *spacing* – by distorting the chessboard. History, including the history of technologies, is thus for him at the origin of practices of spacing. Since this is also the essential point that we recognized in Lovelock,[24] with the same suspicion regarding the global, which has to be composed organism by organism, you will understand why I was struck when I read Schmitt's book. Moreover, what is so surprising about turning toward a recognized master of geopolitics and international law to reopen the questions raised by Gaia-politics and the New Climate Regime? Between the *nomos* of an Earth conceived as a Globe and the *nomos* of an Earth conceived as Gaia – that is, as the anti-Globe – Schmitt will allow us to choose.

*

Like many readers, I put off reading *Nomos* as long as possible. But one day I opened it and came across the following paragraph, the last one in the preface:

> The traditional Eurocentric order of international law is foundering today, as is the old *nomos* of the earth. This order arose from *a legendary and unforeseen discovery* of a new world, from an *unrepeatable* historical event. Only in fantastic parallels can one imagine a modern recurrence, such as men on their way to the moon discovering *a new and hitherto unknown plane*t that could be exploited freely and utilized effectively to *relieve their struggles* on earth. The question of a new *nomos* of the earth will not be answered with such fantasies, any more than it will be with further scientific discoveries.[25]

"Fantastic" is obviously not the term one would use today to speak of the carnage experienced by those who have been "discovered" this way! Let us recall the year 1610, used as a golden spike for the discovery of the Anthropocene because of the elimination of the Native Americans and the reforestation that followed.[26] What interests Schmitt is not the fate of the indigenous peoples but the connection between the rivalry of the European states and the claiming of empty land – that is, land emptied in advance of its empires and its nations. Now, this question, in a different form, has concerned us from the outset: can humans spread themselves still further, toward new lands? Schmitt's answer is negative. We will find no more "new

---

[24] See the end of the third lecture.
[25] Schmitt, 2003, p. 39, emphasis added.
[26] Simon L. Lewis and Mark A. Maslin, "Defining the Anthropocene" (2015), discussed at the beginning of the sixth lecture.

The States (of Nature) between war and peace 233

and hitherto unknown" celestial bodies except in science fiction. Here is the Great Enclosure! Neither the conquest of space nor "scientific inventions" will offer us, any longer, the opportunity to diminish the rivalries among nation-states. We are once again confined to sublunary space alone. Our dreams of conquest henceforth resemble the Concorde, the supersonic airplane now suspended at the end of a runway at Charles de Gaulle airport, a sort of involuntary memorial to past futurisms. The old *nomos* of the Earth – I'm restoring its capital letter – depended on discoveries of worlds *in extension*, whereas the future *nomos* depends on the discovery of a New Earth *in intensity*.

Schmitt was obviously mistaken in saying that humans have not found new earths. Those that they have exploited with the same frenzy, the same violence as the New World, were not found between the Earth and the Moon, and they were not approached by rockets: they were found *under the surface of the Earth*, and, if the States were able to reach down deep into them to attenuate their rivalries even while exacerbating them, it was through pit mining, exploration, foraging, extraction, and hydrofracking. We might even say that coal, petroleum, and gas indeed constitute new celestial bodies, if we remember that we are dealing with the sun captured by living entities whose remains were eventually sedimented in layers of rock.[27] Here is their *new* New World. And this new continent has really been appropriated as a *res nullius*, and without the smallest scruple: "Drill, baby, drill!"[28] Until we reach the current situation by crossing the CO_2 threshold of 400 ppm.

However, Schmitt was right on one point; this new land grab, as fantastic as it is unexpected, is also unrepeatable. Since the publication of his book, the enclosure has been locked up for good, imprisoning us in the unforeseen effects of such extractions. The powers that be have limited themselves by getting all tangled up in the consequences of their action of conquest. The conclusion is definitive: nothing more can come along to attenuate the rivalries among the nation-states imprisoned in this Great *Enclosure*.[29] We are thus

---

[27] The infinitization of the economy is linked by Timothy Mitchell, in *Carbon Democracy: Political Power in the Age of Oil* (2011), to the "new earth" of petroleum that seems accessible in unlimited quantities – and this corresponds moreover to the beginning of the "great acceleration."
[28] The rallying cry at Republican meetings in the United States, expressing virtually cosmic enthusiasm for indefinite access to petroleum and radical opposition to any restrictions.
[29] In the unforeseen form of the theme of "planetary limits" proposed by Will Steffen et al., "Planetary Boundaries: Guiding Human Development on a Changing Planet" (2015b).

Seventh Lecture

headed again toward war of all against all, with no way to delay the conflicts by diminishing rivalries among the powers through the occupation of new lands.[30]

But what astonished me most was the end of the paragraph. Schmitt concludes with an invocation that is totally different both in direction and in tone:

> Human thinking again must be directed to the elemental orders of its terrestrial being here and now. We seek to understand the normative order of the earth. That is the hazardous undertaking of this book and the fervent hope of our work. The earth has been promised to the peacemakers. The idea of a new *nomos* of the earth [*Sinnreich der Erde*] belongs only to them.[31]

Whereas Schmitt had been directing our attention toward a war without end, here he is speaking about the "peacemakers" seeking what might be better translated as "the reign of the order of the earth." And he does this by citing – astonishingly, for the legal expert of the Third Reich – the Sermon on the Mount! It is true that Schmitt manipulates it a bit.[32] But we understand that the bellicose Carl Schmitt cannot go so far as to entrust such a revelation to "the gentle"! It is thus "the peacemakers" whom he charges with discovering the "new *nomos* of the earth," a "hazardous undertaking" and the "fervent hope" of his work.

The unusual term *nomos* – the configuration in which "the orders and orientations of human social life become apparent"[33] – should not trouble us. Even if Schmitt calls on a treasure trove of erudition to produce its etymology,[34] he is attached to it, fundamentally, for other reasons. He is looking for a term that can adequately dignify a concept that would allow his readers to situate themselves at a

---

[30] There will never be a new way to "abolish the land constraints," as Kenneth Pomeranz showed in his classic book *The Great Divergence: China, Europe, and the Making of the Modern World Economy* (2000).

[31] Schmitt, 2003, p. 9; *Sinnreich der Erde* means the empire/kingdom/realm of the sense of the earth.

[32] Matthew's Gospel reads: "Blessed are the peacemakers [or the peaceful ones], for they shall be called children of God" (Matt. 5: 9), whereas those who will "inherit the earth" (who will have the earth as their share) are "the meek" or "the gentle" (Matt. 5: 5). The French ecumenical translation avoids the possessive, saying "*Heureux les doux car ils auront la terre en partage*" (Blessed are the gentle, for they will have the earth available to them).

[33] Schmitt, 2003, p. 42.

[34] See Emmanuel Laroche, *Histoire de la racine "nem" en grec ancien: nemo, nemesis, nomos, nomizo* (1949).

point prior to the invention of the nature/politics distinction.[35] And, as always, when one seeks to go backward in time, one must rely on mythology – Greek, if possible. In practice, the term *nomos* technically fulfills the same function as the much more austere term I have used in these lectures: *redistribution* of agency. Through this concept, I, too, have tried to situate myself in a position prior to the distinction between nature and culture, between primary and secondary qualities, between science and politics. If *nomos* comes across as an element of a mythical history of international law, its real conceptual role is to render the collectives comparable once again. In other words, *nomos* is a more juridical and more erudite version of the term *cosmogram*, which I have used to imagine the diplomatic assembly of the peoples struggling for the Earth.

Must we agree to take seriously the astonishing injunction to "reveal" the cosmogram (or the *nomos*) of the Earth to the "peacemakers," and to them alone? How can we believe that a thinker mixed up with so many horrors can speak this way of peace, revelation, and sharing the Earth? It is on this point that we ought to judge for ourselves. Schmitt saw that one could never speak of peace if one did not first decide to see in the present situation a state of war – and thus agree to have enemies. I maintain that, on this point at least, we have to decide in his favor. "*Hic Rhodus, hic saltus.*"[36]

*

Before taking an interest in what is going to allow the territories to make their front lines explicit, let us try to understand why access to peace negotiations requires prior recognition of a state of war. Everything depends on the distinction introduced by Schmitt in a much better known work, *The Concept of the Political*, between *police operations* and the state of war. This concept is based, as we know, on the friend/enemy distinction. A true enemy must not be confused with an adversary whom one detests for moral, religious,

---

[35] The Canadian legal scholar Richard Janda has clearly grasped this connection: "This is to say that Schmitt was hiding the fact that what he would ultimately call nomos, associated with the appropriation of land, was not so much an original root relationship to the earth but rather the earlier relationship to the earth that had, for him, the greatest energy and majesty to it" (personal communication, March 22, 2013).
[36] *Trans.*: In one of Aesop's fables, an athlete boasts about a spectacular long jump he made in a contest on the island of Rhodes. A listener challenges him to show his prowess on the spot, saying "Here is Rhodes; jump now!"

commercial, or aesthetic reasons. The legitimate opponent would become a mere scoundrel, or, to put it in Latin, *hostis* would be mistaken for *inimicus*.

> The political enemy need not be morally *evil* or aesthetically *ugly*; he need not appear as an economic *competitor*, and it may even be *advantageous* to engage with him in business *transactions*. But he is, nevertheless, the other, the *stranger*;[37] and it is sufficient for his nature that he is, *in a specially intense way, existentially something different and alien*, so that in the extreme case *conflicts* with him are *possible*. These can neither be decided by a previously determined general norm nor by the judgment of a disinterested and therefore neutral third party.[38]

As long as there exists a "third," "disinterested and therefore neutral" party capable of applying a "previously determined general norm" to judge who is wrong and who is right, there is no enemy, no state of war. And so, according to Schmitt, no politics, either. As long as there exists an arbiter recognized by all, a judge, a Providence, a supreme distributor – that is, a State – the thousands of inevitable battles between divided humans are nothing but internal struggles that can be resolved by the application of simple organizational rules. "If there's trouble, call the police!" But there is no war when conflicts can be resolved by calling the police, since even those involved in the dispute agree on the fact that the State has the right to define the situation this way. There is no war in situations where management, positive law, the police, and accounting suffice. All these operations are *deemed legitimate a priori* and can be *calculated* in advance; all the risks one runs in putting them to work have to do with execution, not principles.

War begins when there is no sovereign arbiter, when there are no "general norms" that can be applied in order to render a judgment. This is where we reach the "limit" and "conflicts with foreigners" become possible.

> The friend, enemy, and combat concepts receive their real meaning precisely because they refer to *the real possibility of physical killing*. War follows from enmity. War is the existential *negation* of the enemy. It is the *most extreme consequence* of enmity. It does not have to be common, normal, something ideal, or desirable. But it must nevertheless remain a *real possibility* for as long as the concept of the enemy remains valid.[39]

---

[37] Let us not forget that the notion of stranger has been considerably broadened during the Anthropocene epoch with the insertion of nonhumans.
[38] Schmitt, 1976, p. 27, emphasis added.
[39] Ibid., p. 33, emphasis added.

The States (of Nature) between war and peace 237

Schmitt is obviously thinking only about wars between humans, as they have been triggered, unleashed, exacerbated by the absence of a higher third party, or on the contrary reined in, slowed down, calmed by the presence of an arbiter. As a historian of interstate law, he identifies this arbiter in the old power of the Church or in the modern European law of the nation-states – the *jus publicum europaeum* – for which he has nothing but praise. Politics appears or disappears according to the presence or absence of this third-party arbiter. Even though the argument is well known, it still has not made it possible up to now to slow the dissolution of politics into management, *ethics*, and governance.

What happens when we also acknowledge the absence of an external, disinterested third-party arbiter in conflicts between humans and *other beings*, nonhumans, who may become, "in a specially intense way" – there can be no doubt on this point – "strangers"? If you carry out your ecological conflicts as though they are taking place under the aegis of an impartial arbiter, is it not self-evident that they will be reduced to simple policing operations, without bringing into play the friend/enemy distinction in any form? We will be dealing only with rational beings seeking to bring irrational people *back* to reason or to *indisputable knowledge* of deanimated objects. Here we have the source of the depoliticization of ecological questions: the naturalists have no enemy, since, in the proper sense, *the case has been made and won*, in legal as well as scientific terms. Isn't there an adage that says "a closed case should never be reopened"?

If the key concept is *the presence or absence of a disinterested and neutral third party*, we understand, if we want to repoliticize *ecology*, that we must not hesitate to extend Schmitt's argument to all conflicts, including those that bring heretofore "natural" agents into play. Even though, on first reading, that "other, the stranger" designated an anthropomorphic entity, eighty years later, the number of those who have descended into the arena has dramatically increased. We contemporaries of the Anthropocene are compelled to recognize what Schmitt could only glimpse: every time we find ourselves facing a situation in which "the existential negation" of another being is at stake – and thus, today, everywhere – enmity turns out to be infinitely broadened. This means not that we are necessarily going to fight – war is not "common, normal," or even "something ideal, or desirable" – but, rather, that the Dome of Nature, under which all the old conflicts took place, has disappeared. It is this disappearance that obliges each of us to take seriously the "real possibility" of hostilities, even when we are dealing with "strange" beings whose existence, in the proper sense, we deny, and who can *in their turn* – this is the novelty – deny ours.

We have reached the point when we should make no mistake about the role of Gaia in the return to the situation of war. Gaia no longer occupies in any sense the position of arbiter that Nature occupied during the modern period. Such is the tipping point between unified, indifferent, impartial, global "nature" whose laws are determined in advance by the principle of causality, and Gaia, which is not unified, whose feedback loops have to be discovered one by one, and which can no longer be said to be *neutral toward our actions*, now that we are obliged to define the Anthropocene as the multiform reaction of the Earth to our enterprises. Gaia is no longer "unconcerned" by what we do. Far from being "disinterested" with respect to our actions, it now has *interests* in ours. Gaia is indeed a third party in all our conflicts – especially since the emergence of the Anthropocene – but at no point does it play the role of a *higher* third party capable of dominating situations. The whole, here again, here as always, is inferior to the parts.[40]

We can understand that the Spirit of the Laws in the two regimes differs to this extent: in the Old Climate Regime every conflict is prejudged by the simple application of the laws of "nature," while in the New Climate Regime there is no longer a sovereign arbiter; we have to fight point by point to discover – and no longer to apply – the reactions of the agents, one after another. In the first regime, objects are deanimated – only subjects have souls; in the second, we find ourselves truly in a state of war. In the first regime, Peace is given in advance; in the second, it has to be invented, through the establishment of a specific diplomacy. The first is a naturalist regime; the second is, let us say, a compositionist one.[41]

This is why we have to be skeptical of the concept of the Globe and why it is so essential not to confuse Gaia with the Sphere, the System, or the Earth taken as a Whole. The Globe offers a geometric way, as it were, of representing the supreme arbiter that reigns over all conflicts – and that consequently depoliticizes them at once. Gaia, in contrast, can be defined as the multiplication of the sites in which radically foreign entities practice mutual "existential negation." Never again will the complex set of sciences of nature that constitutes climatology be capable of playing the role of ultimate, indisputable arbiter. Not

---

[40] One can never underline too much the tension that exists between globalism and the concept of Gaia. See the fourth lecture.

[41] In the sense of my modest initiative: see Bruno Latour, "Steps toward the Writing of a Compositionist Manifesto" (2010c).

because of the artificially maintained controversy over the anthropic origin of climate change, but because of the quality of loops that the sciences have to put in place, one after another, to make us sensitive to the sensitivity of Gaia. "Nature," or at least the sublunary Earth, has been placed in a situation that obliges everyone to make decisions about the "extremes" of life and death in the face of strangers who purport to deny their own existential condition. Gaia and the sciences of the Earth System are fully engaged in a geohistory that is as "full of sound and fury" as the history of the earlier age, it too "told by an idiot"!

This is why, in earlier periods, when we invoked Nature, we put ourselves, without even thinking about it, under the protection of a State of Nature, a State with a capital S, a monstrous Leviathan, half politics, half Science. If that monstrous State managed to subsist, for what it was worth, with half of its body in nature and the other half in politics, it was because it was necessary to put an end, as we saw with Toulmin,[42] to the religious wars, through a cult of indisputable certainty. And yet the armistice proposed by Hobbes never managed to achieve, through a formally adopted treaty, a situation of lasting peace between the contradictory imperatives of the various forms of counter-religion. Hence the construction of the dubious Constitution that feigned to be offering peace to the nations even as it led a war against "nature," an all the more limitless war in that it never appeared to be a war at all.

As we know, a major part of Schmitt's work is devoted to the question of what makes a war become limitless. His answer is that it is always for want of a clear recognition of what characterizes the enemy. It is precisely this denial of a state of war and the dissimulation of the friend/enemy relation in the guise of simple policing operations that leads, in Schmitt's eyes, to the transformation of limited wars into *wars of extermination*.[43] Any reader of the contemporary ecological conflicts can only agree with him on this point: the conflicts would never have gone so far toward radical extermination if they had been considered as wars in which *the other side*, in its turn, could endanger the existence of those who were attacking it. The possibility of extermination, of what has to be called a *war of annihilation*, came from the illusion that we were carrying out, in

---

[42] See the sixth lecture.
[43] Hence his critique of the Treaty of Versailles, which had considered Germany not only as the party that had lost but as the criminal party, and hence the way Schmitt revisited the history of the war; see the fine introduction by Céline Jouin to Carl Schmitt, *La guerre civile mondiale: essais (1943–1978)* (2007).

the name of civilization, only a simple operation of *pacification*! As Schmitt writes: "A world in which the possibility of war is utterly eliminated, a completely pacified globe, would be a world without the distinction of friend and enemy and hence a world without politics."[44]

Schmitt obviously was not aiming at ecology as it has developed up to now, but he accurately targeted the ideal of those who want a definitively pacified planet. Is this not the ideal of the naturalists, the utopia of the deep, superficial, or in-between ecologists; the horizon of those who hope to become the managers or the engineers or the re-engineers of the planet; of those who hope to get beyond the crisis with "sustainable development," the ideal of the ecomodernists,[45] of those who claim to be the good caretakers, the serious stewards, the astute gardeners, the attentive quartermasters of the Earth? In short, is it not in fact the dream of those who want so badly, when they are dealing with "simple material questions," to get along without politics altogether?

The choice Schmitt sets before us is terribly clear: either you agree to distinguish the enemy from the friend, and then you are engaging in politics, strictly defining the boundaries of very real wars, "wars over what the world is made of," or else you scrupulously avoid waging wars and having enemies, but then you are *renouncing* politics, which means that you are giving yourselves over to the protection of a State of Nature that encompasses everything and that has *already* unified the world in a single whole, in a Globe that is supposed to be capable of solving all conflicts from its own disinterested, neutral, all-encompassing point of view. A stupefying amalgam of religious, scientific, and political powers: "*Sub specie aeternitatis, sub specie Dei, sive Spherae, sive Naturae.*"

The second solution would be preferable, I readily acknowledge, since it would allow us at least to postpone the conflicts: "Let us all be brothers on the same blue planet, aligning ourselves under the

---

[44] Schmitt, 1976, p. 35.

[45] See the site of the Breakthrough Institute (http://thebreakthrough.org/), established after the publication of a book by Ted Nordhaus and Michael Shellenberger, *Break Through: From the Death of Environmentalism to the Politics of Possibility* (2007). Because I wrote a piece for the journal of this institute which has been misunderstood by people who haven't read Shelley's *Frankenstein*, I have been associated with their political line. However, it is fair to say of the two founders that they were open-minded enough to have me for a while as a fellow even though I disagreed with much of what they tried to do! See the original unpublished pieces, including a review of Nordhaus and Shellenberger's book: Bruno Latour, " 'It's Development, Stupid!' or: How to Modernize Modernization" (2007a); see also "Fifty Shades of Green" (2015).

The States (of Nature) between war and peace 241

same politico-scientific authority in order to escape from more serious conflicts." As I am not particularly bellicose, that would suit me just fine. But only on condition that *such a State could exist*. If there is no such State, then what might have been accepted as a useful last resort becomes quite simply criminal, since we would be agreeing to place our security and that of all the other entities with which we share the Earth *under the protection of a political body incapable of defending us*. When it is a matter of ensuring one's security, pacificists are dangerous people.

The perilous virtue of reactionary thinkers such as Schmitt is that they force us to make a more radical choice than the one proposed by so many ecologists, who are still driven by the hope of getting out of the crisis without ever politicizing the questions of "nature." It is a difficult choice, I admit: either "nature" puts an end to politics, or else politics obliges us to give up "nature" – and thus finally to agree to face up to Gaia. Remember the passage from the Gospel I cited earlier, and that Schmitt would have understood only too well: "Do not suppose that I have come to bring peace to the earth. I did not come to bring peace, but a sword" (Matt. 10: 34). Between the pacifiers and the "peacemakers" to whom, alone, the *"nomos* of the Earth" has been promised, we are going to have to choose.

*

Agreeing to go through a state of war in order to search, afterward, through diplomatic transactions, for peaceful solutions requires significant transformations in the way the collectives present themselves to one another. They have to agree to specify the epoch in which they live and the name they give their people, and above all they have to be able to mark off the space that is theirs so that the others understand what territory they are ready to defend. The spatial limits – and among Schmitt's innovations this is the one that matters most to us – are traced by the identification of strangers recognized as others "in a specially intense way" (*hostis*), "so that in the extreme case conflicts with [them] are possible." Bringing out these limits is the only way to repoliticize ecology and to put an end, consequently, to the simple operations of conquest, land grabs, or pacification.

Let's start with the epoch. To stand up to the threat, we first have to understand why we feel that it is *coming toward us*, and why it is so hard to *face up* to it head on.[46] As I recalled in the introduction, I

---

[46] See the short film by the dancer Stéphanie Ganachaud, whose movement continues to nag at me: www.vimeo.com/60064456, cited in the introduction.

began the strange project of turning toward Gaia with a mental image of the silhouette of a dancer, fleeing backward at first, as if she were escaping from something so frightful that she was indifferent to the destruction she was leaving behind her by pulling back blindly, a little like the "angel of history" made famous by Walter Benjamin.[47] This "angel of geohistory," as I have named her, glances behind her with more and more anxiety, then slows down as though she were getting caught in thorny brush; she finally turns around and suddenly grasps the full horror of what she must now face, until she stops completely, her eyes wide open, incredulous, before she starts to pull back, terrified by what is coming toward her.

Contrary to what is often said about them, the Moderns are creatures who look not *forward* but almost exclusively *backward* and, curiously, *up in the air*. This is why the emergence of Gaia surprises them so much. Since they don't have eyes in the back of their heads, they completely *deny* that it is coming toward them, as if they were too busy fleeing the horrors of the old days. Their vision of the future would seem to have blinded them to the direction in which they are headed; or, rather, it is as though what they mean by "future" were entirely constituted by the rejection of their past, without any realistic content attached to the "things to come." The children of the Enlightenment are in the habit either of rejecting with terror the threatening past from which they have had the courage to escape or, conversely, of endowing that past with magnificent qualities to which they aspire with nostalgia, but they remain quite taciturn as to the shape of the *things to come*.

As we have learned with Voegelin, the future of the Moderns is not in front of them, entrusted to a realistic, hesitant vision of the time that is passing; rather, it consists in an inaccessible transcendence that they nevertheless seek to situate in time in order to *replace* the course of time. The future, for them, is what is to come, but deprived of the means of *becoming*, since they never look at it straight on and never take it in its ordinary humble form. Hence their striking lack of realism, their susceptibility to "hype," their constant resumption of a futuristic vision of the future. Because of the phenomenon that Voegelin calls immanentization,[48] the Moderns are never *of their time* but always on the other side of the Apocalypse, suspended between senseless hope and senseless despair. Moreover, as they have completely forgotten the sources of the counter-religion they have inherited without realizing it, they are incapable of treating themselves for

[47] Walter Benjamin, "Theses on the Philosophy of History" ([1940] 1969).
[48] See the discussion in the sixth lecture.

that illusion by returning to the texts that would have made them aware once again of the demands of the counter-religion. In short, *the Moderns' time is strangely atemporal.*[49]

They see the future only in the form of futuristic fiction. This is hardly surprising: they have never paid enough attention to the *direction* in which they are heading, obsessed as they have been by the idea of *escaping* their attachments to the old Earth. Ready for detachment, they seem excessively naïve when they encounter the prospect of reattachment to a new residence, of the delineation of a new *nomos*. They resemble astronauts preparing to take off without space suits. The Moderns are extraordinarily clever at freeing themselves from the chains of their archaic, provincial, enclosed, local, and territorial past, but when their task is to designate the new localities, new territories, new provinces, the narrow networks toward which they are emigrating, they settle for utopia, dystopia, advertising, and great heavings of breasts, as if they really had lungs suited for breathing in the subtle, toxic air of globalization.[50]

But then toward what horizon do we turn when we face up to Gaia? We have to choose between two opposing conceptions of progress, because Gaia is simultaneously what was there, something that has been abandoned and forgotten along the way – Ge, the ancient goddess – and what is coming toward us, our future. The irony of geohistory is that it lies between two goddesses, one from the most remote past, the other from the nearest future, and they bear the same name. Thus as soon as we begin to concern ourselves with the climate, with what belongs to the land, with territory, we don't know whether we are enjoined to head backward or forward, whether we have to look up, down, behind, or ahead. It is hardly astonishing that we are divided and that ecology drives us crazy!

If the future and what is to come lead us in different directions, the same thing is true of the word *land*. Depending on whether you are speaking of feudal land and soil or of land as Earth, the orientation of the arrow of time changes immediately. You tip from a reactionary attitude to a progressive attitude. To insist on the land and the soil in the feudal sense is to be reactionary in the old way – by invoking "the land that tells no lies," *Blut und Boden*. And it is quite true that reactionaries of all stripes, Schmitt included, have always insisted on the criminal aspect of the will to leave the old land, to abandon the old soil, to forget the limits of the old *nomos*, to be emancipated and

---

[49] Restoring temporality to the Moderns sums up the project of Henri Bergson's inspired disciple Charles Péguy, especially in his "Clio I" ([1917] 1958).
[50] Peter Sloterdijk, *Globes: Macrospherology* ([1999] 2014).

cosmopolitan. Against these calls to remain "behind," the revolutionaries have always appealed to emancipation. And yet what they could not imagine was that there could be another meaning in the attachment to the old soil, in the sense, this time, of "the good old Earth." As soon as you say this, things are reversed, and the earth which was formerly something that one was supposed to leave behind in order to profit from modernization becomes the new Earth that is coming to you. Contrary to what the nostalgics say, coming back down to Earth has nothing to do with some longing for Arcadian rurality.

This may be surprising, but, in the epoch of the Anthropocene, the Great Emancipation Narrative has made us unsuited for finding the path of the Earth to which we belong. As if the very notions of "belonging" and "territory" gave off a whiff of something reactionary! Still, one might think that, after several centuries of critiques of religion, we wouldn't have much trouble recognizing that we are "of this Earth." How strange it is that, after having heard so many appeals in favor of materialism, we find ourselves totally unequipped to approach the *material conditions* of our atmospheric existence! After so much sarcasm toward those who preached to the masses that they had to escape into "the world beyond" in order to flee from the harsh conditions of this world here below, we now find ourselves taken aback by the notion that there can be *limits* to our objectives; we are incapable of defining a behavior that would be down-to-earth, terrestrial, embodied. Although the "death of God" ought to have brought us back to a human, all too human, condition, we find ourselves hesitating, mumbling in the dark, in the "vale of tears," asking ourselves with surprise how it happens that we have so much trouble feeling the ground beneath our feet! Whereas for several centuries we had been congratulating ourselves in the certainty that we were solid realists surrounded by matters of fact, we are now astonished to be *from* here. We are obliged to ask the materialists: "Please give us back our materiality." It is as though underneath the vale of tears there were another vale of tears!

What is coming, Gaia, has to *appear* as a threat, because this is the only way to make us *sensitive* to mortality, finitude, "existential negation" – to the simple difficulty of being of this Earth. This is the only way to make us conscious, tragically conscious, of the New Climate Regime. Only tragedy can allow us to measure up to this event. As we saw in the previous lecture, the fireworks of the Apocalypse are not there to prepare us for an ecstatic elevation toward the Heavens; on the contrary, they are there to *keep us* from being driven from the Earth as it reacts to our efforts to dominate it. We have misunderstood the injunction: we weren't supposed to bring Heaven onto the Earth but, first, to take care of the Earth, thanks to the Heavens. This is the only way to oblige us to change the direction of our attention after so many years spent neglecting what was taking place behind our backs. If the "angel of geohistory" is starting to look ahead with horror and incredulity, it is because she has become aware that there is a threat and that she has waged a war that will never cease if she denies it! To put it baldly: in the face of what is to come, we cannot continue to believe in the old future if we want to have a future at all. This is what I mean by "facing Gaia."

*

We understand nothing about the ecological questions if we don't agree to be divided over them. To resist the desire to empty ecology of its politics, we have to suspend these unanimous, universal, and global visions. Without first recognizing that humans are divided into so many war parties, no peace will be possible; no Republic will ever be built. I beg you not to conclude that I am disdaining the ideal of universality: I recognize, I share, I cherish this ideal. But I am seeking a *realistic* way to achieve it. And, to do so, we have to act as though we were certain that it has not *already* been realized. Our situation is thus at once the same as and the opposite of that of Hobbes: the same, because it is imperative to seek peace; the opposite, because we cannot go from the state of nature to the State; we can only go from the State of Nature to the recognition of a state of war. Whereas Hobbes needed the state of nature to generate the concept of social contract, we need to acknowledge a new state of war before seeking new forms of sovereignty. This is why it was so important, in the earlier lectures, to combat the curse of the Globe and to introduce multiple and dispersed peoples, distributing their powers to act, their agency, in relation to specific cosmograms and to the various deities that convoked them. Let us agree for the moment to raise the question in the following form. Instead of imagining that you have no enemies because you live under the protection of Nature (supposedly depoliticized), can you *designate your enemies and delineate the territory you are prepared to defend*?

This amounts, I am afraid, to doubting the solidity of the social contract. In fact, what makes the designation of the enemy even more urgent is that it makes hardly any sense to speak of the "human species" as if it were a party in conflict with another – for example, with "nature."[51] The front line does not merely divide each of our

---

[51] The title of James Lovelock's *The Revenge of Gaia: Earth's Climate in Crisis and the Fate of Humanity* (2006) is thus misleading: there are not two parties in conflict.

souls; it also divides all the collectives on the topic of all the cosmopolitical problems we have confronted. The *Anthropos* of the Anthropocene is nothing but the dangerous fiction of a universalized agent capable of acting like a single humanity.[52] For such a Humanity to be viable, there would have to be a worldwide State already in place behind it. The Human (with a capital letter) as agent of history has been demobilized and disbanded.[53] As we saw in the fourth lecture, the advantage of the Anthropocene is that it brings to an end not only anthropocentrism but also any premature unification of the human species, even as it makes it possible to imagine a new understanding of the notion of species – but not right away, above all not right away.

Whether you take the worldwide controversy over GMOs, the calculation of fish stocks, the development of wind power, the modification of coastal features, the manufacture of clothing, food, medications, or cars, the reconfiguration of cities, the transformation of agricultural techniques, the protection of wildlife, the change in the carbon cycle, the role of water vapor, the influence of sunspots, or the creeping of icebergs – in every case, you find yourselves facing stakes that bring together those who oppose one another on the topic.[54] Now that there is an acknowledged state of war, it is possible for each of the warring parties to be explicit about its *war aims*.

Apart from tactical reasons, it is no longer necessary to hide behind some appeal to the objectivity of knowledge, to the incontrovertible values of human development, to the Public Good or to the well-being of common humanity.[55] Tell us, rather, who you are, who are your friends and your enemies, *whom* you are ready to sacrifice to your own happiness, *which* foreigners can put you in a situation such that your existence will be denied – and, in addition, please tell us clearly, finally, *by what deity* you feel convoked and protected. If you find this argument too cruel, remind yourselves that the ecological crises

---

[52]This is the thrust of the criticism addressed to the notion of the Anthropocene by Christophe Bonneuil and Pierre de Jouvancourt, "En finir avec l'épopée: récit, géopouvoir et sujets de l'anthropocène" (2014), as well as Isabelle Stengers, "Penser à partir du ravage écologique" (2014).

[53]See Dipesh Chakrabarty, "Postcolonial Studies and the Challenge of Climate Change" (2012). I was reassured to hear Anna Tsing respond calmly to an objector who asked her what new actor was going to replace the revolutionary proletariat: "Perhaps we have already had too many of these heroic actors...!" (Utrecht, April 18, 2015).

[54]See Noortje Marres, *Material Participation: Technology, the Environment and Everyday Publics* (2012).

[55]That everyone will manage to fight under his own colors is the sole democratic hope of Walter Lippmann, and the only one he deems realistic; see *The Phantom Public* (1925).

have not deprived us of a disinterested third party capable of serving as arbiter in our conflicts but that, on the contrary, they have revealed *that this third party has never existed* and that the seventeenth-century solution has never been anything but a provisional armistice. This is the state of exception opened up by the New Climate Regime. This is what compels us to take up politics once again.

I quake at the idea of defending a thesis that is so easy to misinterpret, but I have to go ahead and draw the consequences from these seven lectures: if we want to have a political ecology, we have to begin by acknowledging the *division* of a human species that has been prematurely unified. We have to make room for collectives in conflict with one another, and not only for cultures known through a science such as physical or cultural anthropology. We have to call back into question not only the idea of a Nature conceived as indifferent to our misery – Gaia is exceptionally touchy[56] – but also *the notion of humans prematurely unified*. This is why it may be preferable to say that the "people of Gaia" come together, assemble, behave in a manner that is not easily reconcilable, for example, with those who call themselves "people of Nature," "people of the Creation," or with those who take pride in being simply "Humans." Remember the strange "Game of Thrones" that I tried to have you play in the fifth lecture? These diverse peoples could come together in the future, but only once they were able to understand in what respects they differed.[57] Too many preoccupations divide "us" – and this "us," to begin with, has borders that it would be good to try to redraw.

With the Anthropocene, the Humans are now at war not with Nature but with...in fact, *with whom*? I have had a lot of trouble settling on a name for them. We would need a title that divides those who have been called Humans while making it possible to specify their supreme authorities, their epochs, their grounds – in short, their cosmogram – instead of melding them all into a shapeless mass.[58] Science fiction often uses the term "Earthlings," but that would be too evocative of *Star Trek*, and in any case it would designate the whole of the human species considered from another planet, on the

---

[56] This is a property attributed to Gaia by Isabelle Stengers, in *In Catastrophic Times: Resisting the Coming Barbarism* ([2009] 2015).
[57] See Richard White, *The Middle Ground: Indians, Empires, and Republics in the Great Lakes Region, 1650–1815* ([1991] 2011). We shall focus on creating such a common space – a fiction, a simulacrum – in the next lecture.
[58] Before becoming indignant at a loss of humanism, it is fitting to recall to what extent the properties of the human that one would want to save have been restricted, since, for fear of sinking into "naturalism," they incorporated neither world nor bodies nor materiality.

occasion of an "encounter of the third kind" with little green men. Can we speak of "Gaians"? That would be too weird. Call them "country bumpkins"? That would be too pejorative. I prefer the term "Earthbound."

I know it is risky to state the problem so bluntly, but I am obliged to say that in the epoch of the Anthropocene the Humans and the Earthbound would have to agree to go to war. To put it in the style of a geohistorical fiction, the *Humans* living in the epoch of the *Holocene* are in conflict with the *Earthbound* of the *Anthropocene*.

*

The Earthbound have to be able to map the territories on which they depend for their existence. This last point is the one I want to broach in concluding the current lecture, before exploring, in the next one, the geopolitics of the New Climate Regime. Hobbes – the somewhat simplified Hobbes that I am taking as a convenient reference point in order to move ahead on these questions – had managed to achieve a semblance of peace by entrusting full sovereignty to the State, by entrusting an indisputable form of certainty to the Sciences of Nature, by granting a strictly moral and personal interpretation to biblical exegesis, and, finally, by making sure that the objects of the natural world were totally deanimated and that human agents were limited to the calculation of their interests alone, excluding any other values.[59] The cosmogram of the great Leviathan, while it may have made it possible to delay the declared ecological state of war, had the gross defect of depriving politics of any territorial anchorage. The Leviathan could move around anywhere, indiscriminately, since the limits that designated its enclosure came only from the State and the State's designation of friends and enemies. Hence the division between physical geography – the grid of the chessboard – and human geography – the societies that represented the pawns.

What was *above* these States? The rules of economic calculations, the phantom of the pre-Reformation Church,[60] the laws of human nature, the war of all against all among the sovereign States? Nothing that could ensure lasting peace. The drama of this provisional solution is that the narrow limits of sovereignty allowed and still

[59] On the figure of Hobbes as depicted by Voegelin, see the decisive commentary by Bruno Karsenti, "La représentation selon Voegelin, ou les deux visages de Hobbes" (2012a).
[60] Carl Schmitt, *The Leviathan in the State Theory of Thomas Hobbes: Meaning and Failure of a Political Symbol* ([1938] 1996).

The States (of Nature) between war and peace 249

allow – this is the essential point – *unlimited land grabbing*. Civil peace among States has been achieved at the price of an invisible, total war against the territories. Hence the strange abstraction of a geopolitics that is fundamentally without Earth, without any "geo" but the two-dimensional form of maps taken to be territories. What political ecology has allowed us to understand is the extent to which this Realpolitik was, at bottom, unrealistic.

When Schmitt made the Earth the principal agent that defined the concrete forms of politics, he had not anticipated that the role attributed to that Earth could change so quickly. He had indeed seen that the nation-states were not simply localizable in undifferentiated space and that they located themselves by defining as many spacings as there were decisions about friends and enemies. This was self-evident for the geopolitical borders: the boundary lines also mark the difference between allies and strangers. He had well understood that every new technology had opened up additional opportunities to place and space oneself: the caravels of the first explorers, as well as warplanes or submarines, defined new land grabs every time.[61] (We can easily imagine how attentively he would have followed political theory on the topic of drones.)[62] And yet, if he succeeded in spatializing politics, he obviously did not manage to historicize the Earth's agency. Whereas the whole intent of his book is to put the Earth back at the beginning of the reflection, this Earth finally remains stable from end to end.

> In mythical language, the *earth* became known as the *mother of law*.... In this way, the *earth* is bound to *law* in three ways. She contains law within herself, as a reward of labor; she manifests law upon herself, as fixed boundaries; and she sustains law above herself, as a public sign of order. *Law is bound to the earth and related to the earth.* This is what the poet means when he speaks of the *infinitely just* earth: *justissima tellus*.[63]

With such statements, Schmitt reinvents the long-lost path between positive law and nature, a path that the modernist solution had definitively cut off, since "nature" had been entrusted to deanimated objects that could not engender any law or any politics whatsoever. As long as the Earth was confused with "nature," it was no longer possible for anyone to qualify it as "infinitely just." And yet, one senses very quickly that something is wrong, and that a possibility

---

[61] One has no choice but to resort to mythology in speaking of these matters, as Schmitt illustrates in *Land and Sea* ([1954] 1997).
[62] As Grégoire Chamayou does with talent in *Théorie du drone* (2013).
[63] Schmitt, 2003, p. 42, emphasis added.

of thought has been finally closed off. For "earthbound," the French translation of *The Nomos of the Earth* uses the adjective *terrien* (*Le droit est terrien et se rapporte à la terre*), which is not quite the same as *terrestre* ("terrestrial"). The world envisaged by the earthbound mind is not necessarily of a scale comparable to that of the Earth. Schmitt, in other words, projects onto his theory of law the prejudices of an old man looking out of his window onto an old European agricultural landscape. In his vision of the land, there is neither anthropology nor ecology. This traditional land-based, earthbound distribution of roles between man and soil is clearly visible in one of the many definitions he gives of *nomos*:

> *Nomos* comes from *nemein* – a [Greek] word that means both "to divide" and "to pasture." Thus, *nomos* is *the immediate form in which the political and social order of a people becomes spatially visible* – the initial measure and division of pastureland, i.e., the land-appropriation as well as *the concrete order contained in it and following from it*.... *Nomos* is the *measure* by which the land in a particular order is divided and situated; it *is also the form of political, social, and religious order determined by this process. Here, measure, order, and form* constitute a spatially concrete unity.

And he adds:

> The *nomos* by which a tribe, *a retinue or a people becomes settled*, i.e., *by which it becomes historically situated* and turns *a part of the earth's surface into the force-field of a particular order*, becomes visible in *the appropriation of land* and in the founding of a city or colony.[64]

Here is indeed the limit, Schmitt's, not that of cultivated plots: even though the concrete order is drawn from the earth instead of being simply imposed on the soil, it is nevertheless still man who measures the land and takes it. The actor still remains humanity.[65] Humans are the ones who found, who measure, who settle, who turn "a part of the earth's surface into the force-field of a particular order." Schmitt did not imagine for a moment – and how could he, at the time he was writing? – that the Earth could occupy a position other than that of *what is taken*!

The paradox with Schmitt is that he makes the Earth the "mother of law" in mythic language, but without being able to grant it any power except that of making the "political and social order of a people"

---

[64] Schmitt, 2003, p. 70, emphasis added.
[65] It is this ambiguity that Heinz explores in "La terre comme l'impensé du Léviathan" (2015).

"spatially visible" by giving it "immediate form." What Schmitt could not imagine was that the expression "land-appropriation" – *Landnahmen* – could begin to mean *"appropriation by the land"* – that is, by the Earth. Whereas Humans are defined as those who take the Earth, the Earthbound are *taken by it*. In both cases, the Earth is still the Mother of their law, but it is *not the same mother*, it is *not the same law*, and thus *these are not the same humans* – they are no longer drawn from the same feudal plot, made of the same humus, taken from the same compost – in short, they do not have the same composition. That the mother of law, basically maternal and benevolent, in any case sympathetic, could become the wicked stepmother, the witch, or even the virago of law: this role wasn't anticipated in the stupefying idea of putting the ancient Ge, right in the middle of the twentieth century, at the beginning of the mythic history of the concrete order.

It is this radical reversal in the direction of taking possession that we are going to have to consider. Unlike the Earthbound, Humans are not trustworthy, because you never know where they are headed or what principle marks off the borders of their people. Thus it is impossible to draw a precise map of their geopolitical conflicts. Either they tell you that they belong to no place in particular, that they are defined only by the fact that, thanks to their spiritual or moral qualities, they have been capable of freeing themselves from the harsh "necessities of Nature"; or else they tell you that they belong wholly to Nature and to its realm of material necessity, although what they mean by materiality has so little relation to the agents they have previously deanimated that the "kingdom of necessity" – *phusis* – seems just as outside-of-the-Earth, out-of-this-world, as the realm of freedom – *nomos*. In both cases, they seem incapable of belonging to any cosmos, of drawing any *cosmogram*. Because of this lack of localization, they seem to remain indifferent to the consequences of their actions, postponing the payment of their debts, indifferent to the feedback loops that might make them aware of what they are doing and responsible for what they have done. The Moderns pride themselves on being rational and critical, even while being resolutely non-reflective. Paradoxically, what they call "being oriented toward the future" amounts to saying, like King Louis XV: *"Après moi le déluge!"*

The Earthbound, in contrast, can call themselves sensitive and responsive, not because they possess superior qualities, but because they belong to a *territory* and because the *delimitation* of this people is made explicit by the state of exception in which they agree to be placed by those whom they dare to call their enemies. Of course, the territory in question does not resemble the geographical maps of our classrooms. It is made up not of nation-states enclosed within their borders – the only actors that Schmitt took into account – but, rather, of networks that intermingle, oppose one another, become mutually entangled, contradict one another, and that no harmony, no system, no "third party," no supreme Providence can unify in advance. The ecological conflicts do not bear upon the nationalist *Lebensraum* of the past; and yet they bear, in spite of everything, on "space" and on "life." The territory of an agent is the series of other agents with which it has to come to terms and that it cannot get along without if they are to survive in the long run.

Of course, such a division between inside and outside is as fragile as it is variable, since the series of agents on which each of us depends and to which we belong cannot be summed up without the installation of instruments capable of tracing the loops that make the least of our actions react in response to its causes. At the slightest weakening in the sensitivity of the instruments, the slightest reduction of bandwidth in the sensors, the agent suddenly becomes less sensitive, less reactive, less responsible; it becomes incapable of defining what it belongs to; it literally begins to *lose its territory* along with its bearings. As we shall see in the next lecture, this is what makes these geopolitical maps so hard to stabilize.

If the Humans and the Earthbound are at war, this could also happen to "their" scientists in conflict. Naturalist scientists – those who proudly assert that they are "of Nature" – are unfortunate figures, bound to disappear, disembodied, behind their Knowledge, or to have souls, voices, and places, but at the risk of losing their authority.[66] In contrast, earthbound scientists are embodied creatures. They form a people. They have enemies. They belong to the territory outlined by their instruments. Their knowledge extends as far as their ability to finance, to control, to maintain the sensors that make the consequences of their actions visible. They have no scruples about acknowledging the existential drama in which they are engaged. They dare to say how afraid they are, and from their viewpoint such fear *increases* the quality of their science rather than *diminishing* it. They appear clearly as a new form of *non-national power that is explicitly participating as such in geopolitical conflicts*. If their territory knows no national boundaries, this is not because they have access to the universal, but because they keep on bringing in new agents to be full participants in the subsistence of the other

---

[66] See the fifth lecture.

agents. Their authority is fully political, because they represent agents who have no other voice and who intervene in the lives of many other agents. They do not hesitate to outline the shape of the world, the *nomos*, the cosmos in which they prefer to live.

The earthbound scientists no longer try to be the third party with an overview in all discussions. They are just one party; sometimes they win, sometimes they lose. They are of this world. For them, there is no shame in having allies. They are not afraid to embark on what Schmitt calls, in his brusque language, *Raumordnungskriege*, wars for spatial order. Freed from the terrible obligation of being priests of a divinity in which they do not believe, they could almost say proudly, "We are of Gaia." Not because they trust in the ultimate wisdom of a super-entity, but because they have finally given up the dream of living in the shadow of any super-entity whatsoever. If Gaia weighs on them, it is because they have understood that it is with Gaia, rather than with Nature, that they will have to share every form of sovereignty from now on. They are profane in the sense of secular, not because they take credit for profaning the values of others, like the old-style rationalists, but in the much more banal sense that they accept being ordinary and of this world. What probably appears to most people, scientists included, as a catastrophe – the fact that researchers are now engaged in geopolitics – is what I take as the only tiny source of hope arriving to enlighten us in the current situation. Finally we know what we are facing and with whom we are going to have to face up to it.

*

If only I were wrong! How I would love to be able to end this lecture by telling you that you can now wake up from a bad dream, that the expression "war and peace" applied to Nature was just a figure of speech. How nice it would be to go back to the Old Climate Regime. To turn away once again from the tragicomedy and stop facing Gaia. We would lie back down cozily, our heads on the soft pillow of climate skepticism.

I don't know whether you remember this or not, but, once upon a time, when we looked at the sky in the morning, we could contemplate the spectacle of a landscape indifferent to our cares, or quite simply observe the changing weather, without it looking back at us in any sense. Nature was outside. How restful it was! But today, instead of finding enchantment in the clouds, it is our actions, in part, and every day a slightly less infinitesimal part, that those clouds are transporting. Whether it is rainy or beautiful outside, from now on, we can no longer avoid telling ourselves that it is partly our fault! Instead of enjoying the spectacle of jet trails in the blue sky, we shudder to think that those planes are modifying the sky they are crossing, that they are dragging it in their wake the way we are dragging the atmosphere behind us every time we heat our homes, eat meat, or get ready to travel to the other side of the world. No, unquestionably, unless we contemplate the celestial bodies in the supralunary world, there is nothing outside on which we can meditate calmly.

Here below, in the sublunary world, the feeling of the sublime – that too! – has escaped us. To experience it, we had to feel our smallness before the grandeurs of nature, as well as the grandeur of our souls in the face of the brutality of that same nature. But how can we keep on experiencing the sublime, in the Anthropocene, since we are henceforth a geological force with a grandeur comparable to that of mountain chains, volcanos, erosion? As for brutality, we Moderns are the ones who have stuffed our souls with it to the point that, here too, we rival nature – we who henceforth share the same prospect of becoming rocks. Never again will we be able to tamp down our hubris simply by contemplating the spectacle of grandiose landscapes. In the Great Enclosure where we are now confined, an eye is fixed on us, but it is not the eye of God fixed on Cain crouching down in the tomb; it is the eye of Gaia looking straight at us, in broad daylight. Impossible, from now on, to remain indifferent. From now on, *everything is looking at us*.

Expelled from the bend in the Elba, the eye of the virtual spectator was forced to hesitate about the proper angle that would make it possible to grasp Caspar David Friedrich's painting, obliging the visitor to direct his attention inward. When we come back to this painting, two centuries later, we notice that we have indeed been expelled from Nature, no longer because it is external, indifferent, inhuman, eternal, but because we ourselves are so mixed up with it that it has become internal, human, all too human, provisional perhaps, in any case sensitive to everything we do, a third party in all our actions. A third party that demands its share. According to what distribution rules are we to give it its due, this Nature that the poet greeted with the invocation *justissima tellus*?

EIGHTH LECTURE

How to govern struggling (natural) territories?

> In the Theater of Negotiations, Les Amandiers, May 2015 • Learning to meet without a higher arbiter • Extension of the Conference of the Parties to Nonhumans • Multiplication of the parties involved • Mapping the critical zones • Rediscovering the meaning of the State • *Laudato Sí* • Finally, facing Gaia • "Land ho!"

I was afraid they wouldn't come. When they began to climb up on stage, delegation after delegation, "Forest" after "France," "India" next to "Indigenous Peoples," the "Atmosphere" delegation before "Australia," "Oceans" after "Maldives," each one introducing itself with pride, equal in sovereignty to all the others, I began to believe it. When after three days and one sleepless night the delegations came back on stage to present the result of their work to the public, exhausted but fully in command of their performance, I understood that these young people from some thirty countries had surpassed all my expectations. At the Théâtre des Amandiers, that weekend in May 2015, I really think I sometimes glimpsed, coming out of the smoke in which the director, Philippe Quesne, likes to cloak his productions, something like the "new *nomos* of the Earth," that *nomos* promised by Schmitt to the "peacemakers." Something that, in my enthusiasm, I would characterize as *constitutive*. To begin this final lecture, I

Figure 8.1 The stage of the theate Les Amandiers, May 31, 2015, on the last day of the simulation of the COP21 conference "Making it Work" (author's photograph).

would like to share with you some elements of the constitutional law of the Earth that these student delegations explored.[1]

How could you give any sort of credence, you'll ask me, to a game some young people are playing on the stage of a theater? I grant it the same credibility I give to the equally fragile, equally provisional, equally awkward activity of philosophizing. The scenario staged by Frédérique Aït-Touati to mobilize a simulated negotiation over the climate is no more and no less enlightening than readings on political philosophy or my own very hesitant writing of these lectures. When it is a matter of measuring up to the Gaia event, one has to use any materials at hand. If I am the last to be astonished that two hundred students can solve an insoluble problem of geopolitics, it is

---

[1] "Theater of Negotiations," a simulation carried out in the context of "Make it Work," was produced in Paris at the Théâtre des Amandiers, May 26–31, 2015, under the direction of Philippe Quesne and Frédérique Aït-Touati, with the participation of SPEAP, the school of the political arts at Sciences Po (www.cop21makeitwork.com/simulation/), at the initiative of Laurence Tubiana and myself. See the film by David Bronstein and Les Films de l'Air in French and English, *Climate: Make It Work*, Theater of Negotiations (2015), www.lesfilmsdelair.com/film/climat.

How to govern struggling (natural) territories? 257

because a dancer's steps first warned me that I had better get to work. Moreover, I learned more from the actors in "Gaia Global Circus" improvising scenes in the brightly lit monks' cells of the Chartreuse at Villeneuve-lès-Avignon than from many works of literature labeled "ecological."[2] What have I been doing, in these pages, except commenting by way of further improvisations on the "stage writing" that commented on mine? Conceptual characters relocate themselves as they see fit, breaking through all the walls.

In any case, the concept of a new *nomos* of the Earth cannot appear as anything other than a fiction. Do you remember the work of invention that was required, once upon a time, to bring to light the improbable being called *the people* or, later, the *social question*? How could we imagine that anyone could discover all at once, simply by thinking very hard about it, what peace negotiations between warring territories might look like? If, as the old maxim maintains, "politics is the art of the possible," there still need to be arts to multiply the possibles.[3]

There is a fascinating link, moreover, between the principle of political simulation and that of scientific modeling.[4] Our knowledge about the ecological mutation is based on long-term measuring campaigns but also on models, which offer the only way to approach phenomena whose complexity outstrips our capacities for analysis. As for the loops that are beginning to be added to our existence, one after another, making us more aware every day of the reciprocal feedback among agents of the terrestrial world, we need to make models of them – fictions – long before they can be verified in reality. Fiction anticipates what we hope to observe soon. To each generation of models we can add new variables, further complicating an image of the world that is gradually becoming more and more realistic – and harder and harder to measure! Similarly, to each political simulation we can add new delegations, new representatives, further complicating an image of the *res publica* that is becoming more and more realistic – and whose aberrations are harder and harder to control! *Complicating* the models and *implicating* in them those whom they

---

[2] The project called "Gaia Global Circus" was developed at the Chartreuse in 2011, 2012, and 2013 from a text by Pierre Daubigny (unpublished) thanks to the unwavering support of François Debanne, and in Reims in 2013 thanks to the unwavering support of Ludovic Lagarde. It was performed one last time in Calgary in September 2016 in the festival "Under Western Skies."
[3] This is the maxim of the experimental program in the political arts (SPEAP) created in 2010 at Sciences Po with Valérie Pihet and now directed by Frédérique Aït-Touati.
[4] See Amy Dahan and Michel Armatte, "Modèles et modélisations, 1950–2000: nouveaux pratiques, nouveaux enjeux" (2004).

concern in order eventually to *compose*: this strikes me as a definition common to the sciences, the arts, and politics.

This is exactly what happened in the Theater of Negotiations in May 2015, and what gave this seemingly pedagogical episode a constitutive dimension. Indeed, I maintain that this reduced model – 41 delegations, 208 delegates – is *more realistic* than the real world at full scale, and especially in comparison with the famous Conference of the Parties (COP) in Paris in December 2015, whose twenty-first edition we were prefiguring. Watching the delegates, in the "transformable" room they preferred to a larger room they found overly formal, while they decided that they would sit wherever they wanted and for as long a time as they needed, I couldn't help thinking – you'll have to forgive me – about the room in the Jeu de Paume and the extraordinarily decisive moment on June 20, 1789, when the Estates General decided no longer to seat themselves by orders – nobility, clergy, and the Third Estate – but to meet in a Constitutive Assembly![5]

*

Before transforming themselves into something entirely different, the Estates General had been brought together, as we know, to resolve a simple matter of taxation. Similarly, keeping everything in proportion, starting from the question of the climate, the simulation set itself quite different goals. If the model is more realistic, it is first of all because those who developed it had decided not to concentrate on the impossible question of reducing CO_2 emissions in order to try to remain below the fateful limit of 2° C of warming. Indeed, an excellent book by Stefan Aykut and Amy Dahan[6] had convinced them that the Climate Regime could only lead to an impasse. How could one claim to solve the remote problem – the action of CO_2 on the mechanisms of the climate – without attacking its proximate causes – the multiple decisions regarding ways of life made by the participating nations? Rather like trying to limit the use of guns after having

---

[5] Just like the expressions "Old Regime" and "New Regime," this episode has taken on a mythical dimension in French political philosophy. Until June 20, 1789, the Estates General (the closest that France had come to a system of representation) had been divided by orders – nobility, clerics, and the much more numerous Third Estate. This was still the way the king expected the three bodies to assemble. The refusal to vote in this fashion marked the beginning of the Revolution, a month before Bastille Day.

[6] It was a question of getting out of the impasses brought to light by Aykut and Dahan's book *Gouverner le climat? Vingt ans de négociations internationales* (2014).

How to govern struggling (natural) territories? 259

encouraged their free distribution. For the negotiation to be realistic, it was necessary to concentrate, unlike the real Conference of the Parties, on the various ways of occupying territories, and not solely on the allocation of quotas for CO_2. This was a way of taking advance precautions against a possible failure of COP 21, by anticipating the procedural reforms that will eventually have to be carried out.

Above all, it was necessary to consider that entrusting to nation-states alone the task of solving the problems created by their very utopian – or at least not very earthbound – ways of occupying their lands was not an achievable goal. National borders, as we saw in the two earlier lectures, solve a four-century-old problem, having been put in place on the one hand to impose peace among religions that had run amuck, on the other to ensure unlimited grabbing of lands that had previously been cleared of the other collectives that had possessed them. After four centuries, after imperial expansions, colonization, decolonization, globalization, there is no longer anything realistic in an assembly of one hundred ninety-five nation-states. Even if they managed to reach agreement, all the problems that assail them would escape them nevertheless, since they are intertwined in the most inextricable way, to the point where all these problems have become, as it were, *transversal*.

Ah, you will say, but, naturally, we have to treat all these problems in a "global way"! And yet this utopia should have been resisted. The members of the COP are not *parts* of a higher Whole that would allow them to be unified by attributing to each a role, a function and limits; rather, they are "parties" in the diplomatic sense, in a negotiation that can begin precisely only because there is *no longer a higher arbiter* – neither power, nor law, nor nature. Against the deluge of good feelings that too often accompanies the ecological question, there should have been an agreement not to come together under a common higher principle. Here we return to the figure of the Globe, a figure that we have come to see, in the course of these lectures, as not only impossible but morally, religiously, scientifically, and politically deleterious. Such was the point of departure of the students in May 2015: neither God nor Nature – and thus no Master!

Let us list the higher common principles that they agreed not to invoke. They understood, first, that they must not count on the mirage of a world government that could, by a miracle of coordination and good governance, attribute to each party its share of CO_2 or financial compensation, under the threat of sanctions. While we have the right to dream of such a thing, the absence of a planetary government is all too obvious. One has to ask about the United Nations what Stalin asked about the Vatican: "How many divisions?" The slim workings of the COP are there neither to prefigure a worldwide government nor to replace it, but simply to inhibit, when possible, to slow down the run-up to war.

But, secondly, there is no longer a global Nature, either, that would be capable – if only the world turned to it – of silencing all disagreements. We have not yet seen a single case in which the appeal to the Laws of Nature would have permitted an *automatic* alignment of interests. As one message among the graffiti on the walls of the Théâtre des Amandiers noted: "The blue planet doesn't unify!" And, thirdly, the Science of Nature does not have the capacity to bring everyone together. Even without the pseudo-controversy instigated by the climate skeptics, if there is one thing it is always healthy to avoid, it is government by the learned. Unanimity is not their strong point, and it's a good thing, too.[7]

What is interesting about the experiment is that the students also understood, even if it was harder for them to admit, that the Laws of the Market known to Economics could not serve as a substitute Dome, a Globe, an Absolute, a Mammon-God capable of imposing indisputable decrees on everything that consumes, produces, buys, and sells. Even if, in a paradox that has never stopped surprising me, good sense tends to attribute more indisputable certainty to the laws of the capitalist economy than to those of nature (the two being fused moreover in the common theme of naturalization),[8] it nevertheless seems hard to forget that from ten economists we can get fifteen contradictory pieces of advice about the policy to be implemented. For all its assemblages of useful technologies, economics cannot offer the Great Unification of the Laws of the Planet any more than the other sciences can. By seeking to economize ecology, you are adding to a dizzying multiplicity yet another multiplicity.

If there were a worldwide government, a unified Nature, a universal Science, or an Economy functioning according to unbreakable laws, the delegates would have met, as we saw in the preceding lecture, under the aegis of what has to be called a (quasi-)State of Nature. It

---

[7] Climate skeptics take this unanimity as a proof that there is something fishy in this part of science, though it should actually reassure them: the case is so rare that it must be taken as the signal of a truly exceptional situation. In *A Vast Machine: Computer Models, Climate Data, and the Politics of Global Warming* (2010), Paul Edwards makes the even more troubling suggestion that the certainties will never be greater than they are now, since, by modifying the system so much, we are making it less and less predictable.

[8] The second nature – the Economy – being always more difficult to doubt than the first (see Karl Polanyi, *The Great Transformation: The Political and Economic Origins of Our Time*, [1944] 2001).

hardly matters whether that State appeared secular rather than religious; it would have been apolitical in the sense that it would have maintained the fiction of a sovereign arbiter to whom the delegates could appeal in order to bring disagreements to an end. The delegates would have fulfilled a function, played a role, followed a script. They would have done no more than mimic simple police operations. Their delegations would have been parties, in both the legal and the organizational senses of the term, since it would have sufficed for them to obey rules. The young delegates would have had a good time, perhaps, but in the same way they know so well from games of Risk or Dungeons and Dragons. No political invention would have been required. There would have been nothing constitutive.

What made the simulation in May 2015 at the Amandiers realistic was that the delegations met *in the absence of* any escape valve, without an elsewhere, without a court of appeals, without an external sovereign, without reference to a Dome, a Tent, a Dais capable of sheltering them. Moreover, when the delegations introduced themselves to one another on the first day, allusions to the good of Nature, Humanity, the Planet, or the Globe were rare. Each delegation spoke only about itself. Each one knew it was alone. Each one knew that the others were alone. Nothing unified them in advance. Their common higher "power" was only the fictional frame proposed by the student secretariat that had brought them together and that they had provisionally accepted. Nothing more than a middle ground, a clearing between two suspensions of hostilities.[9] Only the tiny fiction of finding themselves on stage in a theater for four days, surrounded by a minimum of furnishings expressly tailored to the occasion,[10] defined limits that were totally artificial and recognized as such. It was because there was *nothing natural* in the exercise that it was realistic! As nothing was spelled out in advance, *it could fail*. And, indeed, it never stopped almost failing.

*

Still, those who conceived of this event had to give some plausibility to this inside without an outside arbiter. If I stress certain of the decisive innovations that were introduced, it is because I am convinced that

---

[9] See Richard White, *The Middle Ground: Indians, Empires, and Republics in the Great Lakes Region, 1650–1815* ([1991] 2011).
[10] The material transformation of the space, known to be important in every diplomatic undertaking, had been entrusted to Raum Labor, a group of German designers.

they will be useful in the future when real peace negotiations will have to be undertaken.[11]

The first and most radical innovation seemed to be self-evident: we can no longer let the nation-states occupy the stage all by themselves. It is precisely to avoid this utopia that we have to add *non-state* delegations. No longer because they would represent interests *higher* than those of Humanity, but quite simply because they are other powers, possessed by other interests,[12] which exert continual pressure on the interests of Humanity and consequently form other territories, other *topoi*. The crucial point is that the delegations whose names recall ancient elements said to be "of nature" – "Land," "Oceans," "Atmosphere," "Endangered Species" – are there not to naturalize the discussion by reminding humans of what their "environment" requires but to repoliticize the negotiation, by preventing coalitions from forming too quickly at the expense of the others.

This is why it was important for these unconventional delegations to present themselves in the same apparatus and according to the same protocol as those of the old or new nation-states: each delegation was formed in the same way, expressed itself in the same language (in this instance, English), and all were represented by exactly the same young people wearing dresses or suits and ties. No extravagances would have been appropriate. The "Ocean" delegation didn't pretend to be speaking by way of storms and tsunamis; "Atmosphere" didn't take on the guise of Boreas, nor did "Land" purport to be a clump of soil crawling with worms.[13] Represented on stage were only powerful interests capable of designating the other interested parties as their enemies. For example, the actions of a country that acidifies oceans to the point of turning them into deserts certainly constitute evidence that that country weighs on the quasi-domain "Ocean," leading to the following response by the latter's delegation: "We consider unacceptable for our sovereignty what you, the delegation representing 'United States' or 'Australia,' are inflicting on our domain. By opposing you, we are defining the limit of our territory and *we are redefining the shape of yours*."

---

[11] Ever since *We Have Never Been Modern* ([1991] 1993), I have been stubbornly seeking the precise form and the practical feasibility of what I called in that early text the "Parliament of Things."
[12] "Interest" has to be understood in the sense in which it was used in the second and third lectures, as a general property of the agents that overlap and interpenetrate one another.
[13] Each delegation was required to have five delegates – or entities: a governmental or quasi-governmental representative, an economic actor, a representative of civil society, someone with scientific knowledge, and a fifth freely chosen.

It is a fiction, of course, but the fiction bears on more than a technique of personification that gives equal sovereignty to all the interests represented. It is not hard to understand the surprise of a sovereign peacefully surveying his domain who suddenly hears the virulent *response* of territories that start to shout: "This isn't yours any longer!" The direction of land grabbing is immediately reversed and, with this, the very definition of what it means, for any power whatsoever, to possess land. Up to now, these interests, these entanglements, had no presence in the debate except that of data summarized in reports sketching the general framework under which the national delegations were operating. The data were there, of course, but mute and deanimated – or at least de-dramatized. They formed a framework; they were not agents. They were numbers, not a voice, not a drama, not a role in a developing plot. In other words, we were still in the Holocene: the land was not reacting to human actions. Everything changes when agents are given a voice *compatible* with those of the other agents. *Redistribution* can then begin.

If you agree to define a territory not as a two-dimensional segment of a map but as *something on which an entity depends for its subsistence, something that can be made explicit or visualized, something that an entity is prepared to defend*, then any dramatization of the actors that compose it, even a fictitious one, will modify the composition of the scenario.[14] The form of representation you start with hardly matters; what counts is the reactivity of the parties involved. If you are surprised to see "Forest" given a voice, then you have to be just as surprised that a president speaks as the representative of "France." Each corporate body has a good deal to say, and each can express itself only through a dizzying series of indispensable intermediaries. It took many decades to agree that the definition of democracy as the will of a sovereign people corresponds, even vaguely, to a reality, and it was necessary to start with a fiction. "What? The people, sovereign? You must be crazy!" "What, a delegation representing forests? That's unthinkable!" But the students thought it, and it didn't seem to pose any problems.

I very much enjoyed observing that the negotiations were never impeded by that sort of objection. The tireless president Jennifer Ching addressed "Lands" or "Amazonia" just as politely and straightforwardly as she addressed "Canada" or "Europe." If the fiction appeared so plausible, it was because each delegation was presumed

---

[14]See Michel Lussault, *L'avènement du monde: essai sur l'habitation humaine de la Terre* (2013), and Bruno Latour, "*Onus Orbis Terrarum*: About a Possible Shift in the Definition of Sovereignty" (2016a).

capable of speaking; this is obviously easier in a theater accustomed to hearing the voices of choruses, divinities, monsters, and fairies echoing under its rafters. But it was also because all arrangements for speech have the same strangeness, whether it is a matter of representing humans (who do not speak) or nonhumans (who are made to speak). For the Earthbound, the question no longer arises: they are moved by too many articulated agents to believe they themselves are the only ones who speak. This is perhaps the only advantage of living in the Anthropocene epoch.

In any case, to speak with some authority is always to interpret what mute actors would say if only they could speak – and to be interrupted by another who asserts that those mute parties are saying something else! Doubt about representation appears only at the moment of conflict, when a dispute becomes tense and when someone opposes what an elected official, a scientist, an expert, a citizen, is saying about some particular state of the world, going so far as to ask "How do you know this? What is the evidence?" The time is past when humans spoke with one another in front of an audience of inert things. If humans speak an articulated language, it is because the world is articulated as well.[15] What is cast into doubt in the negotiation is the *quality* of the representation, and no longer the *principle* of representation itself.[16] The New Climate Regime has come to remind the Moderns of what they had forgotten.

Moreover, it is hardly surprising that this principle of representation was developed by scientists with respect to the things of this world, before becoming a principle of political representation of these same things, which have now become subjects of controversy and concern. Without the sciences, the ecological mutations would have remained invisible. In some sense, scientists are the *activists* on behalf of this new "social" issue. They were the first to politicize the mutations (in the good sense of the term "politicize") by becoming their representatives and introducing them into the old question of democracy and representative government. It was scientists who put the acidification of the ocean and the stripping of the land on the political *agenda* of representative assemblies. All we have to do now is extend what they have begun.

The objection on principle that so obsesses journalists ("How can you claim to 'represent' the oceans or the atmosphere?") was all the

---

[15] This essential element of *An Inquiry into Modes of Existence: An Anthropology of the Moderns* (2013b) was addressed at the end of the second lecture.
[16] See Bruno Latour and Peter Weibel, eds, *Making Things Public: The Atmospheres of Democracy* (2005).

less bothersome to the delegates in that they had all included scientists in their delegations, but without giving them elevated status; scientists were simply *added* as spokespersons. The sciences were neither excluded nor marginalized nor elevated to a position of superiority in relation to the other players. This was another astute innovation. Each delegation mobilized the inputs of research, technology, instrumentation, and expertise in its own way, so its members could respond to questions about the quality of the representation of a given interest or a given state of the world.[17] In any case, Science was not there to dictate the general framework in which the negotiation was obliged to take place. The objectivity of the sciences was not in doubt, only their unification. Here, too, we must no longer expect to appeal to some ultimate outside authority. This first post-natural assembly was also post-epistemological.

If this distribution of the sciences seems to weaken the authority that they have never had in any case, in exchange it secures a privileged place for researchers who are led to find themselves *everywhere*. They become capable at last of defending the originality, the power, the interests of the beings on whose behalf they are speaking and that they can embody – represent, interpret – with their contradictions and their controversies in all the negotiations, in order to try to redraw the lines. *Situated* knowledge is much more realistic than knowledge from nowhere or knowledge that claims to remain above the concerned parties. We all had these views confirmed when we saw Jan Zalasiewicz – Mr Anthropocene himself! – share a frenzied night among the delegates, without being in any way shocked by this innovation. For he knows better than anyone how hard it is to create a consensus among scientists, and in how many delicate negotiations the geologists in the working group he heads for the Subcommission on Nomenclature of the Quaternary are embroiled![18]

So it was very important that no one claimed to represent Nature conceived in its globality and that no delegation purported to be, for example, the "voice of Gaia." Such claims would have emptied out all the politics at once. This is the point at which it becomes politically, and no longer scientifically, crucial *not* to consider Gaia to be a unified System. If Lovelock's astute approach, as I have made clear,

---

[17] Despite the presence of many students who had had both scientific and "literary" training, access to the sciences was inadequate. The innovation, however, consisted in distributing researchers among all the delegations and not keeping them apart from and above the others, as is the current case with the Intergovernmental Panel on Climate Change (IPCC).

[18] I introduced Jan Zalasiewicz at the beginning of the fourth lecture.

consisted in disaggregating the system into a multiplicity of actors, each capable of *encroaching* on the action of the others, the political translation of a similar disaggregation of agents has to be achieved in order for the *encroachments* of territories on one another to become clearly visible at last.[19] Hence the importance of multiplying (in the of course limited framework of the reduced model) the erstwhile beings of nature. It is at this point that, in place of the ancient relation between the order of a society and the natural order that would serve as its framework, the order of a human geography layered on top of a physical geography, we begin to define the borders between friend and enemy and thus to trace the front lines of the territories in conflict.

*

Little by little, we are slipping from traditional conflicts between nation-states to conflicts between territories. The pluralism of the delegations, all equally legitimate, gave a clear sense that the relations between the different ways of interweaving interests are finally going to become truly conflictual, since there is no longer any way out. The students were not trying to establish a new version of the *Whole Earth Catalog*.[20] What interested them, on the contrary, was something like *land redistribution*, the fictional equivalent of an immense agrarian reform! From that point on, the concerned parties really got caught up in the game. Even if, in the language of governance, the term "stakeholders" seems rather feeble, to rediscover its virulence it suffices to stress the *stake*, the portion, the land that each tries to *hold to*, and to remember how many others try to *grab it away* from those who *hold* it. If territory-holders proliferate, it becomes harder and harder to remain in the position of *stakeholder*. Indeed, this was the experience of the delegations from the nation-states; they found other stakeholders interrupting them at every stage. Here we see the parallel with the revolutionary situation that I couldn't help evoking earlier: the moment when the traditional orders refused to meet *separately*.

---

[19] Superimposition, penetrability, overlap: this is the essential point of the reterritorialization of the New Climate Regime. Without this, we fall back into identities separated by borders while continuing to dream of a global world. We fall back into the schema of the parts and the Whole.

[20] See Diedrich Diederichsen and Anselm Franke, eds, *The Whole Earth Catalog: California and the Disappearance of the Outside* (2013): the book offers a fascinating review of the history of this catalog, which played such an important role in the 1980s, starting of course from the Whole taken as a unifying *a priori* principle.

Still, the scene of conflict constructed this way would have been of little interest if those who developed the concept had limited the non-state delegations to the traditional "material" objects. We would have inevitably returned to the oppositions between Humans and Nature, falling back on the old Nature/Culture dualism that would have paralyzed the whole discussion. It would have been impossible to struggle against this scheme – we know how powerful it is – without the introduction of non-state delegations that did not define themselves as heirs of the "material" objects endowed with speech at last. Thus it was crucial that delegations such as "Cities," "Indigenous Peoples," or "Non-Governmental Organizations" come in to defend their own stakes in their own voice.[21] This is when we begin to understand that what is contributed by the non-state delegations is not "concern for nature" but corrosive action against the delimitation of territories that nation-states continue to believe are theirs alone. If "Lands," "Atmosphere," or "Oceans" can still appear as the (ex-natural) framework of a government of men, the claims of "Cities," "NGOs," or "Indigenous Peoples" to govern likewise intervene directly to erode the very logic of the exercise of power, as well as its administrative projection onto a two-dimensional map.

And yet we remain aware that even these innovations would not suffice to make the simulation realistic. There are in fact certain powers that always act in obscure or devious ways and that seem to toy with the political activity of the unfortunate nation-states, which have become mere marionettes in their hands. These are powers that are brought together as a whole when someone talks about "globalization." These are the powers said to act surreptitiously and that are belittled by being called lobbies – or even mafias. "Well," the organizers tell themselves, "if these powers act, if they oppose one another, if they are concerned parties or, better, *land-grabbing* parties, then they must not stay outside, they should come inside, with equal sovereignty, so that we can find out at last how they define their territory, who are their friends and their enemies, and for what cause they are ready to fight, to the death if necessary – which generally means the death of the other concerned parties." Thus the inclusion on the list of delegations representing "Economic Powers," "International Organizations," and even one of the strangest but also one of the most effective delegations, representing "Stranded

---

[21] Certain delegations occupied an intermediate position between a classic geographic definition and a plurinational definition, such as the Arctic, the Sahara, and the Amazon. This corresponds more or less to reality as shown by François Gemenne in *Géopolitique du changement climatique* (2009).

Petroleum Assets," capable of ruining the other countries by reducing their petroleum-based wealth to zero.²²

You can understand, now, what is constitutive about these innovations. In the real COP, all these interests, these position-takers, have a place, but it is located outside the main negotiating room, and it takes the form of countless campaigns for influence: lobbying, publicity, side events. In the negotiating room, on the other hand, there are only nation-states, supposedly all equal. Inside, according to a strict protocol, the countries try to reduce the impact of remote consequences – what carbon dioxide emissions are doing to the machinery of the climate – by seeking a consensus; outside, the other parties, all of which have become pressure groups, fight in great disorder about the proximate causes. At the Théâtre des Amandiers, the organizers decided to place all the parties inside, so that there would be no more "outside," and so that the position-takers could be seen exercising their pressures *all together*, so that all could fight under their own colors.²³

The rule for composition is of the utmost simplicity: every time someone characterizes a problem posed for governments as *transversal*, the organizers will try to insert it into the simulation by giving it power, representation, and a voice. In other words, if you want to take one position away from another, then participate in the redistribution, but *show your hand*.²⁴ Following this principle, it is necessary to decide on the delegations not according to the plausibility of their more or less conventional representation – "Land" or "City," "Atmosphere" or "Congo," "NGO" or "Arctic" – but according to their capacity to *oppose the others by making explicit what territory they occupy*. If one party is capable of taking the territory of another because that other is already occupying, invading, or restricting it, then that party will be granted equal sovereignty. It will not have to act surreptitiously; it will have to introduce itself and state its interest, indicate its war aims, specify its friends and its enemies – in short, say *where it is*, what allows it to *distance itself* from the others. In so

---

²²This delegation was inspired by the Territorial Agency project developed by John Palmesino and Ann-Sofi Rönnskog, "Oil Left in the Ground." See Palmesino and Rönnskog, "Radical Conservation: The Museum of Oil" (2016). The same data were used for the Museum of Oil project in the *Reset Modernity!* Exhibition (2016).
²³See Walter Lippman, *The Phantom Public* (1925).
²⁴This is what has been done so effectively by Richard Heede in "Tracing Anthropogenic Carbon Dioxide and Methane Emissions to Fossil Fuel and Cement Producers, 1854–2010" (2014); Heede has managed to define the "entities" that are actually most responsible for CO_2 emissions.

doing, it will make visible to the others the territory that it occupies or that preoccupies it.

What seems to me to justify the connection with a constitutive episode is this reorganization of the lighting system, so to speak, that renders visible front lines between territories that were invisible before. This is what allowed the students to discover that they were indeed in a state of war and that the negotiation had nothing to do with the mere distribution of CO_2 quotas under the implicit arbitration of a State of Nature. Whereas Hobbes had to invent a politics after decades of dreadful civil wars, the paradox of climate negotiations is that the protagonists have to be made to understand that they are indeed at war, while they believe that they are in a situation of peace!

What does that change, you ask? Everything. Any geopolitical manual will confirm this: every time one great power has seen the emergence of another, the rest have had to recalculate their interests from scratch (as Spain once had to adjust to the emergence of the Netherlands or, today, as the United States has to adjust to the rise of China). This is what textbooks call the balance of power or the discordant concert of nations.[25] Imagine how this balance wobbles, when "Cities" and "Lands" start to claim their due; imagine the powerful music that has them stamping their feet! Isn't there something here that might warm up the State, that "cold monster," by getting it to dance?

What the simulation allowed us to test is the idea that there are two possible directions for governing in a period of ecological mutation: up or down. Up, by appealing to a higher common principle, to the State of Nature. Unfortunately, not only does this latter not exist, but the appeal depoliticizes the entire negotiation, turning it into the simple application of distribution rules. Down, by agreeing to have no sovereign arbiter but by treating all the stakeholders as having an equal degree of sovereignty. The first direction is utopian, in the etymological sense of "no place"; the second consists in giving oneself a ground. Such a situation does not exist either, you say? This is true, but at least it makes it possible to repoliticize the negotiation through what is most essential about it: belonging to a territory, to a land, to a soil. If democracy has to start over, it will have to begin at the bottom. The soil is a good starting point: there is nothing lower! You wanted to work from the bottom up? Well then, here you are!

You may remember General de Gaulle's words: "We found occupying the comfortable chairs in the club of the great powers as many

[25] See Frédéric Ramel, *Philosophie des relations internationales* (2011).

hallowed egotisms as there were charter members."²⁶ Realism in geopolitics requires never believing that one will be able to demand that the "registered members" give up their "hallowed egotisms," their "sacrosanct self-interests," for the higher good of all. Realism in Gaia-politics makes it possible to demand at least that the stakeholders define in different terms what that self-interest is supposed to defend to the death, by modifying precisely *the territory that it is a matter of defending*. After all, the same General de Gaulle well knew that, to defend his fatherland by choosing to remain with his weapon at his feet, immobile behind the Maginot Line, or by mobilizing divisions of armored tanks, was not at all to remain faithful to the same "hallowed egotism" – or to the same fatherland.

This is the major innovation of the May 2015 simulation: if one cannot abandon the narrow defense of one's self-interests, is it feasible to *lengthen* the list of entities in which one is directly interested? If the nation-states find themselves affected by other delegations who claim to be exercising their authority over the same ground, or over portions of the same ground, how are they going to react? How are they going to modify the definition of what they value most of all? You enter into the negotiation with one idea of your interests, you come away with a different idea. Responding to Realpolitik with Realpolitik squared: isn't this, in essence, how the "brilliant art of diplomacy" is learned?²⁷

*

For the simulation at the Théâtre des Amandiers to make it possible to institute or inaugurate Gaia, the delegates would have had to achieve two other goals set by the planners; unfortunately, they didn't manage to reach either one. The delegates were to have been asked to find appropriate ways to *visualize* the new forms of overlapping sovereignty that they were exploring. And, during a final ceremony, the old nation-states were to have *redefined* their sovereignty in front of the other delegations. If the new *nomos* of the Earth is to be more than a fleeting vision, these are the tasks that must be tackled in order to complete the exercise.

You may recall that we have already come up against the difficulty of giving precise limits to "hallowed egotism." In the third lecture, I tried to show you how Lovelock had made fun of the strange idea

---

[26] *The Complete War Memoirs of Charles de Gaulle* ([1959] 1998), p. 729.
[27] "The brilliant art of diplomacy," cited in mid-crisis by president Jennifer Ching.

of the selfish gene, not because he doubted that living beings took an avid interest in their fate – how could it be otherwise? – but because he doubted that anyone could assign assured limits to their own interests. It is the very distinction between an organism and its environment that the Gaia theory calls into question. Here again we have the problem of gauging the selfishness – still just as "sacrosanct" – not of organisms, now, but of the Great Powers. In the context of the Theater of Negotiations, what the emergence of Gaia obliges us to reconsider is the distinction between a nation-state and its environment. That nation-states and genes have something in common can no longer surprise us since, in each case, we are still borrowing the notions of limits and calculations from organizational theory, from economics, from accounting formats. Tracing the limits of interests is the most directly political activity there is.[28] Here is where the question of the distribution of agency (which is basically the only subject of these lectures) is always settled.

Contrary to what is generally believed, the famous tragedy of the commons does not arise from the inability of individuals to forget their selfish interests because they are unable to devote themselves over time to the "good of all."[29] The tragedy comes from the recent belief that the interest of the individual – nation-state, animal, human, it hardly matters – can be calculated in only one way, by placing the entity on a territory that belongs to it exclusively and over which it reigns with sovereignty, and by shunting to the "outside" everything that *must not* be taken into account. The novelty as well as the artificiality of this type of calculation is well brought out by the technical term "externalization" – a precise synonym for *calculated negligence*, and consequently for irreligion.[30] To get back to the common world, and perhaps also to the sense of the common (that is, to common sense!), the solution is not to appeal to Totality, which in any case does not exist, but to learn to represent differently the territory to which one belongs. This would then make it possible to modify what

---

[28] Just as localization in space and time is the most formal of the operations that nevertheless purport to define matter (as Whitehead demonstrates), the formatting of individual interests apart from their "context" is the most political operation there is, even though it purports to define the somehow autochthonous self-evidence of human interests. The problem is the same in physics and social science alike, and the two procedures arose at the same time, in the seventeenth century. See Latour 2016a.
[29] See Elinor Ostrom, *Governing the Commons: The Evolution of Institutions for Collective Action* (1990).
[30] See Michel Callon, "An Essay on Framing and Overflowing: Economic Externalities Revisited by Sociology" (1998a). On negligence as the antonym for religion, see the citation from Michel Serres in the fifth lecture.

one is claiming to defend in the name of hallowed egotism. It is finally a matter of *internalizing* the countless encroachments of the entities on which we depend – to an extent that we are gradually discovering – for our own subsistence.[31]

In geopolitical terms, the question then comes down to visualizing *several overlapping authorities* on the same ground. The Dutch, for example, have proved able to choose, at the same moment, ever since the thirteenth century, deputies called to represent human subjects, but also representatives to serve on the National Water Authority (*Rijkswaterstaat*), whose decisions are followed attentively by dairy and poultry farmers as well as tulip-growers.[32] You will object that there is nothing astonishing in the fact that a country built artificially by means of dikes and polders should give the powers of seas and rivers a degree of representation worthy of their sovereignty. After all, if the Masters of Water make errors in their calculations, all Holland will disappear, swallowed up under the North Sea as surely as Atlantis. Where it is a question of life and death, it is normal for Water to exercise acknowledged domination, and for it thus to be represented by the intermediary of a power that is added to, opposed to, superimposed on, that of monarch and parliament. This is proof, in any case, that there is no obstacle to imagining on a single plot of land sovereignties that encroach upon one another as surely as those of the pope and the emperor in the Middle Ages.[33]

There is obviously nothing natural in such an arrangement. To be convinced of this, it suffices to compare the situation with that of almond growers in California's Central Valley. They too depend so totally on the powers of water that their green valley ought to be nothing but a sandy desert scorched by the sun.[34] But as there is no one to represent the aquifer from which they blithely pump deeper and deeper in periods of drought, all farmers steal their

---

[31] Let us remember that the difficulty of defining an individual is the same in biology, ecology, economics, and politics: see Scott Gilbert, Jan Sapp, and Alfred Tauber, "A Symbiotic View of Life: We Have Never Been Individuals" (2012).

[32] Wiebe E. Bijker, "The Politics of Water: The Oosterschelde Storm Surge Barrier: A Dutch Thing to Keep the Water Out or Not" (2005).

[33] This is one of Schmitt's essential points in *The Nomos of the Earth in the International Law of the Jus Publicum Europaeum* ([1950] 2003): it is by no means a question of separate domains – contrary to what has happened starting with Hobbes – but one of a principle of overlapping influence over the same affairs by distinct forms of power. This same principle presides over the "constitutional revision" I proposed in *Politics of Nature: How to Bring the Sciences into Democracy* ([1999] 2004b).

[34] It is an artificially produced desert, since the area was a vast humid zone systematically destroyed after colonization.

neighbors' water, to such an extent that the ground level is literally sinking beneath their feet, offering the best caricature there is of the tragedy of the commons.[35] Those who have seen the film *Chinatown* know that tracing the entangled interests is not risk-free.[36] Unlike the Dutch, the Central Valley farmers have been economized[37] – modernized, naturalized, materialized, the adjective hardly matters – to the point of finding themselves helpless as they face the phenomenon of a calamity said, quite wrongly, to be "natural": they have neither enough water nor enough skill to take charge of the situation.[38] It is odd that Californians are still ignoring the procedures of the ancient commons, which over millennia had invented clever arrangements for distributing water to all interested parties, taking droughts in their stride. Or, rather, it is tragic, actually, to note that people can intentionally lose a competence so essential to their own survival – which is proof enough that, however "sacrosanct" self-interest may be, that does not make it lucid!

In the case of the Central Valley, the difficulty of representation is twofold: for a geologist, there is nothing harder to map than an aquifer whose limits never correspond neatly to official land surveys.[39] But even if one could produce an accurate map, how could the water be represented *without the fiction* of a representative, a public servant, an officer, an intermediary who would speak in its name, and especially who could speak *face to face* with the rugged California farmers? The fiction resides not in giving water a voice but in believing that one *could get along without* representing it *by a human voice* capable of making itself understood by other humans. The error does not lie in claiming to represent nonhumans; we do that in any case all the time when we talk about rivers, voyages, the future, the past, States, the Law, or God. The error would lie in believing it possible to take such interests into account without a human who embodies, *personifies, authorizes, represents* their interests. This personification, so necessary to the Leviathan if it is to exit from the state of

---

[35] See Matt Richtel, "California Farmers Dig Deeper for Water, Sipping Their Neighbors Dry" (2015). On the geohistorical context of the current crisis, see John McPhee, *Assembling California* (1993).
[36] Roman Polanski (1974).
[37] The performative powers of economization are what allow us to shed the idea that *Homo oeconomicus* is a "native." See the now classic book by Donald MacKenzie, *An Engine, Not a Camera: How Financial Models Shape Markets* (2006).
[38] A general point in time of artificial droughts, as shown by Mike Davis, *Late Victorian Holocausts: El Niño Famines and the Making of the Third World* (2002).
[39] Thanks to Professor Roger Banes, who heads the critical zone observatory of South Sierra, for allowing me to visit his site in July 2015.

nature, is even more indispensable for the territories in conflict that are trying to put an end to the State of Nature.[40]

Now you understand why I have insisted so much on the continuity to be established among agents in what I have called a metamorphic zone. There is not an objective aquifer as defined by geology, then a legal aquifer as defined by the complex laws related to the land, and, over and above it, still another, a political aquifer governing California water. There are no levels; the world is not a layered puff pastry. The water of the Central Valley aquifer loses or gains its properties, its attributes, according to the way it is associated with other agents. The water externalized by each drilling event decided on "freely" by each independent property owner is not at all the same as the water patiently surveyed by the *Rijkswaterstaat* in the Netherlands. Because it is not well represented, it does not have the same *properties*, either, and consequently not the same *proprietors*; thus the water cannot be *appropriated* by the interested parties, treated as a substance over which they can claim ownership, and be seized as *property*. It is in a sense rejected, deanimated water – and it soon fades away like water in a mirage. This water is in the literal sense utopian.

Here we can see all the practical consequences of what we studied in the sixth lecture under the term *immanentization*, that curious way of simultaneously escaping immanence through a misplaced appeal to transcendence, and escaping transcendence by a too-hasty short-circuit into immanence.[41] It is this very strange, very modern, also very perverse mix that gives humans the impression that they are receiving a good that they are due in an *infinite* quantity for an *infinite* time – as if it had fallen from Heaven – and that at the same time *is going to disappear* – as if, literally, it had sunk down under the earth. It is this mix that makes those who believed they had the right to possess it forever fall from infinite enthusiasm for the future into deep despair over the errors of the past. The exact opposite, consequently, of Dutch water, which is *well governed* and thus *delimited*, or, as we say, *appropriated*. "Good government" of water, lands, air, cities, or economies requires a *representative government*, and thus spokespersons, emblems, figures, to whom one can speak face to face. With "bad government," this is impossible. Ever since Lorenzetti

---

[40] This play of personification is the topic of chapter 16 in Hobbes's *Leviathan* ([1651] 1998, p. 107): "From hence it followeth, that when the actor maketh a covenant by authority, he bindeth thereby the author, no less than if he had made it himself."
[41] See Eric Voegelin, *The New Science of Politics: An Introduction* ([1952] 2000a), and my summary in the sixth lecture.

painted his fresco in Siena, we have known that only by erecting such figures can we "conjure away fear."[42] Why has what people knew how to paint in the fourteenth century been so completely forgotten in the twenty-first?

The problem with "ecological questions," to use an outdated term, is that they seem to speak of objects that have been beamed into utopia as well as into uchronia. Neither water nor land nor air nor living beings are in the time or in the space of those who make them the framework for their actions. We're familiar with the debate, as old as the very idea of geopolitics, over the existence or non-existence of "natural boundaries" – the Rhine, the Urals, or the Rubicon. After all we have put (the notion of) "nature" through, it goes without saying that this kind of limit can no longer enable us to stabilize relations among agents. Yet we still have to face the task of tracing the limits of these agents. These limits cannot be dictated from the outside simply because they are deemed to have been "objectively determined by the Laws of Nature." These limits have to be felt, they have to be generated, they have to be discovered, they have to be *decided on from within* the peoples themselves. Without decisions, as we know, there is no body politic, no freedom, and no autonomy.

This is what is interesting about the terms "planetary limits"[43] and "critical zones,"[44] these notions invented, like the Anthropocene, by scientists becoming aware that the notion of limit entails law, politics, science – and perhaps also religion and the arts. Everything that allows us to become sensitive to the retroaction of beings. With these hybrid terms, scientists are inventing a *geo-tracing* activity, which only reminds us, after all, of the old meanings of geo*graphy*, geo*logy*, geo*morphology* – that is, the writing, the inscribing, the tracing, the mapping, and the inventory of a territory. No one can belong to a land without these activities of tracking space, marking plots, tracing lines, activities identified by all those Greek terms – *nomos, graphos, morphos, logos* – that are rooted in the same *Ge, Geo,* or *Gaia.*

Unfortunately, if there is a crisis of representation, it is not only because we hesitate to give voice to the things that concern us. It is also because we are limited to the imaginary realm of the two-dimensional maps, the highlighted borders that are very useful, as

---

[42]See Patrick Boucheron, *Conjurer la peur: Sienne 1338: essai sur la force politique des images* (2013).
[43]See Will Steffen et al., "Planetary Boundaries: Guiding Human Development on a Changing Planet" (2015b).
[44]See Susan L. Brantley, Martin B. Goldhaber, and K. Vala Ragnarsdottir, "Crossing Disciplines and Scales to Understand the Critical Zone" (2007).

we know, for "making war,"[45] but very inadequate for finding our way around in the geopolitics of territories in conflict. Were we to give ourselves at last a realistic vision of our belongings, we would need a geography that we lack, a geography of the discontinuous and overlapping territories – something like a geological map with a three-dimensional view, its multiple layers embedded in one another, its dislocations, its breaks, its sinuous movements, all the complexity that geologists have been able to master for the long history of soils and rocks, but of which geopolitics unfortunately remains deprived.[46] We don't know how to represent the encroachments that are nevertheless the only way to reopen, at new costs, the question of sovereignty. Networks, alas (it's my job to know this), remain hard to read.[47] When they are projected onto the background of a map, we find ourselves once again within the limits of the old cartography, without having progressed very far.

Geohistory would need a visual representation as good as the old representations of geography and history, finally fused. It is as though every limit, every border, every boundary marker, every encroachment – in short, every feedback loop – has to be simultaneously and collectively recounted, traced, replayed, and ritualized. Each of the loops registers the unanticipated actions of some external agent that comes in to complicate human action. Owing to this reactivity, what a "territory" signifies has been totally disrupted: it is no longer the old pastoral landscape of well-marked fields on which harvests ripen slowly and reliably – "*Et in Arcadia ego.*" Far from being "land-appropriation," the *Landnahme* celebrated by Carl Schmitt, it is rather the violent *reappropriation* of all human claims *by the Earth itself* – as though "territory" and "terror" had a common root.

The Earthbound have to trace and retrace these loops endlessly by all available means, as if the old distinctions among scientific instrumentation, the emergence of a public, the political arts, and indeed the definition of civic space were in the process of disappearing. Such distinctions are far less important than this powerful injunction: act in such a way that a loop is traceable and publicly visible; if we fail to do this, we'll end up blind and destitute, with no land on which to

---

[45] An allusion to the title of Yves Lacoste's essay *La géographie, ça sert, d'abord, à faire la guerre* ([1982] 2014).
[46] Jan Zalasiewicz, personal communication, May 30, 2015.
[47] Despite the numerous efforts of Science Po's medialab to make it easier to follow the logic of networks. See the attempt by the Bureau d'Études to represent the influence of capital through networks: *An Atlas of Agendas: Mapping the Power, Mapping the Commons* (2015).

settle.⁴⁸ We'll become foreigners in our own country. Everything takes place through such loops: it is as though the threads of tragedy were woven not just by the Olympian gods of long ago but by all the agents from the beginning of time. This is the story of the Anthropocene: a truly Oedipal myth. And, unlike Oedipus, who was blind to his own actions for so long, as we face the revelation of past errors we must resist the temptation to blind ourselves anew: we must agree to look at them head on, in order to be able to face what is coming toward us with our eyes wide open.

*

The designers of the simulation had imagined a last scene, before the final signing ceremony, that would have brought together the delegates representing the governments of the nation-states, the only parties recognized by the official COP. Such an assembly would not have had the goal of finally making decisions based on what the other delegations had simply proposed; their goal would rather have been to discern what legal forms, in conformity with international law, would have to be given to the decisions taken by the other delegations. Such an innovation would have reversed the direction of sovereignty.⁴⁹ Instead of occupying the entire space, the States would have found themselves in the position of servants, facilitators, organizers, logisticians, or legal experts. The only competence for which they are truly indispensable would have been recognized – that of creating, signing, and maintaining international agreements. All the rest would have stayed in other hands. We would have had the surprise of seeing the emergence of the equivalent of a civil society encompassing the territories in conflict, which would have made the nation-state apparatus not an organ of command, any longer, but one of *service*!

---

⁴⁸Like the tsunami markers that show the extent of past cataclysms and that have been ignored or forgotten; see Martin Fackler, "Tsunami Warnings, Written in Stone" (2011). Reiko Hasegawa was kind enough to translate for me the text of one of those stones, erected in 1933: "Houses on the higher ground, happiness and joy of children and descendants / Memory of the tragedy of great tsunamis / Must not build houses below this stone / The tsunami came until here in 1896 as well as in 1933 / The district was completely destroyed, survivors counts only two for the first and four the other / Be warned no matter how many years go by" (personal communication, July 1, 2015).
⁴⁹Thus we would have reversed the scene of the 2009 COP meeting in Copenhagen, where heads of state, after unraveling all the work of negotiation, sat around a table and drafted on a blank piece of paper what seemed acceptable to them, in just a few lines. See the astonishing video of this secret negotiation captured by *Der Spiegel*: "Secret Copenhagen Climate Recording Reveals Resistance from China and India" (2010).

Would the institution of nation-states have been reduced, for all that? Not necessarily. It would have experienced a powerful shock, of course, but at bottom, starting with the opening session of the simulation at the Amandiers, the spectators watching "Cities" or "Lands" negotiate on an equal footing with "Russia" or "Brazil" had already felt the extent to which the nation-states were showing their age. In fact, these states would have liked to be freed from the impossible task of holding onto a territory protected from all encroachments, a task they have always handled very poorly, and one that hardly makes sense in the epoch of ecological mutation. In the last analysis, the nation-states would have come out actually rejuvenated. Following the historic parallel, it would have been as important as the passage from monarchy by divine right to constitutional monarchy. Who can deny the gain in civilization that made it possible to pass from the power of kings to the power of constitutional states? What an advance it would be if we could finally pass from nation-states reigning without counterforces, on land delimited by borders, to a constitutional order finally endowed with a complex system of counterforces exercised by the other delegations – those famous checks and balances so celebrated by the Humans, but that the Earthbound are still trying to find?

If it is true that the modern conception of sovereignty stems from the need to find a solution to the impossible question of the double power of religion and politics, we understand how much the state would benefit if it could get rid of a sovereignty that got off to such a bad start. A solution that was imagined in order to solve the religious problem and to seize foreign lands emptied in advance of the multiform collectives that had learned to inhabit them, the nation-state has been suffocating ever since under the burden of having to take responsibility for the whole Earth. All the more so given that, since the wars of religion, the question of sovereignty has been made still more complicated by the authority of Science with a capital S, which has had to be understood, most often, for several decades now, as that of Economics. Under the authority of this apparently worldwide but curiously deterritorialized power, nation-states *have lost the capacity to ensure the defense of their subjects*. "Globalization" means that *no one knows where to live any longer*.[50] The failure of

---

[50] Hence the astonishing reaction, visible everywhere, of falling back on identity, just at the moment when the ecological mutation is imposing the overlapping and interweaving of all agents. It is basically this crisis that Aykut and Dahan explore in *Gouverner le climat?* (2014). The question mark implies a negative response: "No, the climate cannot be governed," not only because there is no rudder (French

How to govern struggling (natural) territories?   279

the nation-states' struggles against successive globalizations has left them completely unprepared to take into account this new form of globalization *by the Earth itself*.⁵¹ In the Anthropocene epoch, the sovereign State thus turns out to be afflicted with obsolescence, just at the moment when planetary globalization is becoming, literally and not just figuratively, the planet. How can the State maintain that it has "monopoly on legitimate physical violence" in the face of the geohistorical violence of the climate?

Soon, the nation-state's claim to represent total sovereignty over a territory that in any case is escaping it will appear as strange as the claim of a king to exercise absolute power. Inevitably, nation-states will have to learn to share power. Just as inevitably, then, they will have to prepare for a reinforcement – or, let's say, a rearticulation – of what is called sovereignty. There is no reason why the same term should continue to designate the amalgam of religious, scientific, and political authorities that purports to fill, completely, a continuous space bounded by a border. The scene that I imagined at the end of the simulation was one in which sovereignty would shed that burden in order to redistribute its limits in a different way. It might end up being reinforced, provided that everything that surrounded it, everything that it had been externalizing, were included inside – as the simulation supposed.⁵² Not only the old states of nature but also what are called, quite wrongly, the supranational forces, all of which in the last analysis occupy a territory, however discontinuous it may be, that we also have to learn to map. If we are going to claim to govern what happens offshore, we are going to have to redefine the shore, the borders, the limits that will finally *contain* all the powers, in the literal sense of limiting their expansion. Can you imagine the scene? "Today, May 31, 2015: got rid of States." We would have finally made it into the twenty-first century!⁵³

---

gouvernail from the Greek *kubernētēs*, "steersman"; cf. *kubernan*, "to steer"), but because there is no governing State. This is what it means to pass from the Old to the New Climate Regime.
⁵¹This is why it would be useful to establish a new compass for politics, one that would make it possible to register positions not simply along a line going from Land to Globe but also through a third attractor, the Earth. We have attempted to make such a "triangulation" visible in Bruno Latour and Christophe Leclercq, eds, *Reset Modernity!* (2016). For a recent presentation of the argument, see www.bruno-latour.fr/node/684 (unpublished).
⁵²It is significant that both Naomi Klein, in *This Changes Everything: Capitalism vs. the Climate*, (2015), and Aykut and Dahan (2014) end with a vibrant appeal for the return of the State.
⁵³An allusion to the scene imagined by Brecht and cited in the sixth lecture, p. 185.

And it was at this point that the figure of Gaia, now less enigmatic, was to have come on stage. Unlike Nature, Gaia emerges not in order to reign in the place of all the States forced to submit to its laws but *as that which requires that sovereignty be shared*. It is as though Nature had been confused with the local, historical, sublunary *oikos* called Gaia. In an earlier epoch, when we mentioned the presence of a "natural phenomenon," as soon as someone crossed the invisible threshold of society, culture, and subjectivity, it was as though all the rest, from the innards of our bodies to the Big Bang, from the ground under our feet to the infinite expanses of the galaxies, were made of the *same matter*, belonged to the same domain, and obeyed the same intangible laws. But Gaia is not Nature. Gaia is the localized, historical, and secular avatars of Nature; or, rather, Nature appears retrospectively as the epistemological, politicized, (counter-)religious and legendary extension of Gaia. Hence the surprising reversal that results in the complete consternation of the Moderns. Nature may have been able to provide us with the hope of unifying and pacifying politics, or at least of offering a solid background for the vicissitudes of human history; Gaia does nothing of the sort. Gaia does not promise peace and does not guarantee a stable background.

Contrary to the old nature, Gaia does not play either the role of inert object that could be appropriated or the role of higher arbiter on which, in the end, one could rely. It was the old Nature that could serve as a general framework for our actions even as She remained *indifferent* to our fate. It was Mother Nature who served as nurse-maid to humans capable of neglecting her as a mere inert and mute object even as they celebrated in her the *ultima ratio*. As the proverb says, "You can't do better than *Mother Nature*!" This supposedly maternal figure found itself at once below – as an object that could be manipulated and scorned – and above – as final arbiter and last judgment. All humans could do was play the role of the good child, the reasonable guardian, the rebel sure of being punished, or the respectful gardener. We can see why the offspring of this cruel and bloody stepmother have rushed straight to the psychoanalyst's couch – and why feminists have constantly challenged the myth.[54] We now understand even more clearly that Nature has no power except the power to drive her children crazy. With Nature, ecology, whether scientific or political, didn't stand a chance.

---

[54]See Charis Thompson, *Making Parents: The Ontological Choreography of Reproductive Technologies* (2005); Giovanna Di Chiro, "Ramener l'écologie à la maison" (2014); and especially Silvia Federici, *Caliban and the Witch: Women, the Body and Primitive Accumulation* (2004).

Every conception of the new geopolitics has to take into account the fact that the way the Earthbound are attached to Gaia is totally different from the way humans were attached to Nature. Gaia is no longer *indifferent* to our actions. Unlike the Humans in Nature, the Earthbound know that they are contending with Gaia. They can neither treat it as an inert and mute object nor as supreme judge and final arbiter. It is in this sense that they no longer enter into an infantile mother–child relation with Gaia. The Earthbound and the Earth have grown up. Both parties share the same fragility, the same cruelty, the same uncertainty about their fate. They are powers that cannot be dominated and cannot dominate. As Gaia is neither external nor indisputable, it cannot remain indifferent to politics. Gaia can treat us as enemies. We can respond in kind.

While Nature could reign over humans as a religious power to which a paradoxical cult, civic and secular, had to be devoted, Gaia only requires that power be shared *as secular and not religious powers*. It is useless to hope for a new *translatio imperii* that would go from God to Nature, then from Nature to Gaia. No "law of the three estates" is at work here.[55] Gaia is content to recall the more modest traditions of a body politic that finally recognizes in the Earth that through which this assembled body *solemnly agrees to be definitively bounded*. Even though, up to now, there has been no civic cult for such an outlining of the "planetary borders" that a political body would impose on itself, what we did in the simulation was offer a glimpse of such a ritual. Limits that nothing was imposing – in the sense of the old Nature – were decided on collectively – in the face of the new Gaia. This does not mean that humans have to feel guilty – guilt would paralyze them, and that would be futile – but that they have to learn to become capable of *responding*.[56] It is by making themselves capable of response, by endowing themselves with a new sensitivity, that Humans in Nature become Earthbound with and against Gaia. Here are the checks and balances, that strange technical metaphor used by constitutional law, newly repurposed as a principle of composition for agents.[57]

---

[55] An allusion to the familiar triad, invoked especially by Auguste Comte, that purports to divide the pace and evolution of history into three stages: theological, metaphysical, and positive. See Auguste Comte, *The Catechism of Positive Religion: Or Summary Exposition of the Universal Religion in Thirteen Systematic Conversations between a Woman and a Priest* ([1891] 2009).
[56] See the first lecture for Donna Haraway's use of "response-able."
[57] The technical metaphor of the regulator has always been a source of fascination in political theory. See Otto Mayr, *Authority, Liberty, & Automatic Machinery in Early Modern Europe* (1986).

This is what will allow us finally to understand the highly unsettling metaphor of feedback loops and the highly unstable use of the notion of *cybernetics*. In the very etymology of the word *cybernetics*, there is a whole *government* that purports to be holding the tiller! The question is whether the metaphor tilts toward technology, with a proliferation of server commands and *control* centers, or toward politics, with a proliferation of opportunities to hear *protests* by those who insist on reacting in response to the commands! On one side, the modern ambition par excellence is extended further and further, all the way to the nightmarish dream of geo-engineering;[58] on the other, the situation is turned to advantage, allowing for demodernization and a return back to Earth.

It all depends on what is meant by *responding* to commands. Everything that reacts to our actions is beginning to take on a consistency, a solidity, a cohesiveness that can be treated either as inert objects having the predictability of a cybernetic system, in the technical sense of the term, or else as agents that are all called to make their voices heard. How do you do react, for example, when you listen to the climate specialists, who keep on adding to their models the "response" of the ice sheets to the warming of the waters, the "response" of micro-organisms to the acidity of the oceans, the "response" of the Gulf Stream to thermohaline circulation, the "response" of the land to the loss of biodiversity? Do you think in terms of a more and more naturalized system or as a political body to be composed, one agent after another? If you make it a global system, you overanimate and you depoliticize just as surely. Can we become capable of limiting ourselves to the animation proper to the Earth, which would make it possible to redefine politics as well as nature? Is this an extension of politics? Yes, in fact, it is. Isn't it strange that we could once have thought that only humans were "political animals"? What about the animals, then, and all the animated agents?

Gaia does not possess, must not possess, the legal quality of the *res publica*, of the State, of the great artificial Leviathan invented by Hobbes. *It is from the State as well as from the State of Nature that it comes, as it were, to set us free.* If we have long pretended that we had to exit from Nature in order to be emancipated as Humans, it is in the face of Gaia that the Earthbound seek emancipation. When we begin to come together as Earthbound beings, we realize that we are convoked by a power that is fully political, because it reverses

---

[58] See Clive Hamilton, *Earthmasters: The Dawn of Climate Engineering* (2013), but see also the subtle plea from Oliver Morton to repoliticize the question: *The Planet Remade: How Geoengineering Could Change the World* (2015).

How to govern struggling (natural) territories?   283

all titles, all legal rights to occupy land and to claim to be its owner. Confronted with such a reversal of property titles, the Earthbound understand that, contrary to what the Humans have never stopped imagining, they will never play the roles of Atlas or Earth Gardener; they will never be able to serve as Master Engineers of Space Ship Earth or even as modest and faithful Guardians of the Blue Planet. It is as simple as that: *they are not alone in the command post*. Some other entity has preceded them, although they have become aware very belatedly of its presence, its precedence, and its priority. The expression *power-sharing* means just that.

So far Gaia has no legal form beyond that of addressee. While it has no sovereignty, it may at least have what the Romans called *majesty*.[59] One can address Gaia, not *as one addressed* Nature, as an impersonal but nonetheless personalized entity, but rather directly, naming it as a configuration of new political entities. To live in the epoch of the Anthropocene is to acknowledge a strange and difficult *limitation of powers* in favor of Gaia, considered as the secular aggregation of all the agents that can be recognized thanks to the tracing of feedback loops. Here, just as with the earlier invention of the political personification of the State,[60] both thought and practice need fiction: "Gaia, I name you as that which I am addressing and that which I am prepared to face."[61]

If it is always appropriate, retrospectively, to mull over the question "How would I have behaved if I had found myself among the criminals of the past century?," it is still more useful, it seems to me, to avoid finding oneself among the criminals of the present century when we are going to have to confront, to build on one of Carl Schmitt's sentences, "struggles for the ordering, appropriation, and

---

[59] I thank Pierre-Yves Condé for calling to my attention Yan Thomas's discussion of majesty and the associated concept of plenitude: "It was not yet the plenitude of a sum of competences in action, such as monarchic law must have conceived of it at the end of the Middle Ages and the beginning of the modern epoch. It was a plenitude affirmed only as untransgressible, through a prohibition. An empty place of Majesty, which projected its sanctified circle around power....The history of the Roman state, if one means by the word 'state' something more than a vague descriptive approximation, that is, if one wants to understand it in the very terms in which it was formulated in Rome, the problematics...and even more the practice of the legal construction of the One have to include the history of the crime of *lèse-majesté*. This crime is not an incident along the way, an accidental anomaly. It is on the contrary the event presupposed by the political institution built around the defense of an ultimate point of reference" (Yan Thomas, "L'institution de la majesté," 1991).
[60] See note 40 on the fiction of the person and Hobbes's quote.
[61] Hence the importance of exploring such fictions through plays, exhibitions, art forms, poetry, and maybe also rituals.

distribution of spaces and climates." Schmitt credits the *jus publicum europeanum* with limiting intra-European wars over the span of two centuries by exporting them elsewhere, before they exploded in the twentieth century, breaking through all boundaries to become worldwide. Will the Earthbound be capable of inventing a successor to this *jus publicum*, in view of limiting the wars to come? Will we be capable of placing this new law under the same ancient invocation, that of the "Earth, mother of law," an entity that Virgil saluted with the name *"justissima tellus"*? Such a shift would lead to a different mode of action for the old "laws of nature." These laws would become something like a *"jus publicum telluris,"* still to be invented, in view of limiting what Schmitt, in his terribly precise language, called the *Raumordnungskriege*, the "wars over spatial order" – an expression that, once purged of its associations with the conflicts of the twentieth century, offers a radical definition of earthly life, but an earthly life finally capable of taking the presence of Gaia into account, so that we shall be able to *limit* the extent of wars to come.

The alternatives can be presented in concise terms as follows: do we extend the hegemony of the nation-states over the Earth by giving the Moderns a new horizon of mastery – a form of eco-modernization that would be even more imperious and much more violent than all previous land-appropriations – or do we agree to bow before the majesty of Gaia while making the distribution of agency the political question par excellence – a renewal of the great question of democracy? The latter course would presumably mean giving up on the expressions "modern," "nature," and even "ecology," a relinquishment I have summed up by proposing to pass from the Old Climate Regime to the New.

*

The outcome of this battle necessarily depends on the way we make ourselves capable of inheriting religion. If it is true, as I believe along with many others, that what is called "secularization" has only reappropriated the principal feature of the counter-religions – living in the end of times – while shifting the end of times into the utopia of modernization, this means that access to the earthly has been made impossible. Even if we managed to restore a place for the sciences and to revitalize politics once again, the fact would still remain that the heirs of modernism – that is, today, the entire planet, to the extent that it is globalized – are situated in an impossible time, the time that has forever torn them away from the past and hurled them into a futureless future. Exactly the temporal situation whose obsolescence is marked by the Anthropocene.

How to govern struggling (natural) territories? 285

If we miss this fork in the path, the battle between the religious and the secular will continue. Instead of discovering materiality, the earthbound, the ordinary, the mundane, we will find ourselves in endless wars over the utopian foundations of existence – with, in addition, under the new name of fundamentalism, the return of the wars of religion from which the State was supposed to protect us. One can even imagine the worst, wars of religion waged in the name of protecting Nature! Let us recall Schmitt's argument: wars waged in the name of reason, morality, and calculations – the "just" wars – are the ones that lead to limitless extermination. Global wars waged in the name of the survival of the Globe would be much worse than the ones called "world wars." The extent, the duration, and the intensity of such wars can be *limited* only if we agree that the composition of the common world has *not yet been achieved*, that there is no Globe. How can we decide on these limits? By accepting finitude: that of politics and of the sciences, but also of religions.

I know that the usual solution consists in saying "leave religions behind and move on." But how can we cope if, in this move, we bring with us the worst religions have to offer and leave aside the antidote that they have also been able to develop? With our strange idea of the secular, we can neither return to the religious nor extricate ourselves from it. The only solution is to make a new effort to consider what the expression "counter-religion" means. If there is nothing to be done with the rump religion that has become the salvation of souls and a morality police, we really need to domesticate that ferocious invention of a time that does not pass, since in any event we have inherited it. Around the somewhat obscure questions of the end, goals, finitude, infinity, meaning, absurdity of life, and so on, there is always the religious question. To rediscover meaning in the question of emancipation, *we have to free ourselves from the infinite.*

The only way to do this, it seems to me, is to take seriously the apocalyptic dimension of which we are the descendants – the apocalypse that we have imposed on the other collectives and that is falling back on us today – but whose meaning we have lost the ability to comprehend. The question then becomes the following: can we relearn to live in the time of the end without tipping thereby into utopia, the utopia that has beamed us into the beyond, as well as the one that has caused us to lose the here below? In other words, can we return to humility three times in a row – for sciences, for politics, and for religion – instead of the deadly amalgam that has mixed up their virtues but has succeeded only in poisoning us? If you find the word "humility" shocking, remember that there is humus, and compost, in it. The Ash Wednesday phrase "Remember that you are dust and to dust you will return!" is not a curse but a blessing:

what is worth more than anything else lasts only through that which does not last.

To *live in the time of the end* is first of all to accept the finitude of the time that passes and to put an end to negligence. Before being blown up into grandiose big-budget cosmic scenarios, the radical rupture of eschatology has to be acknowledged first in a lighter, more humble, and more economical tonality. The end of time is not the Final Globe that encircles all the other globes, the final answer to the question of the meaning of existence; it is, rather, a new difference, a new line traced *inside* all the other lines, traversing them everywhere, a line that gives a different meaning to all events, a line that is a goal, a final and radical presence, a completeness. Not another world, but the same world grasped *in a radically new spirit*.

Tragically, this twist in the flow of time, this event in the event, this *eschaton* situated within the movement of history, has been metamorphosed into an escape outside of time, a leap into eternity, into that which knows no time. The Incarnation has been changed into a vanishing point far from all flesh, pointing toward the disembodied realm of a remote spiritual domain. As if the calamity of the *natural* were not enough, generations of priests, pastors, preachers, and theologians have started mistreating the Holy Gospels in order to add, above Nature, a domain of the *supernatural*. As if the (non-)existence of Nature could serve as a solid foundation to the (non-)existence of the Supernatural! The whole of religion, or at least of Christianity and its multiple avatars, has gradually been displaced toward the project of saving the disembodied souls of humans from their sinful attachment to the Earth. With eyes always turned upward, in a gaze made ecstatic by the expectation of the final event! It is in large part the belief that a pitiless battle against materialism must be waged that has led Christianity astray, forcing the faithful to disdain the path of the sciences, at the very moment when the sciences were showing the path of the Earth more clearly than the column of smoke that led the Hebrews into the desert.

The idea was not futile. Creation as an alternative to Nature made it possible to assure oneself that the power of conversion of the Incarnation was not limited to the intimate reaches of the soul, and that it could extend little by little, I ought to say neighbor by neighbor, to the entire cosmos. But only on condition that Creation *not become* another name for Nature, distinguished only by the presence of overanimated agents and governed by a providential Grand Design.[62] The

---

[62]This superficial opposition between overanimation and deanimation has been taken up in the fifth lecture.

Holy Spirit may "renew the surface of the earth," but it is powerless when it confronts a faceless Nature. It is because Gaia offers such secular, worldly, terrestrial figures that it can allow the dynamic of the Incarnation to recapture its momentum in a space freed from the limits of Nature. If we truly "know that the whole creation has been groaning as in the pains of childbirth right up to the present time" (Rom. 8: 22), this means that creation has not been completed and that it therefore must be composed, step by step, soul by soul, agent by agent.

How strange it is that the theologians who combat materialism have taken so long to understand that they are the ones who have constructed, over the centuries, a veritable cult of Nature – that is, the search for an external, immutable, universal, and indisputable entity, in contrast to the changing, local, entangled, and disputable story that we Earthbound beings inhabit. To save the treasure of Faith, they abandoned it to Eternity. In seeking to emigrate toward this supernatural world, they did not notice that what was "set aside" was not sin but that for which, according to their own story, their own God had had his own Son die, namely, the Earth of His Creation. They must have forgotten that another definition of the word "ecology" – to go back to the lovely fictional etymology proposed by Jürgen Moltmann – could be *oikos logou* – that is, the House of the Logos, that Logos which, as John's Gospel says, "has many rooms" (John 14: 2).[63] I hope you have understood that, in order to occupy the Earth, or, rather, to *be occupied and preoccupied by* the Earth, we have to inhabit all these rooms at the same time. The cosmos doesn't need us to spread out the Glory of God in it; on the contrary, it needs to see religion limiting itself in order to learn to conspire with the sciences and with politics, to restore meaning to the notion of limit.

I was without hope on this point, I confess, when I had the happy surprise of reading the encyclical of Pope Francis, someone who is capable of taking up the Canticle of the Creatures again while addressing the Earth as "mother" and "sister." I had sworn never to cite St Francis: too much sentimentality, too many good feelings. And yet, when I read "Praise be to you, my Lord, through our Sister, Mother Earth, who sustains and governs us, and who produces various fruit with colored flowers and herbs," I told myself that, between the terrifying genealogy of Gaia and the family tree set up by Pope Francis, there were perhaps links to be established that the

---

[63] Jürgen Moltmann, *God in Creation: A New Theology of Creation and the Spirit of God* (1993), p. xiv.

old quarrel over paganism seemed to have cut off forever.[64] All the more so in that the author, full of verve, made it a new version of the Communist Party Manifesto by reconnecting ecology to politics, and without belittling the sciences in the process. I then began to wonder whether Voegelin's wish might be realized at last:[65] those who had passed through all the avatars of the successive counter-religions were perhaps going to become capable of opening their souls, as Voegelin says, to a supreme authority without having to give up the others. Would it be possible, I asked myself as I read Pope Francis's call to conversion, that the intrusion of Gaia might bring us closer to all the gods? That the poet's overly celebrated statement – "Only a God can save us now!" – could be reworded: "Only the assembly of all the gods can save us now..."?

*

If, to conclude, I wanted to pull together in a lively sketch everything I've said about Gaia, I would say that nothing has been played out. The worst may happen; in particular, Gaia may be taken as a reincarnation of the old State of Nature. Imagine the catastrophe: political, scientific, and religious elites would make Gaia the power that must be obeyed, in the name of the indisputable truths of the State, Science, and Religion combined. "Gaia requires! Gaia wants! Gaia demands!" All the powers of the Globe fused in the most toxic of amalgams. The Empire of the Globe would be back! With all the totalitarianisms acting in concert, a government by Gaia would be an absolute horror. If you have followed me to this point, you will have understood that Gaia is neither a Globe nor a global figure but, rather, the *impossibility* of limiting oneself to a figure of the Globe. Gaia is historical through and through. Gaia is neither a nurturing Mother nor a cruel stepmother, indifferent or remote. It is not maternal at all! If you still have doubts about this, go back to the Gaia of Greek mythology: it is the most ambiguous, the most complex, the least stable of the past powers.[66] The contemporary Gaia that we have to face is no more a salutary divinity than the old Ge was. It obliges all divinities to reopen the question of their mode of presence. It is no more the heir to political forces than it is the heir to any forms of cosmic religion.

---

[64] Francis, *Laudato Sí: On Care for Our Common Home* (2015).
[65] See the fifth lecture, on the impossible pluralism of the Western tradition that has never been able to keep the three forms of religion together.
[66] After all, it was Lynn Margulis herself who famously exclaimed "Gaia Is a Tough Bitch" (1995).

How to govern struggling (natural) territories? 289

It is shaped by too many sciences, instrumentations, models, sensors, to resemble the old forms of access to the world. In this sense, it is as remote from the Inca earth mother goddess Pachamama as from the ancient Ge. And yet it metamorphizes the sciences effectively and will change them forever: it anthropologizes them, brings them back down to Earth, encourages their multiplicity, welcomes their instrumentation, conspires with their rediscovered modesty. Gaia requires the sciences to say where they are situated and what portion of Earth they inhabit. Gaia is no more scientific in the old style than it is an ersatz pagan of the Creation. It mistrusts paganism – that pejorative version of the old way of belonging to the world – as much as it mistrusts the notion of letting itself be transformed by the Christian religion into the providential design of a transcendent God. It mistrusts all transcendence. It does not reject design, but it wants there to be as many designs as there are actors on its Earth. It objects to any flight into the beyond. Gaia is the great figure opposed to utopia and uchronia. Gaia is the great huntress of Gnostics. Gaia is the third party in everything done by men, divinities, organisms, and gods; it is another name for Third Estate. Gaia can welcome the present, but it mistrusts the Apocalypse and everything that claims to jump to the end of time. It belittles the exaggerations of religion along with those of the sciences and politics. It wants the present to be celebrated first of all for what it is: the time that makes things last, through what does not last. Gaia is finitude, a very just and very worldly finitude. Once this is understood, you're quite free, you adherents to (counter-) religion, to add to it the time of salvation, finally realized; but *let such a fulfillment be within time*. Far from being the frog swollen with air who believes that it is bigger than an ox, Gaia is the great power of *deflation*. It is the thorn that deflates all the obsessions of the Globe. It requires of the Moderns that they stop believing that they are on the other side of the Apocalypse. It is a great figure of exegesis: reread your sacred texts, you scientists, you religious types, you politicians. With its finger, quite simply, Gaia designates the Earth.

*

I am sure you have often contemplated the admirable maps said to be in the shape of a capital T through which monks in the Middle Ages represented the world, with Jerusalem in the middle, before the maps went out of fashion with the stupefying discovery of an infinitely larger world whose shores the monks had to learn to draw. As I prepared these lectures, I often thought about the extent to which the present situation resembles that of our learned predecessors at the moment when the news reached them that Christopher Columbus, against all expectations, had returned from his travels toward China. We too draw our maps in the form of a capital T, with Man at the center and circular, global Nature surrounding, threatening, or protecting him. And we too are going to have to redraw them entirely, in order to absorb newly discovered lands that oblige us to exit completely from Nature and from Humanity while redistributing the sciences, religion, and politics – in short, while remapping our entire cosmology. What a surprise for the people of the sixteenth century, to discover how much more vast nature turned out to be than their little Mediterranean world. What a surprise for the people of the twenty-first, to discover how narrow (the notion of) nature is compared to the behavior of the Earth that is suddenly opening up under their feet.

There is no point soothing ourselves with illusions: we are as ill-prepared for the upheavals to come in the image of the world as was Europe in 1492. All the more so in that, this time, it is not the expansion of space we have to prepare for, not the discovery of new lands emptied in advance of their inhabitants, that gigantic land grab that made possible what has long been called the "Western expansion." We are still dealing with space, with the earth, with discovery, but it is the discovery of a new Earth considered in its *intensity* and no longer in its *extension*. We are not stunned spectators witnessing the discovery of a New World at our disposal; we are rather witnessing the obligation to relearn completely the way we are going to have to inhabit the Old World![67] The novelty is all the greater and our surprise all the more complete in that we are no longer the ones chasing the earlier peoples from their lands; it is our own land, ours too, that is being taken. Or, rather, it seems that all the formerly human peoples are finding themselves simultaneously the object of a reverse appropriation of land, by the Earth itself. Moreover, all these reversals are still so obscure that we know what has befallen us no better than Columbus did when he returned from Hispaniola, which he had mistaken for the shores of China! As I end these lectures, I am not even sure of the quality of the news I have relayed in telling you that the Anthropocene was going to modify our ways of life – could it be just a rumor?

What is certain is that, while humans of the modern species could be defined as those who always emancipated themselves from the constraints of the past, who were always trying to pass through the impassable Pillars of Hercules, conversely, the Earthbound have to

---

[67] See the citation from Carl Schmitt discussed in the seventh lecture, p. 232.

How to govern struggling (natural) territories? 291

explore the question of their limits. Whereas the Humans had *"Plus ultra"* as their motto, the Earthbound have no motto but *"Plus intra."* They cannot rely on any other, older version of what the land, the earth, the terrain represented. Not because they fear being reactionary and retrograde (retrogressing is what they *stopped* doing when they stopped believing that they were modern!),[68] but because there is no way to shrink their ways of life, their technologies, their values, their multitudes, their cities, to fit within the narrow limits of what "belonging to a country" means. Paradoxically, in view of determining their limits, the Earthbound have to pull themselves away from the limits of what they used to consider space: the narrow countryside they were so eager to leave behind, as well as the utopia of indefinite space they were so eager to reach. Geohistory requires a change in the very definition of what it means to have, hold, or occupy a space, of what it means to be appropriated by an earth.

The transformative power of billions of people might be able to discover the problem that the politics of Nation-States could not envisage. Where might we discover the "four planets" necessary to our progress and our development, if not in the curves and crevices of Gaia itself:[69] namely, the *interior* of the planetary borders, enveloped *in* their multiple worlds, and *because* we shall learn to maintain our activity within limits agreed on deliberately and politically? Here is where the transcendence of religion lies, deep within human souls; it is here that the sciences and technology reside, deep within the countless narratives intermingled with *all* the events of *all* the agents in *all* the deviations and folds of Earth's *natural history*; here is where the resources of politics are found, underlying the indignation and revolt of those who cry out as they see the ground slipping away beneath their feet. What the motto *Plus intra* designates is also, in a way, a path for progress and invention, a path that connects the natural history of the planet with the sacred history of the Incarnation, and with the revolt of those who are going to learn never to hold still on the pretext that one has to obey the laws of nature.

It is always the proud old injunction: "Go on! Go on!," not toward a new earth, but toward an earth whose face must be renewed. You know that Christopher Columbus took his first name, "bearer of Christ," very seriously, and that he was convinced that he was

---

[68] It is the dancer's turning around that has served as our index from the beginning; see the first lecture and the seventh.
[69] According to the rough calculations of the World Wildlife Fund's 2014 *Living Planet* report, it would take more than four planets, calculated in terms of "global acreage," to allow all human beings to enjoy the lifestyle of North Americans.

helping his God cross the Atlantic in the same way that the legendary Christopher had allowed the child Jesus to cross the river. No one can believe any longer that we have solid enough shoulders to bear such a weight. Rather, we should agree to put less weight on the back of what is bearing us across the ford of time, namely, Gaia.

As far as we may be from Captain Columbus's spirit of conquest, perhaps we are nevertheless still like the thirsty sailors aboard his caravel, waiting day after day for the cry that the lookout will surely end up shouting some morning, from up in the crow's nest: "Land ho! Land ho!"

References

Abram, David (1996) *The Spell of the Sensuous: Perception and Language in a More-than-Human World*. New York: Pantheon Books.
Acot, Pascal, ed. ([1988] 1998) *The European Origins of Scientific Ecology*, trans. B. P. Hamm, 2 vols. Amsterdam: Gordon & Breach.
Aït-Touati, Frédérique (2012) *Fictions of the Cosmos: Science and Literature in the Seventeenth Century*, trans. Susan Emanuel. Chicago: University of Chicago Press.
Allègre, Claude, et al. (2012) "No Need to Panic about Global Warming: There's No Compelling Scientific Argument for Drastic Action to 'Decarbonize' the World's Economy," *Wall Street Journal*, January 27.
Anders, Günther ([1981] 2006) *La menace nucléaire: considérations radicales sur l'âge atomique*, trans. Christophe David. Paris: Le Serpent à plumes.
—— (2007) *Le temps de la fin*. Paris: L'Herne.
"The Anthropocene Curriculum," www.anthropocene-curriculum.org/.
Archer, David (2010a) *The Global Carbon Cycle*. Princeton, NJ: Princeton University Press.
—— (2010b) *The Long Thaw: How Humans Are Changing the Next 100,000 Years of Earth's Climate*. Princeton, NJ: Princeton University Press.
Ashmore, Malcolm, Derek Edwards, and Jonathan Potter (1994) "The Bottom Line: The Rhetoric of Reality Demonstrations," *Configurations* 2(1): 1–14.
Assmann, Jan (1998) *Moses the Egyptian: The Memory of Egypt in Western Monotheism*. Cambridge, MA: Harvard University Press.
—— (2009) *Violence et monothéisme*. Paris: Bayard.

—— (2010) *The Price of Monotheism*, trans. Robert Savage. Stanford, CA: Stanford University Press.
Attali, Jacques (2015) "À quoi peut encore servir la COP 21?" *L'Express*, March 16.
Auerbach, Erich (1959) *Figura*. New York: Meridian Books.
Austin, J. L. ([1955] 1962) *How to Do Things with Words*. Cambridge, MA: Harvard University Press.
Aykut, Stefan, and Amy Dahan (2014) *Gouverner le climat? Vingt ans de négociations internationales*. Paris: Presses de Sciences Po.
Bachelard, Gaston (1998) *Le rationalisme appliqué*. Paris: Presses Universitaires de France.
Banwart, S. A., J. Chorover, and J. Gaillardet (2013) *Sustaining Earth's Critical Zone: Basic Science and Interdisciplinary Solutions for Global Challenges*. Sheffield: University of Sheffield.
Barnes, Barry, and Steven Shapin, eds (1979) *Natural Order: Historical Studies of Scientific Culture*. Beverly Hills, CA: Sage.
Bastaire, Hélène, and Jean Bastaire (2010) *La terre de gloire: essai d'écologie parousiaque*. Paris: Le Cerf.
Bastide, Françoise (2001) *Una notte con Saturno: scritti semotici sul discorso scientifico*, ed. Bruno Latour. Rome: Meltemi.
Beck, Ulrich ([1986] 1992) *The Risk Society: Towards a New Modernity*, trans. Mark Ritter. Newbury Park, CA: Sage.
Benjamin, Walter ([1940] 1969) "Theses on the Philosophy of History," in Benjamin, *Illuminations*, ed. Hannah Arendt, trans. Harry Zohn. New York: Schocken Books, pp. 253–64.
Bensaude-Vincent, Bernadette, and Isabelle Stengers ([1992] 1996) *A History of Chemistry*, trans. Deborah van Dam. Cambridge, MA: Harvard University Press.
Biagioli, Mario (1993) *Galileo, Courtier: The Practice of Science in the Culture of Absolutism*. Chicago: University of Chicago Press.
—— (2006) *Galileo's Instruments of Credit: Telescopes, Images, Secrecy*. Chicago: University of Chicago Press.
Bijker, Wiebe E. (2005) "The Politics of Water: The Oosterschelde Storm Surge Barrier: A Dutch Thing to Keep the Water Out or Not," in Bruno Latour and Peter Weibel, eds, *Making Things Public*. Cambridge, MA: MIT Press, pp. 512–29.
Blumenberg, Hans ([1976] 1983) *The Legitimacy of the Modern Age*, trans. (from 2nd edn) Robert M. Wallace. Cambridge, MA: MIT Press.
—— ([1979] 1997) *Shipwreck with Spectator: Paradigm of a Metaphor for Existence*, trans. Stephen Rendall. Cambridge, MA: MIT Press.
Bonneuil, Christophe, and Jean-Baptiste Fressoz, eds (2016) *The Shock of the Anthropocene: The Earth, History, and Us*, trans. David Fernbach. Brooklyn, NY: Verso.
Bonneuil, Christophe, and Pierre de Jouvancourt (2014) "En finir avec l'épopée: récit, géopouvoir et sujets de l'anthropocène," in Émilie Hache, ed., *De l'univers clos au monde infini*. Paris: Dehors, pp. 57–108.

Bonneuil, Christophe, and Dominique Pestre, eds (2015) *Histoire des sciences et des savoirs*, vol. 3: *Le siècle des technosciences*. Paris: Seuil.
Boucheron, Patrick (2013) *Conjurer la peur: Sienne 1338: essai sur la force politique des images*. Paris: Seuil.
Boureux, Christophe (2014) *Dieu est aussi jardinier*. Paris: Le Cerf.
Bourg, Dominique (2010) *Vers une démocratie écologique: le citoyen, le savant et le politique*. Paris: Seuil.
Brahami, Frédéric (2001) *Le travail du scepticisme: Montaigne, Bayle, Hume*. Paris: Presses Universitaires de France.
Brantley, Susan L., Martin B. Goldhaber, and K. Vala Ragnarsdottir (2007) "Crossing Disciplines and Scales to Understand the Critical Zone," *Elements* 2: 307–14.
Brecht, Bertolt ([1945] 2001) *The Life of Galileo*, trans. John Willett, ed. John Willett and Ralph Manheim. London: Methuen.
Bredekamp, Horst ([1992] 1995) *The Lure of Antiquity and the Cult of the Machine: The Kunstkammer and the Evolution of Nature, Art, and Technology*, trans. Allison Brown. Princeton, NJ: Markus Wiener.
—— (2003) *Stratégies visuelles de Thomas Hobbes: Le Léviathan archétype de l'État moderne*. Paris: Maison des Sciences de l'Homme.
Broecker, Wallace S. (1995) "Ice Cores – Cooling the Tropics," *Nature*, July 20: 212–13.
Bronstein, David (2015) *Climate: Make It Work*. Les Films de l'Air.
Brotton, Jerry (2012) *A History of the World in Twelve Maps*. London: Allen Lane.
Bureau d'Études (2015) *An Atlas of Agendas: Mapping the Power, Mapping the Commons*. London: Anagram Books.
Callon, Michel (1986) "Some Elements of a Sociology of Translation: Domestication of the Scallops and the Fishermen of St Brieux Bay," in John Law, ed., *Power, Action, and Belief: A New Sociology of Knowledge?* London : Routledge & Kegan Paul, pp. 196–229.
—— (1998a) "An Essay on Framing and Overflowing: Economic Externalities Revisited by Sociology," in Michel Callon, ed., *The Laws of the Market*. Oxford: Blackwell, pp. 245–69.
——, ed. (1998b) *The Laws of the Markets*. Oxford: Blackwell.
——, ed. (2013) *Sociologie des agencements marchands: textes choisis*. Paris: Presses de l'École nationale des Mines.
Cartwright, Nancy (1983) *How the Laws of Physics Lie*. Oxford: Clarendon Press.
—— (1999) *The Dappled World: A Study of the Boundaries of Science*. Cambridge: Cambridge University Press.
Chabard, Pierre (2001) "L'Outlook Tower, anamorphose du monde," *Le Visiteur* 7: 64–89.
Chakrabarty, Dipesh (2009) "The Climate of History: Four Theses," *Critical Inquiry* 35(2): 197–222.
—— (2012) "Postcolonial Studies and the Challenge of Climate Change," *New Literary History* 43(1): 1–18.

—— (2014) "Climate and Capital: On Conjoined Histories," *Critical Inquiry* 41(1): 1–23.
Chamayou, Grégoire (2013) *Théorie du drone*. Paris: La Fabrique.
Charbonnier, Pierre (2015) *La fin d'un grand partage: de Durkheim à Descola*. Paris: Presses du CNRS.
Charlson, Robert J., James E. Lovelock, Meinrat O. Andreae, and Stephen G. Warren (1987) "Oceanic Phytoplankton, Atmospheric Sulphur, Cloud Albedo and Climate," *Nature*, April 16: 655–61.
Charvolin, Florian (2003) *L'invention de l'environnement en France: chroniques anthropologiques d'une institutionnalisation*. Paris: La Découverte.
Chen, Angus (2014) "Rocks Made of Plastic Found on Hawaiian Beach," *Science*, June 4, www.sciencemag.org/news/2014/06/rocks-made-plastic-found-hawaiian-beach.
Clark, Christopher M. (2013) *The Sleepwalkers: How Europe Went to War in 1914*. New York: Harper.
Clarke, Bruce (2012) "Gaia Is Not an Organism: Scenes from the Early Scientific Collaboration between Lynn Margulis and James Lovelock," in Dorian Sagan, ed., *Lynn Margulis: The Life and Legacy of a Scientific Rebel*. White River Junction, VT: Chelsea Green, pp. 32–43.
—— (2014) *Neocybernetics and Narrative*. Minneapolis: University of Minnesota Press.
——, ed. (2015) *Earth, Life, and System: Evolution and Ecology on a Gaian Planet*. New York: Fordham University Press.
Cline, Eric H. (2014) *1177 B.C.: The Year Civilization Collapsed*. Princeton, NJ: Princeton University Press.
Commoner, Barry (1971) *The Closing Circle*. New York: Random House.
Comte, Auguste ([1891] 2009) *The Catechism of Positive Religion: Or Summary Exposition of the Universal Religion in Thirteen Systematic Conversations between a Woman and a Priest*, trans. Richard Congreve. Cambridge: Cambridge University Press.
Conant, James Bryant (1952) *Pasteur's Study of Fermentation*. Cambridge, MA: Harvard University Press.
Conway, Philip (2016) "Back Down to Earth: Reassembling Latour's Anthropocenic Geopolitics," *Global Discourse* 6(1–2): 43–71; http://dx.doi.org/10.1080/23269995.2015.1004247.
Crary, Jonathan (1999) *Suspensions of Perception: Attention, Spectacle, and Modern Culture*. Cambridge, MA: MIT Press.
Cronon, William, ed. (1996) *Uncommon Ground: Rethinking the Human Place in Nature*. New York: W. W. Norton.
Cruikshank, Julie (2010) *Do Glaciers Listen? Local Knowledge, Colonial Encounters, and Social Imagination*. Seattle: University of Washington Press.
Crutzen, Paul J., and Eugene F. Stoermer (2000) "The 'Anthropocene,'" *Global Change Newsletter* no. 41: 17–18.
Dahan, Amy, and Michel Armatte (2004) "Modèles et modélisations, 1950–2000: nouveaux pratiques, nouveaux enjeux," *Revue d'histoire des sciences* 57(3): 243–303.

Danowski, Déborah, and Eduardo Viveiros de Castro (2016) *The Ends of the World*, trans. Rodrigo Nunes. Cambridge: Polity.
D'Arcy Wood, Gillen (2015) *Tambora: The Eruption that Changed the World*. Princeton, NJ: Princeton University Press.
Daston, Lorraine (1998) "The Factual Sensibility: An Essay Review on Artifact and Experiment," *Isis* 79: 452–70.
Daston, Lorraine, and Peter Galison (2007) *Objectivity*. New York: Zone Books.
Daston, Lorraine, and Fernando Vidal (2004) *The Moral Authority of Nature*. Chicago: University of Chicago Press.
Daubigny, Pierre (2013) "Gaia Global Circus," unpublished MS.
Davis, Mike (2002) *Late Victorian Holocausts: El Niño Famines and the Making of the Third World*. London: Verso.
Dawkins, Richard (1976) *The Selfish Gene*. New York: Oxford University Press.
—— (1986) *The Blind Watchmaker*. New York: W. W. Norton.
de Gaulle, Charles ([1959] 1998) *The Complete War Memoirs of Charles de Gaulle*, trans. Richard Howard. New York: Carroll & Graf.
De Pryck, Kari (2014) "Le groupe d'experts intergouvernemental sur l'évolution du climat, ou les défis d'un mariage arrangé entre science et politique," *Ceriscope Environnement*, http://ceriscope.sciences-po.fr/environnement/part1/content/le-groupe-d-experts-intergouvernemental-sur-l-evolution-du-climat.
Debaise, Didier (2015) *L'appât des possibles: reprise de Whitehead*. Dijon: Presses du Réel.
Deléage, Jean-Paul (1991) *Histoire de l'écologie: une science de l'homme et de la nature*. Paris: La Découverte.
Deleuze, Gilles, and Félix Guattari ([1980] 1987) *A Thousand Plateaus: Capitalism and Schizophrenia*, trans. Brian Massumi. Minneapolis: University of Minnesota Press.
—— ([1991] 1994) *What Is Philosophy?*, trans. Hugh Tomlinson and Graham Burchell. New York: Columbia University Press.
Descola, Philippe ([1994] 1996) *The Spears of Twilight: Life and Death in the Amazon Jungle*, trans. Janet Lloyd. New York: New Press.
—— (2010) *La fabrique des images*. Paris: Musée du Quai Branly [exhibition catalog].
—— ([2005] 2013) *Beyond Nature and Culture*, trans. Janet Lloyd, foreword by Marshall Sahlins. Chicago: University of Chicago Press.
Despret, Vinciane (2009) *Penser comme un rat*. Versailles: Quae.
—— ([2012] 2016) *What Would Animals Say if We Asked the Right Questions?*, trans. Brett Buchanan. Minneapolis: University of Minnesota Press.
Despret, Vinciane, and Isabelle Despret (2011) *Les faiseuses d'histoires: que font les femmes à la pensée?* Paris: La Découverte.
Detienne, Marcel (2009) *Apollon, le couteau à la main*. Paris: Gallimard.
Dewey, John (1927) *The Public and its Problems*. New York: H. Holt.
—— (1938) *Logic: The Theory of Inquiry*. New York: H. Holt.

Di Chiro, Giovanna (2014) "Ramener l'écologie à la maison," in Émilie Hache, ed., *De l'univers clos au monde infini*. Paris: Dehors, pp. 191–220.

Diederichsen, Diedrich, and Anselm Frank, eds (2013) *The Whole Earth Catalog: California and the Disappearance of the Outside*. Berlin: Haus der Kulturen der Welt.

Doel, Ronald E. (2003) "Constituting the Postwar Earth Sciences: The Military's Influence on the Environmental Sciences in the USA after 1945," *Social Studies of Science* 33: 635–66.

"Doomsday Pending," Canadian Television, *The Hour*, www.youtube.com/watch?v=sRQ-NqaYFzs.

Dörries, Matthias (2011) "The Politics of Atmospheric Sciences: 'Nuclear Winter' and Global Climate Change," *Osiris* 26(1): 198–233.

Drouin, Jean-Marc (1991) *Réinventer la nature*. Paris: Desclée de Brower.

Dubos, René (1950) *Louis Pasteur, Free Lance of Science*. Boston: Little, Brown.

Dupuy, Jean-Pierre (2003) *Pour un catastrophisme éclairé: quand l'impossible est certain*. Paris: Seuil.

—— (2012) "On peut ruser avec le destin catastrophiste," *Critique* 738: 729–37.

Durkheim, Émile ([1912] 1965) *The Elementary Forms of the Religious Life*, trans. Joseph Ward Swain. New York: Free Press.

Edwards, Paul N. (2010) *A Vast Machine: Computer Models, Climate Data, and the Politics of Global Warming*. Cambridge, MA: MIT Press.

—— (2012) "Entangled Histories: Climate Science and Nuclear Weapons Research," *Bulletin of Atomic Scientists* 68(4): 28–40.

Elden, Stuart (2014) *The Birth of Territory*. Chicago: University of Chicago Press.

"Environmental Word Games" (2003) *New York Times*, March 15.

Fackler, Martin (2011) "Tsunami Warnings, Written in Stone," *New York Times*, April 20.

Farinelli, Franco (2009) *De la raison cartographique*, trans. Katia Bienvenu. Paris: CTHS.

Federici, Silvia (2004) *Caliban and the Witch: Women, the Body and Primitive Accumulation*. New York: Automedia.

Fleming, James Rodger (2010) *Fixing the Sky: The Checkered History of Weather and Climate Control*. New York: Columbia University Press.

Foessel, Michaël (2012) *Après la fin du monde: critique de la raison apocalyptique*. Paris: Seuil.

Fontanille, Jacques ([1998] 2006) *The Semiotics of Discourse*, trans. Heidi Bostic. New York: Peter Lang.

Ford, J. R., S. J. Price, A. H. Cooper, and C. N. Waters (2014) "An Assessment of Lithostratigraphy for Anthropogenic Deposits," *Geological Society, London, Special Publications* 395: 55–89.

Foucart, Stéphane (2013) "Le taux de CO_2 dans l'air au plus haut depuis plus de 2,5 millions d'années," *Le Monde*, May 7: 4.

Foucault, Michel ([1966] 1972) *The Archaeology of Knowledge*, trans. A. M. Sheridan Smith. New York: Pantheon Books.

Francis (2015) *Laudato Sí: On Care for Our Common Home*, encyclical letter, http://w2.vatican.va/content/francesco/en/encyclicals/documents/papa-francesco_20150524_enciclica-laudato-si.html.
Fressoz, Jean-Baptiste (2012) *L'apocalypse joyeuse: une histoire du risque technologique*. Paris: Seuil.
Freud, Sigmund ([1917] 1973) "A Difficulty in the Path of Psycho-Analysis," in *The Standard Edition of the Complete Psychological Works of Sigmund Freud*, ed. James Strachey. London: Hogarth Press, vol. 17, pp. 135–44.
Fukuyama, Francis (1992) *The End of History and the Last Man*. New York: Free Press.
Gaddis, John Lewis (2006) *The Cold War: A New History*. Harmondsworth: Penguin.
Gagliardi, Pasquale, Anne-Marie Reijnen, and Philipp Valentini (2013) *Protecting Nature, Saving Creation: Ecological Conflicts, Religious Passions, and Political Quandaries*. New York: Palgrave Macmillan.
Galilei, Galileo ([1588] n.d.) "Two Lectures to the Florentine Academy on the Shape, Location and Size of Dante's Inferno," trans. Mark A. Peterson, www.mtholyoke.edu/courses/mpeterso/galileo/inferno.html.
Galilei, Galileo ([1632] 1967) *Dialogue concerning the Two Chief World Systems – Ptolemaic and Copernican*, trans. Stillman Drake, foreword by Albert Einstein. Berkeley: University of California Press.
Galinier, Jacques, and Antoinette Molinié ([2006] 2013) *The Neo-Indians: A Religion for the Third Millennium*, trans. Lucy Lyall Grant. Boulder: University Press of Colorado.
Galison, Peter (2003) *Einstein's Clocks and Poincaré's Maps: Empires of Time*. New York: W. W. Norton.
Gamboni, Dario (2005) "Composing the Body Politic: Composite Images and Political Representations, 1651–2004," in Bruno Latour and Peter Weibel, eds, *Making Things Public: The Atmospheres of Democracy*. Cambridge, MA: MIT Press, pp. 162–95.
Gardiner, Stephen M. (2013) *A Perfect Moral Storm: The Ethical Tragedy of Climate Change*. Oxford: Oxford University Press.
Garfinkel, Harold, Michael Lynch, and Eric Livingston ([1981] 2011) "The Work of a Discovering Science Constructed with Materials from the Optically Discovered Pulsar," *Philosophy of the Social Sciences* 11(2): 131–58.
Gary, Romain (1956) Interview by Pierre Dumayet, in *Lecture pour tous*, December 19, http://fresques.ina.fr/jalons/fiche-media/InaEdu05408/les-racines-du-ciel-de-romain-gary.html.
Geddes, Patrick (1905) "A Great Geographer: Elisée Reclus, 1830–1905," *Scottish Geographical Magazine* 21(9): 490–6, 548–55.
Geison, Gerald, and James A. Secord (1988) "Pasteur and the Process of Discovery: The Case of Optical Isomerism," *Isis* 9: 6–36.
Gemenne, François (2009) *Géopolitique du changement climatique*. Paris: Armand Colin.
Georgescu-Roegen, Nicholas (2011) *La décroissance: entropie, écologie, économie*. Paris: Sang de la terre.

Gervais, François (2013) *L'innocence du carbone: l'effet de serre remise en question*. Paris: Albin Michel.
Gil, Marie (2011) *Péguy au pied de la lettre: la question du littéralisme dans l'œuvre de Péguy*. Paris: Le Cerf.
Gilbert, Scott F., and David Epel (2009) *Ecological Developmental Biology: Integrating Epigenetics, Medicine, and Evolution*. Sunderland, MA: Sinauer Associates.
Gilbert, Scott, Jan Sapp, and Alfred Tauber (2012) "A Symbiotic View of Life: We Have Never Been Individuals," *Quarterly Review of Biology* 87(4): 325–41.
Golding, William (1954) *Lord of the Flies*. New York: Putnam.
Gontier, Thierry (2011) *Politique, religion et histoire chez Eric Voegelin*. Paris: Le Cerf.
Goodman, Barak, and Rachel Dretzen (2004) *The Persuaders*. Alexandria, VA: PBS Home Video.
Gordon, Deborah (1999) *Ants at Work: How an Insect Society Is Organized*. New York: Free Press.
—— (2010) *Ant Encounters: Interaction Networks and Colony Behavior*. Princeton, NJ: Princeton University Press.
—— (2014) "The Ecology of Collective Behavior," *PLoS Biology* 12(3): 1–4.
Gore, Al (2006) *An Inconvenient Truth: The Planetary Emergency of Global Warming and What We Can Do About It*. Emmaus, PA: Rodale Press.
—— (2007) *The Assault on Reason*. New York: Penguin.
Gould, Stephen Jay (1989) *Wonderful Life: The Burgess Shale and the Nature of History*. New York: W. W. Norton.
Grace, Christy Rani R., Marilyn H. Perrin, Jozsef Gulyas, Michael R. DiGruccio, Jeffrey P. Cantle, Jean E. Rivier, Wylie W. Vale, and Roland Riek (2007) "Structure of the N-Terminal Domain of a Type of B1 G Protein-Coupled Receptor in Complex with a Peptide Ligand," *PNAS* 104(12): 4858–63.
Greimas, Algirdas, and Joseph Courtés, eds ([1979] 1982) *Semiotics and Language: An Analytic Dictionary*. Bloomington: Indiana University Press.
Greimas, Algirdas, and Jacques Fontanille ([1991] 1993) *The Semiotics of Passions: From States of Affairs to States of Feeling*, trans. Paul Perron and Frank Collins. Minneapolis: University of Minnesota Press.
Grevsmühl, Sebastian-Vincent (2014) *La terre vue d'en haut: l'invention de l'environnement global*. Paris: La Découverte.
Gribbin, John, and Mary Gribbin (2009) *James Lovelock: In Search of Gaia*. Princeton, NJ: Princeton University Press.
Grove, Richard H. (1995) *Green Imperialism: Colonial Expansion, Tropical Island Edens, and the Origins of Environmentalism, 1600–1860*. New York: Cambridge University Press.
Hache, Émilie (2014) *De l'univers clos au monde infini*. Paris: Dehors.
Hackett, E., O. Amsterdamska, M. Lynch, and J. Wacjman, eds (2007) *The Handbook of Science and Technology Studies*. 3rd edn, Cambridge, MA: MIT Press.

Haggenmacher, Peter (2001) Introduction, in *Le Nomos de la Terre dans le droit des gens du Jus Publicum Europaeum*, trans. Lilyane Deroche-Gurcel. Paris: Presses Universitaires de France, pp. 1–42.
Hamilton, Clive (2010) *Requiem for a Species: Why We Resist the Truth about Climate Change*. London: Earthscan.
—— (2013) *Earthmasters: The Dawn of Climate Engineering*. New Haven, CT: Yale University Press.
—— (2017) *Defiant Earth: The Fate of Humans in the Anthropocene*. Cambridge: Polity.
Hamilton, Clive, and Jacques Grinevald (2015) "Was the Anthropocene Anticipated?" *Anthropocene Review* 2(1): 33–58.
Hamilton, Clive, Christophe Bonneuil, and François Gemenne, eds (2015) *The Anthropocene and the Global Environmental Crisis: Rethinking Modernity in a New Epoch*. London: Routledge.
Harari, Yuval Noah (2016) *Homo Deus: A Brief History of Tomorrow*. London: Harvill Secker.
Haraway, Donna (1991) "A Cyborg Manifesto: Science, Technology, and Socialist-Feminism in the Late Twentieth Century," in Haraway, *Simians, Cyborgs and Women: The Reinvention of Nature*. New York: Routledge, pp. 149–81.
—— (2016) *Staying with the Trouble: Making Kin in the Chthulucene*. Durham, NC: Duke University Press.
Harding, Stephan, and Lynn Margulis (2009) "Water Gaia: 3.5 Thousand Million Years of Wetness on Planet Earth," in Eileen Crist and H. Bruce Rinker, eds, *Gaia in Turmoil: Climate Change, Biodepletion, and Earth Ethics in an Age of Crisis*. Cambridge, MA: MIT Press, pp. 41–60.
Harnack, Adolf (1990) *Marcion: The Gospel of the Alien God*, trans. John E. Steely and Lyle D. Bierma. Durham, NC: Labyrinth Press.
Heede, Richard (2014) "Tracing Anthropogenic Carbon Dioxide and Methane Emissions to Fossil Fuel and Cement Producers, 1854–2010," *Climatic Change* 122(1): 229–41.
Heinz, Dorothea (2015) "La terre comme l'impensé du Léviathan: une lecture de Carl Schmitt en juriste de l'écologie politique," Master of philosophy thesis, Paris: EHESS.
Hergé ([1953] 1976) *The Adventures of Tintin: Explorers on the Moon*, trans. Leslie Lonsdale-Cooper and Michael Turner. New York: Little, Brown.
Hesiod (1914) *Theogony*, in *The Homeric Hymns and Homerica*, trans. Hugh G. Evelyn-White. Cambridge, MA: Harvard University Press.
Hobbes, Thomas ([1651] 1998) *Leviathan*, ed. J. C. A. Gaskin. Oxford: Oxford University Press.
Hochstrasser, Julie (2007) *Still Life and Trade in the Dutch Golden Age*. New Haven, CT: Yale University Press.
Hoggan, James (2009) *Climate Cover-Up: The Crusade to Deny Global Warming*. Vancouver: Greystone Books.
—— (2016) *I'm Right and You're an Idiot: The Toxic State of Public Discourse and How to Clean it Up*. Gabriola Island, BC: New Society.

Hölldobler, Bert, and Edward O. Wilson (2008) *The Superorganism: The Beauty, Elegance, and Strangeness of Insect Societies*. New York: W. W. Norton.
Hulme, Mike (2009) *Why We Disagree about Climate Change: Understanding Controversy, Inaction and Opportunity*. Cambridge: Cambridge University Press.
Hultén, Pontus, ed. (1987) *The Arcimboldo Effect: Transformations of the Face from the 16th to the 20th Century*. New York: Abbeville Press.
Hume, David ([1779] 1993) *Principal Writings on Religion, including Dialogues concerning Natural Religion; and, The Natural History of Religion*, ed. J. C. A. Gaskin. New York: Oxford University Press.
Husserl, Edmund (1970) *The Crisis of European Sciences and Transcendental Phenomenology: An Introduction to Phenomenological Philosophy*, trans. David Carr. Evanston, IL: Northwestern University Press.
Huzar, Eugène ([1855] 2008) *La fin du monde par la science*, introduction by Jean-Baptiste Fressoz. Alfortville: Ère.
"Hymn to Gaia," trans. Alec Roth, www.alecroth.com/documents/Hymn to Gaia Text.pdf.
IPCC (Intergovernmental Panel on Climate Change) (2010) "Statement on IPCC Principles and Procedures," February 2, www.ipcc.ch/pdf/press/ipcc-statement-principles-procedures-02-2010.pdf.
James, William ([1909] 2012) *A Pluralistic Universe: Hibbert Lectures at Manchester College on the Present Situation in Philosophy*. Auckland: Floating Press.
Jameson, Fredric (2003) "Future City," *New Left Review* 21 (May–June): 65–79.
Jeandel, Catherine, and Remy Mosseri (2011) *Le climat à découvert: outils et méthodes en recherche climatique*. Paris: CNRS.
Jonas, Hans (1962) "Immortality and the Modern Temper: The Ingersoll Lecture, 1961," *Harvard Theological Review* 55(1): 1–20.
—— (1984) *The Imperative of Responsibility: In Search of an Ethics for the Technological Age*, trans. Hans Jonas with David Herr. Chicago: University of Chicago Press.
—— ([1958] 2001) *The Gnostic Religion: The Message of the Alien God and the Beginnings of Christianity*. 3rd edn, Boston: Beacon Press.
Jouin, Céline (2007) Introduction to Carl Schmitt, *La guerre civile mondiale: essais (1943–1978)*, ed. and trans. Céline Jouin. Paris: ERE, pp. 7–27.
Karsenti, Bruno (2012a) "La représentation selon Voegelin, ou les deux visages de Hobbes," *Revue des sciences philosophiques et théologiques* 96(3): 513–40.
—— (2012b) *Moïse et l'idée de peuple: la vérité historique selon Freud*. Paris: Le Cerf.
Keeling, Charles David (1998) "Rewards and Penalties of Recording the Earth," *Annual Review of Energy and Environment* 23: 25–82.
Keller, Evelyn Fox (1983) *A Feeling for the Organism: The Life and Work of Barbara McClintock*. San Francisco: W. H. Freeman.

Kepel, Gilles, and Jean-Pierre Milelli, eds (2008) *Al Qaeda in its Own Words*. Cambridge, MA: Belknap Press.
Kershaw, Ian (2011) *The End: The Defiance and Destruction of Hitler's Germany, 1944–1945*. New York: Penguin.
Kierkegaard, Søren ([1843] 2013) *Fear and Trembling; and, The Sickness unto Death*, trans. Walter Lowrie. Princeton, NJ: Princeton University Press.
Klein, Naomi (2015) *This Changes Everything: Capitalism vs. the Climate*. New York: Simon & Schuster.
Koerner, Joseph Leo (2009) *Caspar David Friedrich and the Subject of Landscape*. London: Reaction Books.
Koestler, Arthur (1959) *The Sleepwalkers: A History of Man's Changing Vision of the Universe*, introduction by Herbert Butterfield. London: Hutchinson.
Kohn, Eduardo (2013) *How Forests Think: Toward an Anthropology beyond the Human*. Berkeley: University of California Press.
Kopenawa, Davi, and Bruce Albert (2013) *The Falling Sky: Words of a Yanomami Shaman*. Cambridge, MA: Harvard University Press.
Koyré, Alexandre (1957) *From the Closed World to the Infinite Universe*. Baltimore: Johns Hopkins University Press.
Kupiec, Jean-Jacques, and Pierre Sonigo (2000) *Ni Dieu ni gène*. Paris: Seuil.
Lacoste, Yves ([1982] 2014) *La géographie, ça sert, d'abord, à faire la guerre*. Paris: La Découverte.
Laroche, Emmanuel (1949) *Histoire de la racine "nem" en grec ancien: nemo, nemesis, nomos, nomizo*. Paris: Klincksieck.
Larrère, Catherine (1997) *Les philosophies de l'environnement*. Paris: Presses Universitaires de France.
Latour, Bruno ([1984] 1988) *The Pasteurization of France*, trans. Alan Sheridan and John Law. Cambridge, MA: Harvard University Press.
—— ([1991] 1993) *We Have Never Been Modern*, trans. Catherine Porter. Cambridge, MA: Harvard University Press.
—— (1995a) "Pasteur and Pouchet: The Heterogenesis of the History of Science," in Michel Serres, ed., *History of Scientific Thought*. Oxford: Blackwell, pp. 526–55.
—— (1995b) "Joliot: History and Physics Mixed Together," in Michel Serres, ed., *History of Scientific Thought*. Oxford: Blackwell, pp. 611–35.
—— ([1992] 1996) *Aramis, or the Love of Technology*, trans. Catherine Porter. Cambridge, MA: Harvard University Press.
—— (1998) "To Modernize or to Ecologize, That Is the Question," in Bruce Braun and Noel Castree, eds, *Remaking Reality: Nature at the Millennium*. London: Routledge, pp. 221–42.
—— (1999) *Pandora's Hope: Essay on the Reality of Science Studies*. Cambridge, MA: Harvard University Press.
—— (2003) "What if We Talked Politics a Little?," *Contemporary Political Theory* 2(2): 143–64.

—— (2004a) "Why Has Critique Run out of Steam? From Matters of Fact to Matters of Concern," *Critical Inquiry* 30(2): 225–48 [special issue on the future of criticism].
—— ([1999] 2004b) *Politics of Nature: How to Bring the Sciences into Democracy*, trans. Catherine Porter. Cambridge, MA: Harvard University Press.
—— (2005) *Reassembling the Social: An Introduction to Actor-Network Theory.* Oxford: Oxford University Press.
—— (2007a) "'It's Development, Stupid!' or: How to Modernize Modernization," www.bruno-latour.fr/node/153 [review of Ted Nordhaus and Michael Shellenberger, *Break Through: From the Death of Environmentalism to the Politics of Possibility*].
—— (2007b) "The Recall of Modernity – Anthropological Approaches," trans. Stephen Muecke. *Cultural Studies Review* [Sydney] 13 (March): 11–30.
—— (2008a) "The Powers of Facsimiles: A Turing Test on Science and Literature," in Stephen J. Burn and Peter Dempsey, eds, *Intersections: Essays on Richard Powers*. Champaign, IL: Dalkey Archive Press, pp. 263–92.
—— (2008b) Introduction to Walter Lippmann, *Le public fantôme*, trans. Laurence Décréau. Paris: Demopolis, pp. 6–44.
—— (2008c) *What Is the Style of Matters of Concern? Two Lectures on Empirical Philosophy.* Amsterdam: Van Gorcum.
—— ([1996] 2010a) *On the Modern Cult of the Factish Gods*, trans. Catherine Porter and Heather MacLean. Durham, NC: Duke University Press.
—— (2010b) "Si tu viens à perdre la Terre, à quoi te sert d'avoir sauvé ton âme?," in Jacques-Noël Perès, ed., *L'avenir de la terre: un défi pour les Églises.* Paris: Desclée de Brouwer, pp. 51–72.
—— (2010c) "Steps toward the Writing of a Compositionist Manifesto," *New Literary History* 41: 471–90.
—— ([2002] 2013a) *Rejoicing: The Torments of Religious Speech*, trans. Julie Rose. Cambridge: Polity.
—— ([2012] 2013b) *An Inquiry into Modes of Existence: An Anthropology of the Moderns*, trans. Catherine Porter. Cambridge, MA: Harvard University Press.
—— (2013c) "Facing Gaia: A New Enquiry into Natural Religion," www.ed.ac.uk/arts-humanities-soc-sci/news-events/lectures/gifford-lectures/archive/series-2012-2013/bruno-latour (video); "Facing Gaia: Six Lectures on Natural Religion," www.bruno-latour.fr/node/700 (text) [University of Edinburgh Gifford Lectures].
—— (2013d) *Kosmokoloss*, www.bruno-latour.fr/node/538 [radio play (in German)].
—— (2014a) "Agency at the Time of the Anthropocene," *New Literary History* 45: 1–18.
—— (2014b) "Formes élémentaires de la sociologie: formes avancées de la théologie," *Archives des sciences sociales des religions* 167 (July–September): 255–77.

—— (2014c) "Nous sommes des vaincus," in Camille Riquier, ed., *Charles Péguy*. Paris: Le Cerf, pp. 11–30.
—— (2014d) "Some Advantages of the Notion of 'Critical Zone' for Geopolitics: Geochemistry of the Earth's Surface, GES-10, Paris, France, Aug. 18–23, 2014," *Procedia Earth and Planetary Science* 10: 3–6.
—— (2015) "Fifty Shades of Green," *Environmental Humanities* 7: 219–25.
—— (2016a) "*Onus Orbis Terrarum*: About a Possible Shift in the Definition of Sovereignty," *Millennium: Journal of International Studies* 44(3): 305–20.
—— (2016b) "How Better to Register the Agency of Things," in Mark Matheson, ed., *The Tanner Lectures on Human Values*, vol. 34. Salt Lake City: University of Utah Press, pp. 79–117.
—— (2016c) "Why Gaia Is Not a God of Totality," *Theory, Culture and Society* (June): 1–21 [special issue: *Geosocial Formations and the Anthropocene*].
Latour, Bruno, and Paolo Fabbri (2000) "The Rhetoric of Science: Authority and Duty in an Article from the Exact Sciences," trans. Sarah Cummins, *Technostyle* 16: 115–34.
Latour, Bruno, and Émilie Hache (2010) "Morality or Moralism? An Exercise in Sensitization," trans. Patrick Camiller, *Common Knowledge* 16(2): 311–30.
Latour, Bruno, and Christophe Leclercq, eds (2016) *Reset Modernity!* Cambridge, MA: MIT Press [exhibition catalog].
Latour, Bruno, and Shirley C. Strum (1986) "Human Social Origins: Oh Please, Tell Us Another Story," *Journal of Social and Biological Structures* 9: 169–87.
Latour, Bruno, and Peter Weibel, eds (2002) *Iconoclash: Beyond the Image Wars in Science, Religion and Art*. Cambridge, MA: MIT Press.
—— (2005) *Making Things Public: The Atmospheres of Democracy*. Cambridge, MA: MIT Press.
Latour, Bruno, and Steve Woolgar (1979) *Laboratory Life: The Social Construction of Scientific Facts*. Beverley Hills, CA: Sage.
Latour, Bruno, Pablo Jensen, Tommaso Venturini, Sebastian Grauwin, and Dominique Boullier (2012) "The Whole Is Always Smaller Than its Parts – a Digital Test of Gabriel Tarde's Monads," *British Journal of Sociology* 63(4): 590–615.
Legg, Stephen, ed. (2011) *Spatiality, Sovereignty and Carl Schmitt: Geographies of the Nomos*. London: Routledge.
Lenton, Timothy (1998) "Gaia and Natural Selection: A Review Article," *Nature*, July 30: 439–47.
—— (2016) *Earth System Science*. Oxford: Oxford University Press.
Lewis, Simon L., and Mark A. Maslin (2015) "Defining the Anthropocene," *Nature*, March 12: 171–80.
Lippmann, Walter (1925) *The Phantom Public*. New York: Harcourt, Brace.
Locher, Fabien (2013) "Les pâturages de la guerre froide: Garrett Hardin et la 'tragédie des communs,'" *Revue d'histoire moderne et contemporaine* 60(1): 7–36.

Locher, Fabien, and Gregory Quenet (2009) "L'histoire environnementale: origines, enjeux et perspectives d'un nouveau chantier," *Revue d'histoire moderne et contemporaine* 56(4): 7–38 [special issue on the history of the environment].

Lovelock, James ([1991] 2000a) *Gaia: The Practical Science of Planetary Medicine*. Oxford: Oxford University Press.

—— (2000b) *Homage to Gaia: The Life of an Independent Scientist*. Oxford: Oxford University Press.

—— (2006) *The Revenge of Gaia: Earth's Climate in Crisis and the Fate of Humanity*. New York: Basic Books.

Lovelock, James, and Dian R. Hitchcock (1967) "Life Detection by Atmospheric Analysis," *Icarus: International Journal of the Solar System* 7(2): 149–59.

Lovelock, James, and Michael Whitfield (1982) "Life Span of the Biosphere," *Nature* (April 8): 561–3.

Löwith, Karl (1949) *Meaning in History: The Theological Implications of the Philosophy of History*. Chicago: University of Chicago Press.

Lubac, Henri de ([1981] 2014) *La postérité spirituelle de Joachim de Flore*. Paris: Le Cerf.

Luisetti, Federico, and Wilson Kaiser, eds (2015) *The Anomie of the Earth: Philosophy, Politics and Autonomy in Europe and the Americas*. Durham, NC: Duke University Press.

Luntz, Frank (2005) *Words That Work*. New York: Hachette.

Lussault, Michel (2013) *L'avènement du monde: essai sur l'habitation humaine de la Terre*. Paris: Seuil.

MacKenzie, Donald (2006) *An Engine, Not a Camera: How Financial Models Shape Markets*. Cambridge, MA: MIT Press.

—— (2009) *Material Markets*. Oxford: Oxford University Press.

Mandeville, Bernard ([1714] 1962) *The Fable of the Bees: Private Vices, Publick Benefits*, ed. Irwin Primer. New York: Capricorn Books.

Mann, Charles C. (2005) *1491: New Revelations of the Americas before Columbus*. New York: Alfred A. Knopf.

—— (2011) *1493: Uncovering the New World Columbus Created*. New York: Alfred A. Knopf.

Mann, Michael (2013) *The Hockey Stick and the Climate Wars: Dispatches from the Front Lines*. New York: Columbia University Press.

—— (2014) "If You See Something, Say Something," *New York Times*, January 17.

Margulis, Lynn (1995) "Gaia Is a Tough Bitch," in John Brockman, ed., *The Third Culture: Beyond the Scientific Revolution*. New York: Simon & Schuster, pp. 129–51.

—— (1998) *Symbiotic Planet: A New Look at Evolution*. New York: Basic Books.

Margulis, Lynn, and Dorian Sagan ([1986] 1997) *Microcosmos: Four Billion Years of Evolution from our Microbial Ancestors*, foreword by Lewis Thomas. Berkeley: University of California Press.

Marres, Noortje (2012) *Material Participation: Technology, the Environment and Everyday Publics.* London: Palgrave.
Martel, Yann (2001) *Life of Pi.* New York: Harcourt.
Martin, Nastassja (2016) *Les âmes sauvages: face à l'Occident, la résistance d'un peuple d'Alaska.* Paris: La Découverte.
Maslin, Mark (2014) "Why I'll Talk Politics with Climate Change Deniers – but Not Science," blog post, December 16, http://theconversation.com/why-ill-talk-politics-with-climate-change-deniers-but-not-science-34949.
Masson-Delmotte, Virginie (2011) *Climat: le vrai et le faux.* Paris: Le Pommier.
Matteucci, Ruggero, Guido Gosso, Silvia Peppoloni, Sandra Placente, and Janutz Wasowksi (2012) "A Hippocratic Oath for Geologists?," *Annals of Geophysics* 55(3): 365–9.
Mayr, Otto (1986) *Authority, Liberty, & Automatic Machinery in Early Modern Europe.* Baltimore: Johns Hopkins University Press.
McNeil, John R. (2000) *Something New under the Sun: An Environmental History of the Twentieth-Century World.* New York: W. W. Norton.
McPhee, John (1989) *The Control of Nature.* New York: Farrar, Straus, & Giroux.
—— (1993) *Assembling California.* New York: Farrar, Straus & Giroux.
Meadows, Donella H., Dennis L. Meadows, Jørgen Randers, and William W. Behrens III (1972) *The Limits to Growth: A Report for the Club of Rome's Project on the Predicament of Mankind.* New York: Universe Books.
Meier, Heinrich (1998) *The Lesson of Carl Schmitt: Four Chapters on the Distinction between Political Theology and Political Philosophy.* Chicago: University of Chicago Press.
Merchant, Carolyn (1980) *The Death of Nature: Women, Ecology and the Scientific Revolution.* London: Wildwood House.
Metz, Christian ([1975] 1982) *Psychoanalysis and Cinema: The Imaginary Signifier*, trans. Ben Brewster, Alfred Guzzetti, Celia Britton, and Annwyl Williams. Blooomington: Indiana University Press.
Mialet, Hélène (2012) *Hawking Incorporated: Stephen Hawking and the Anthropology of the Knowing Subject.* Chicago: University of Chicago Press.
Minca, Claudio, and Rory Rowan (2015) *On Schmitt and Space.* London: Routledge.
Mitchell, Timothy (2011) *Carbon Democracy: Political Power in the Age of Oil.* London: Verso.
Moltmann, Jürgen (1993) *God in Creation: A New Theology of Creation and the Spirit of God.* Minneapolis: Fortress Press.
Monod, Jacques ([1970] 1972) *Chance and Necessity: An Essay on the Natural Philosophy of Modern Biology*, trans. Austryn Wainhouse. New York: Vintage Books.
Montebello, Pierre (2003) *L'autre métaphysique: essai sur Ravaisson, Tarde, Nietzsche et Bergson.* Paris: Desclée de Brouwer.

Montesquieu, Baron de, Charles de Secondat ([1748] 1989) *The Spirit of the Laws*, trans. Anne M. Cohler, Basia C. Miller, and Harold Stone. Cambridge: Cambridge University Press.

Moore, Jason (2015) *Capitalism in the Web of Life: Ecology and the Accumulation of Capital.* New York: Verso.

Morrison, Philip, Phylis Morrison, and the Office of Charles and Ray Eames (1982) *Powers of Ten: A Book about the Relative Size of Things in the Universe and the Effect of Adding Another Zero.* New York: Scientific American Books.

Morton, Oliver (2007) *Eating the Sun: The Everyday Miracle of How Plants Power the Planet.* London: Fourth Estate.

—— (2015) *The Planet Remade: How Geoengineering Could Change the World.* Princeton, NJ: Princeton University Press.

Morton, Timothy (2013) *Hyperobjects: Philosophy and Ecology after the End of the World.* Minneapolis: Minnesota University Press.

Nietzsche, Friedrich ([1882] 1974) *The Gay Science, with a Prelude in Rhymes and an Appendix of Songs*, trans. Walter Kaufman. New York: Vintage Books.

Nordhaus, Ted, and Michael Shellenbeger (2007) *Break Through: From the Death of Environmentalism to the Politics of Possibility.* New York: Houghton Mifflin.

Northcott, Michael S. (2013) *A Political Theology of Climate Change.* Grand Rapids, MI: William B. Eerdmans.

Olwig, Kenneth (2008) "Has 'Geography' Always Been Modern? Choros, (Non)Representation, Performance, and the Landscape," *Environment and Planning, A: Society and Space* 40: 1843–63.

—— (2011) "The Earth Is Not a Globe: Landscape versus the 'Globalist' Agenda," *Landscape Research* 16(4): 401–15.

Oreskes, Naomi (2004) "Beyond the Ivory Tower: The Scientific Consensus on Climate Change," *Science* (December 3): 1686.

Oreskes, Naomi, and Erik M. Conway (2010) *Merchants of Doubt: How a Handful of Scientists Obscured the Truth on Issues from Tobacco Smoke to Global Warming.* New York: Bloomsbury.

—— (2014) *The Collapse of American Civilization: A View from the Future.* New York: Columbia University Press.

Ospovat, Dov (1995) *The Development of Darwin's Theory: Natural History, Natural Theology, and Natural Selection, 1838–1859.* Cambridge: Cambridge University Press.

Ostrom, Elinor (1990) *Governing the Commons: The Evolution of Institutions for Collective Action.* Cambridge: Cambridge University Press.

Otto, Shawn Lawrence (2011) *Fool Me Twice: Fighting the Assault on Science in America.* New York: Rodale.

Palmesino, John, and Ann-Sofi Rönnskog (2016) "Radical Conservation: The Museum of Oil," in Bruno Latour, ed., *Reset Modernity!* Cambridge, MA: MIT Press, pp. 337–53.

Panofsky, Erwin (1954) *Galileo as a Critic of the Arts.* The Hague: Martinus Nijhoff.

—— ([1927] 1991) *Perspective as Symbolic Form*. New York: Zone Books.
Pearce, Fred (2007) *With Speed and Violence: Why Scientists Fear Tipping Points in Climate Change*. Boston: Beacon Press.
Péguy, Charles ([1914] 1992) Note conjointe sur M. Descartes et la philosophie cartésienne, in *Charles Péguy: Œuvres en prose complètes*, vol. III (*Édition présentée, établie et annotée par Robert Burac*). Paris: Gallimard, pp. 1278–477.
—— ([1917] 1958) "Clio I," in *Temporal and Eternal*, trans. Alexander Dru. New York: Harper, pp. 89–159.
Pestre, Dominique (2006) *Introduction aux science studies*. Paris: La Découverte.
—— (2014) "Néolibéralisme et gouvernement: retour sur une catégorie et ses usages," in Pestre, ed., *Le gouvernement des technosciences: gouverner le progrès et ses dégât depuis 1945*. Paris: La Découverte, pp. 261–84.
Pickering, Andy (2011) *The Cybernetic Brain: Sketches of Another Future*. Chicago: University of Chicago Press.
Polanyi, Karl ([1944] 2001) *The Great Transformation: The Political and Economic Origins of Our Time*. Boston: Beacon Press.
Pomeranz, Kenneth (2000) *The Great Divergence: China, Europe, and the Making of the Modern World Economy*. Princeton, NJ: Princeton University Press.
Powers, Richard (1998) *Gain*. New York: Farrar, Straus, & Giroux.
—— (2006) *The Echo Maker*. New York: Farrar, Straus, & Giroux.
Proctor, Robert N. (2011) *Golden Holocaust: Origins of the Cigarette Catastrophe and the Case for Abolition*. Berkeley: University of California Press.
Quenet, Gregory (2015) *Versailles: une histoire naturelle*. Paris: La Découverte.
Ramel, Frédéric (2011) *Philosophie des relations internationales*. Paris: Presses de Sciences Po.
Richtel, Matt (2015) "California Farmers Dig Deeper for Water, Sipping Their Neighbors Dry," *New York Times*, June 5.
Riquier, Camille (2011) "Charles Péguy: Métaphysiques de l'événement," in Didier Debaise, ed., *Philosophie des possessions*. Dijon: Presses du Réel, pp. 197–232.
Rockström, Johan, Will Steffen, et al. (2009) "Planetary Boundaries: Exploring the Safe Operating Space for Humanity," *Ecology and Society* 14(2), www.ecologyandsociety.org/vol14/iss2/art32.
Rudwick, Martin (2014) *Earth's Deep History: How it Was Discovered and Why it Matters*. Chicago: University of Chicago Press.
Ruse, Michael (2013) *The Gaia Hypothesis: Science on a Pagan Planet*. Chicago: University of Chicago Press.
Ruyer, Raymond ([1952] 2016). *Neofinalism*, trans. Alyosha Edlebi. Minneapolis: University of Minnesota Press.
Schaffer, Simon (2005) "Seeing Double: How to Make Up a Phantom Body Politic," in Bruno Latour and Peter Weibel, eds, *Making Things Public*. Cambridge, MA: MIT Press, 196–202.

—— (2008) *The Information Order of Isaac Newton's Principia Mathematica*. Uppsala University, www.vethist.idehist.uu.se/pdf/schaffer. pdf [Hans Rausing Lecture].

—— (2011) "Newtonian Angels," in Joad Raymond, ed., *Conversations with Angels: Essays Towards a History of Spiritual Communication*. London: Palgrave, pp. 90–122.

Schmitt, Carl ([1932] 1976) *The Concept of the Political*, ed. and trans. George Schwab, comments by Leo Strauss. New Brunswick, NJ: Rutgers University Press.

—— ([1938] 1996) *The Leviathan in the State Theory of Thomas Hobbes: Meaning and Failure of a Political Symbol*, trans. George Schwab and Erna Hilfstein. Westport, CT: Greenwood Press.

—— ([1954] 1997) *Land and Sea*, trans. Simona Draghici. Washington, DC: Plutarch Press.

—— ([1950] 2003) *The Nomos of the Earth in the International Law of the Jus Publicum Europaeum*, trans. G. I. Ulmen. New York: Telos.

—— ([1922] 2005) *Political Theology: Four Chapters on the Concept of Sovereignty*, trans. George Schwab. Chicago: University of Chicago Press.

Schneider, Stephen H., James R. Miller, Eileen Crist, and Penelope J. Boston (2008) *Scientists Debate Gaia*. Cambridge, MA: MIT Press.

"Secret Copenhagen Climate Recording Reveals Resistance from China and India" (2010) *Der Spiegel*, May 7, www.youtube.com/watch?v=-ybecKdwj2c.

Serres, Michel (1974) *Hermès III: la traduction*. Paris: Minuit.

—— ([1980] 1982) *The Parasite*, trans. Lawrence R. Schehr. Baltimore: Johns Hopkins University Press.

—— (1995) *The Natural Contract*, trans. Elizabeth MacArthur and William Paulson. Ann Arbor: University of Michigan Press.

—— (2000) *Retour au contrat naturel*. Paris: Bibliothèque nationale de France.

Shapin, Steven (1996) *The Scientific Revolution*. Chicago: University of Chicago Press.

—— (2008) *The Scientific Life: A Moral History of a Late Modern Vocation*. Chicago: University of Chicago Press.

Shapin, Steven, and Simon Schaffer (1985) *Leviathan and the Air-Pump: Hobbes, Boyle, and the Experimental Life: Including a Translation of Thomas Hobbes, Dialogus physicus de natura aeris, by Simon Schaffer*. Princeton, NJ: Princeton University Press.

Shelley, Percy Bysshe (1817) "Mont Blanc – Lines Written in the Vales of Chamouni," www.poetryfoundation.org/poems-and-poets/poems/detail/45130.

Simondon, Gilbert ([1958] 2016) *On the Mode of Existence of Technical Objects*, trans. Cecile Malaspina. Minneapolis: Univocal.

Sloterdijk, Peter ([2005] 2013) *In the World Interior of Capital: For a Philosophical Theory of Globalization*, trans. Wieland Hoban. Cambridge: Polity.

—— ([1999] 2014) *Globes: Macrospherology* (*Spheres*, vol. 2), trans. Wieland Hoban. South Pasadena, CA: Semiotext(e).
Sonigo, Pierre, and Isabelle Stengers (2003) *L'évolution*. Les Ulis: EDP Sciences.
Soudan, Clara (2015) "Théologie politique de la nature: l'ontologie théologique de Hans Jonas au fondement de son éthique environnementale de la responsabilité," Masters thesis, EPHE, Paris.
Squarzoni, Philippe (2014) *Climate Changed: A Personal Journey through the Science*, trans. Ivanka Hahnenberger. New York: Abrams Comic Arts.
Steffen, Will, et al. (2015a) "The Trajectory of the Anthropocene: The Great Acceleration," *Anthropocene Review* 2(1): 1–18.
Steffen, Will, et al. (2015b) "Planetary Boundaries: Guiding Human Development on a Changing Planet," *Science* (February 13): 736–47.
Stengers, Isabelle (1998) "La guerre des sciences: et la paix?," in Jurdant Baudouin, ed., *Impostures scientifiques: les malentendus de l'affaire Sokal*. Paris: La Découverte, pp. 268–92.
—— ([1993] 2000) *The Invention of Modern Science*, trans. Daniel L. Smith. Minneapolis: University of Minnesota Press.
—— (2005) *La vierge et le neutrino*. Paris: La Découverte.
—— (2011a) *Thinking with Whitehead: A Free and Wild Creation of Concepts*, trans. Michael Chase. Cambridge, MA: Harvard University Press.
—— (2011b) "Another Science Is Possible! A Plea for Slow Science," http://we.vub.ac.be/aphy/sites/default/files/stengers2011_pleaslowscience.pdf.
—— (2014) "Penser à partir du ravage écologique," in *De l'univers clos au monde infini*, ed. Émilie Hache. Paris: Éditions Dehors, pp. 187–91.
—— ([2009] 2015) *In Catastrophic Times: Resisting the Coming Barbarism*, trans. Andrew Goffey. London: Open Humanities Press.
Stengers, Isabelle, and Thierry Drumm (2013) *Une autre science est possible! Manifeste pour un ralentissement des sciences*. Paris: La Découverte.
Stern, Nicholas (2007) *The Economics of Climate Change: The Stern Review*. Cambridge: Cambridge University Press.
Strum, Shirley S. (2012) "Darwin's Monkey: Why Baboons Can't Become Human," *Yearbook of Physical Anthropology* 55: 3–23.
Styfhals, Willem (2012) "Gnosis, Modernity and Divine Incarnation: The Voegelin–Blumenberg Debate," *Bijdragen: International Journal in Philosophy and Theology* 73(2): 190–211.
Subcommission on Quaternary Stratigraphy (2011) "What Is the 'Anthropocene'? – Current Definition and Status," report written during the Congress of the International Union for Quaternary Research in Bern, Switzerland, July 21–7, http://quaternary.stratigraphy.org/workinggroups/anthropocene/.
Szerszynski, Bronislaw (2012) "The End of the End of Nature: The Anthropocene and the Fate of the Human," *Oxford Literary Review* 34(2): 165–84.

Tansley, A. G. (1935) "The Use and Abuse of Vegetational Concepts and Terms," *Ecology* 16(3): 284–307.
Tarde, Gabriel de (1899) *Social Laws: An Outline of Society*, trans. Howard C. Warren. New York: Macmillan.
—— ([1893] 2012) *Monadology and Sociology*, ed. and trans. Theo Lorenc. Melbourne: re.press.
Taylor, Bron (2010) *Dark Green Religion: Nature, Spirituality, and the Planetary Future*. Berkeley: University of California Press.
Thomas, Yan (1991) "L'institution de majesté," *Revue de synthèse* 112(3–4): 331–86.
Thompson, Charis (2005) *Making Parents: The Ontological Choreography of Reproductive Technologies*. Cambridge, MA: MIT Press.
Tollefson, Jeff (2013) "Soot a Major Contributor to Climate Change," *Nature* (January 15), www.nature.com/news/soot-a-major-contributor-to-climate-change-1.12225.
Tolstoy, Leo ([1865–6] 1996) *War and Peace*, trans. Aylmer Maude and Louise Maude, ed. George Gibian. New York: W. W. Norton.
Toulmin, Stephen E. (1990) *Cosmopolis: The Hidden Agenda of Modernity*. Chicago: University of Chicago Press.
—— ([1958] 2003) *The Uses of Argument*. New York: Cambridge University Press.
Tresch, John (2005) "Cosmogram," in Melik Ohanian and Jean-Christophe Royoux, eds, *Cosmograms*. New York: Lukas & Sternberg, pp. 67–76.
—— (2012) *The Romantic Machine*. Chicago: University of Chicago Press.
Tsing, Anna L. (2015) *The Mushroom at the End of the World: On the Possibility of Life in Capitalist Ruins*. Princeton, NJ: Princeton University Press.
Twain, Mark ([1883] 1944) *Life on the Mississippi*. New York: Heritage Press.
Tyrrell, Toby (2013) *On Gaia: A Critical Investigation of the Relationship between Life and Earth*. Princeton, NJ: Princeton University Press.
Uexküll, Jakob von ([1940] 2010) *A Foray into the Worlds of Animals and Humans, with A Theory of Meaning*, trans. Joseph D. O'Neill. Minneapolis: University of Minnesota Press.
Van Damme, Stéphane (2002) *Descartes*. Paris: Presses de Sciences Po.
Van Dooren, Thom (2014) *Flight Ways: Life and Loss at the Edge of Extinction*. New York: Columbia University Press.
Vaughan, Diane (1996) *The Challenger Launch Decision: Risky Technology, Culture, and Deviance at NASA*. Chicago: University of Chicago Press.
Venturini, Tommaso (2010) "Diving in Magma: How to Explore Controversies with Actor-Network Theory," *Public Understanding of Science* 19(3): 258–73.
Vernant, Jean-Pierre (1981) Preface to Hesiod, *Théogonie: la naissance des dieux*. Paris: Rivages, pp. 7–50.

Vidal-Naquet, Pierre ([1991] 1992) *Assassins of Memory: Essays on the Denial of the Holocaust*, trans. Jeffrey Mehlman. New York: Columbia University Press.
Vieille-Blanchard, Élodie (2011) "Les limites à la croissance dans un monde global: modélisations, prospectives, réfutations," dissertation, EHESS, Paris.
Viveiros de Castro, Eduardo ([2009] 2014) *Cannibal Metaphysics: For a Post-Structuralist Anthropology*, trans. Peter Skafish. Minneapolis: Univocal.
—— (2016) *The Relative Native: Essays on Indigenous Conceptual Worlds*, trans. Martin Holbraad, David Rodgers, and Julia Saum, afterword by Roy Wagner. Chicago: Hau Books.
Voegelin, Eric ([1952] 2000a) *The New Science of Politics: An Introduction*, in *The Collected Works of Eric Voegelin*, vol. 5: *Modernity without Constraint: The Political Religions, The New Science of Politics, and Science, Politics and Gnosticism*, ed. Manfred Henningsen. Columbia: University of Missouri Press, pp. 75–241.
—— (2000b) "Ersatz Religion: The Gnostic Mass Movements of Our Time," in *The Collected Works of Eric Voegelin*, vol. 5: *Modernity without Constraint: The Political Religions, The New Science of Politics, and Science, Politics and Gnosticism*, ed. Manfred Henningsen. Columbia: University of Missouri Press, pp. 295–313.
Waddington, C. H., ed. ([1972] 2012) *Biological Processes in Living Systems*, vol. 4: *Toward a Theoretical Biology*. New Brunswick, NJ: Aldine Transaction.
Ward, Barbara, and René Dubos (1972) *Only One Earth: An Unofficial Report commissioned by the Secretary General of the United Nations Conference on the Human Environment*. New York: W. W. Norton.
Ward, Peter D. (2009) *The Medea Hypothesis: Is Life on Earth Ultimately Self-Destructive?* Princeton, NJ: Princeton University Press.
Waters, Colin N., Jan Zalaciewicz, et al. (2016) "The Anthropocene Is Functionally and Stratigraphically Distinct from the Holocene," *Science* (January 8): 137–48.
Weart, Spencer (2003) *The Discovery of Global Warming*. Cambridge, MA: Harvard University Press.
"Welcome to the Anthropocene!" (2011) *The Economist*, May 28: 13.
Welzer, Harald ([2008] 2012) *Climate Wars: Why People Will Be Killed in the Twenty-First Century*. Cambridge: Polity.
White, Lynn, Jr (1967) "The Historical Roots of Our Ecological Crisis," *Science* (March 10): 1203–7.
White, Richard ([1991] 2011) *The Middle Ground: Indians, Empires, and Republics in the Great Lakes Region, 1650–1815*. New York: Cambridge University Press.
Whitehead, Alfred North (1920) *The Concept of Nature*. Cambridge: Cambridge University Press.
Williams, Alex, and Nick Srnicek (2013) "#Accelerate Manifesto for an Accelerationist Politics," *Critical Legal Thinking*, http://

criticallegalthinking.com/2013/05/14/accelerate-manifesto-for-an-accelerationist-politics/.

Williams, Mark, Jan Zalasiewicz, Neil Davies, Ilaria Mazzini, Jean-Philippe Goiran, and Stephanie Kane (2014) "Humans as the Third Evolutionary Stage of Biosphere Engineering of Rivers," *Anthropocene* 7 (September): 57–63.

Witham, Larry (2005) *The Measure of God: Our Century-Long Struggle to Reconcile Science & Religion: The Story of the Gifford Lectures*. San Francisco: HarperCollins.

World Wildlife Fund (2014) *Living Planet Report: Species and Spaces, People and Places*, www.wwf.or.jp/activities/lib/lpr/WWF_LPR_2014.pdf.

Yack, Bernard (1992) *The Longing for Total Revolution: Philosophic Sources of Social Discontent from Rousseau to Marx and Nietzsche*. Berkeley: University of California Press.

Zaccai, Edwin, François Gemenne, and Jean-Michel Decroly (2012) *Controverses climatiques, sciences et politiques*. Paris: Presses de Sciences Po.

Zalasiewicz, Jan (2008) *The Earth after Us: What Legacy Will Humans Leave in the Rocks?* Oxford: Oxford University Press.

—— (2010) *The Planet in a Pebble: A Journey into Earth's Deep History*. Oxford: Oxford University Press.

Zalasiewicz, Jan, Mike Walker, Phil Gibbard, and John Lowe (2015) "When Did the Anthropocene Begin? A Mid-Twentieth Century Boundary is Stratigraphically Optimal," *Quaternary International* 383: 196–203.

Index

1610
  assassination of Henry IV 185, 186
  as beginning of the Anthropocene 184–6, 218n, 232
  Galileo's *Sidereus Nuncius* 185
  possible return to 211
  reforestation of America 184–5, 232
  Toulmin's ending of one epoch/beginning of another 186–90

actantial roles 171, 171n
actor-network theory 54n, 95n, 98n, 136
adversaries 224, 235–6
Aesop's fables 235n
agency 4, 223
  anthropomorphic aspects 50–4
  ascribed to living organisms 87
  autonomous 211
  competences 56
  distribution/redistribution of 144, 149, 179–80, 188, 217, 235
  human 50–4
  and intentionality 98–101
  it moves/it is moved 59–60
  and language 67–70
  loss of 170
  multiple 67, 70
  natural 54–8
  neuroscience article 54–8
  novel/newspaper story, examples of 50–4
  sciences as multipliers of 49, 49n
  in scientific report 55–8
  signification 69–70
  subject/object mix 61–3
  tricksters/shapeshifters 66
agents
  accounting procedures 103–4
  chemical reactions 88–90
  limits of 275
  multiplication of 163–4
  outside/inside of 103, 104–5
  territory of 252
Aït-Touati, Frédérique 256
Anders, Günther 218
angel of geohistory 242, 245
Anglocene 138n
animism 70n

Index

Anthropocene 3, 4, 39, 44n, 62, 100n, 110, 183n, 277, 284
   1610 as fateful date 184–6, 218n, 232
   advantage of 246
   beginning of 113, 113n
   criticism of 246, 246n
   definition 113–14
   and disaggregating people/things 120–2, 143
   as golden spike 116
   and Great Emancipation Narrative 244
   interpretation of 118
   not an extension of anthropocentrism 122
   official adoption of 113, 119
   as potential geological epoch 112, 112n
   and redefining political task 143–4
   relocalization of the global 136
   science/Globe confusion 126–30
   and separation of Science/Religion 150
   Sloterdijk's concept of the sphere 122–6
   understanding/knowledge 139–40
anthropomorphism 65, 109–10, 115
   concerning rivers 52–4
   instability of 57–8
   in Tolstoy 50–1
apocalypse 193–4, 196, 206, 212, 230, 285–6
apocalyptic discourse 217–19
Arcimboldo effect 118, 118n
Arendt, Hannah 2, 218
Assmann, Jan 154, 175n, 193, 200
Atchafalaya river 52–4, 57
Atlas's curse 122, 128, 130
Augustine, St 198, 199, 201
Australopithecenes 44
autonomy 62, 188–9, 211, 212, 275
Avatar (film, 2009) 80
axiological neutrality 22, 27
Aykut, Stefan 258, 258n

Benjamin, Walter 242
Bergson, Henri 2
blind Watchmaker 171, 172, 208
Blumenberg, Hans 112n
Branson, Richard 84, 84n
Brecht, Bertolt 60
   The Life of Galileo 79–80
Bush, George H. W. 27

California Central Valley farmers 272–3, 274
capitalocene 138n
carbon footprint 121–2
causalist narrative 68–9, 69n, 71–2
Cavendish Laboratory, Cambridge 126–7
Chakrabarty, Dipesh 39n, 116n, 121n
Christian theology 11, 125, 126, 127, 132, 133–6, 144, 145, 171–2, 182, 197–9, 201–3, 234, 234n, 286, 287
Clark, Christopher, *The Sleepwalkers* 10
climate change 25
   and arguing about science 28, 28n
   attacks/counter-attacks 27–8
   causes of 26
   dramatized debates 29–33
   facts/values 25–7
   and Gaia 239
   and responsibility 26–7
   and scientific uncertainty 26
climate skeptics 11, 226
   apocalyptic origin 206–10
   attacks on scientific certainties 25–6
   attitude toward facts 27–8
   billions spent by 33
   denigration of science of climatologists 32
   dramatized debates 29–33
   and objectivity 48
   pressure groups 25, 26–7
   pseudo-controversy 25–8, 33–4, 142

Index 317

and unanimity 260n
and visibility of science 215–16
climate war 43
Climategate 161, 161n
CO_2 emissions 3n, 8, 8n, 25, 26, 41–2, 43, 44, 59, 59n, 72, 115, 184, 227, 258–9, 268n
Cold War 45
collectives
  epoch 151, 154–5, 157, 165, 169, 241–5
  limits 151, 169
  negotiation 158
  OWWAAB 159, 160–5
  principle of organization 151, 157, 159–60, 166
  shared certainty of being crushed by the falling sky 216
  supreme authority 151, 152–8, 159, 160, 165
  and tables of translation 157
  territory 151, 157, 160–5, 166–9
Columbus, Christopher 185, 185n, 290, 291–2
Comte, Auguste 281n
Conference of the Parties (COP) 258, 259–60, 268, 277, 277n
Conrad, Joseph 206
constative/performative 47–8
Copernican revolution 79, 126
Copernicus, Nicolaus 79
Coppola, Francis Ford 206
Cosmocolossus 2, 2n, 61, 149, 179, 227
cosmogram 151, 235, 245, 248, 251
counter-Copernican revolution 61, 62
Counter-Reformation 187, 189, 199
counter-religion
  and death 176n
  and Design 174
  and disinhibition 192–3, 194
  and end of times 174–7, 174–7nn
  and exercise of terror 198–9
  instability of 200–6
  misunderstanding/indifference to the cosmos 210, 210n
  and the Moderns 217, 242–3
  and Nature 168
  and secularization 284
  turning against divinities, God, nature 209
  two forms of 172–3, 178, 181
  uncertainties 200n
  understanding 157
  see also religion
Counter-Renaissance 187–90
Creation 170–1, 171n, 173, 180–1, 208, 208n, 286
creationists 29, 71
critical zones 3, 60, 61n, 86, 93, 97, 106, 140, 207, 275
cybernetics 282
Cyrano de Bergerac, Savinien de 77

Dahan, Amy 258, 258n
Daisy model 102, 102n
dance 1–2, 242, 257, 291n
Danowski, Déborah 73
Dante Alighieri 73
Darwin, Charles 79
Darwinian revolution 79
data 47, 47n
Daubigny, Pierre, *Gaia Global Circus* 28–33
Dawkins, Richard 171
De Gaulle, Charles 269–70
deanimation 54, 67–70, 70n, 71–2, 85, 86, 87, 97, 100, 169, 170, 172, 207–8, 286n
denial of reality 11n
Descartes, René 66, 77, 85, 90
description/prescription 43, 48–9, 63, 226n
Design 169, 172, 174
Detienne, Marcel 81, 83
Dewey, John 2, 139, 199n
disinhibition 191–2
  Modernity as process of 10n
  religious origin of 192–4
Dobson, Gordon 139
Dome of Nature 226, 237
Dubos, René, *Louis Pasteur, Free Lance of Science* 88n

Dutch National Water Authority 272, 274

Earth
  as active without a soul 86–7, 91–4
  addressed as "mother," "sister" 210n, 287
  and agency 63–6, 70–1, 98–101
  as animate 60–1, 78, 133n
  and appropriation by the land 251
  and chains of causality 63
  compared to billiard ball 84
  different models of 183
  as "good old Earth" 244
  as immanence 202
  importance/similarity to other celestial bodies 76–7, 78
  as a living planet 78
  maintaining inside/outside difference 78–9
  as the mother of law 228
  movement of 59–60, 79, 85, 86
  mythology concerning 81–7
  New World under the surface of 233
  organized agents/principal role 91–4
  as overanimated 94–8
  primary/secondary qualities 85, 109
  as self-regulating system 95–8
  as sensitive to human actions 113
  as stable 249
  as subject rather than object 61–3
  as sublunary world 78, 228
  technological metaphors 96–7
Earth System
  as anti-systematic 97
  connection/totality 85, 85n
Earthbound 248–53, 276–7, 278, 281, 282–3, 284, 290–1
ecological controversies 3
ecological crisis 22
  action/reaction to 8–9
  alternative understanding of 8
  getting used to it 13
  ignoring the warnings 9–10
  madness concerning 10–14
  as reassurance 7–8
  religious origin 210–13
  as a return of the human to nature 14–20
  and third-party arbiter 246–7
ecological questions 8, 36n, 189–90, 191, 206, 230, 237, 245, 259, 275
ecology
  activists in 46
  confusions/instabilities surrounding 34
  economizing 260
  as end of nature 36, 46–7
  and the Gnostics 209
  and Nature 179
  repoliticizing 223–8, 237–8, 241, 249
  scientific/political 34–5
  understanding of 21
ecomodernist manifesto 192n
The Economist 117, 118
Economy 260, 260n
Edwards, Paul 143, 215
end of history 177, 195, 195n, 206, 212
end of time 174–7, 194–9, 206, 206n, 213, 230, 285–6
end of the world 174, 207, 207n
enemies 31, 224, 235–6, 245
environment 144
  historical precedent 137–8
  military link 136, 136n
epoch 139, 169, 189, 200, 208, 278
  age of spheres 181
  Anthropocene 142, 143–4, 150, 236n, 244, 248, 264
  cosmopolitan 154
  end of one/beginning of another 186
  human epoch 115, 119, 147, 149, 184
  Kingdom of the Spirit 197, 199
  of the mushroom cloud 218

naming 113
post-apocalyptic 212
pre-Copernican 125
in which we are living 4, 38, 111–12, 112n, 121, 151, 157, 160, 165, 219, 241
"*Eppur si muove!*" ("And yet the Earth moves!") 59
Estates General 258, 258n

facts/values
and beliefs 33n
definitions 34
distinguishing between 25–7, 230
dramatized debates 29–33
facts as stubborn 27
facts, warnings, decisions continuum 49
gulf between 49
matters of fact as matters of concern 164
and the scientific disciplines 31–2
feedback loops 191, 257, 276–7
following loops to avoid totalizing 142
global vision 140–1
historians of the environment 137–8
knowing/feeling 139–40
material/empirical problem 138–9
sensitivity 141
and supplementary loops 140
Fleming, James Rodger, *Fixing the Sky: The Checkered History of Weather and Climate Control* 43n
Fontenelle, Bernard Le Bovier de 77
Foucault, Michel 189, 189n
The Archaeology of Knowledge 112n
Francis, Pope 195n, 210n, 287, 288
Francis, St 287
Fressoz, Jean-Baptiste 191–2, 192n
Freud, Sigmund 79, 175
Friedrich, Caspar David, *Das grosse Gehege bei Dresden* 220–3, 254
Fukuyama, Francis 195n

Gaia 2, 4, 6, 40, 62, 73, 287
acknowledgment/naming of 283
agents not prematurely unified 87
characteristics 87
and climate change 239
curse attached to theory of 84–5, 86
Darwinian/evolutionists notion 101–4
deanimated/overanimated agents 87
definition 238
description of 288–9
as exceptionally touchy 247, 254
facing 288–9
fleeing backwards 242
founding of 183
and global unification 142
Hesiod's theogony 81–3
holistic conception 96–7
hymn to 153–4
insistence of 138
intrusion of 222, 223, 226, 230
invoking respectfully 183
Lovelock's theory 81, 86, 91–101
as muddle 100, 103
multicellular 142
mythological 81–7, 122, 288
and natural religion 150
neither a Sphere nor global 140
no place in Nature/Culture schema 85
objections to 179
people of 213–17
as political lever 226
and Providence 100
as secular 87, 106–7
as self-regulated System 170
sensitivity of 141
as series of historical events 141
and sharing of sovereignty 280–3
and sovereignty 253
subversion of levels 106
as superorganism 94–8
theory consistent with evolutionary biology 101
Tyrrell's arguments against 131–6

Gaia (cont.)
  and war 238
  waves of action 101, 105
  what it is/is not 106–7
  what was/what will be 243, 244–5
Gaia Global Circus (play) 2, 2n, 149n, 228, 257, 257n
Gaia-politics 152, 232, 270
Galileo Galilei 125
  Brecht's play 79–80
  comment on the planets 75
  Earth in motion 59, 59n, 60, 85, 86
  mythic name 81
Game of Thrones (George R. R. Martin) 151, 159, 169, 213, 247
Ganachaud, Stéphanie 2, 241n
Geddes, Patrick 128
geohistory 3, 39, 40, 45, 45n, 64, 73, 79, 109, 112, 116, 170, 226, 239, 276, 291
geological time 116
geophysiology 88
geopolitics 4, 152, 168, 272, 281
geoscientists 117–18
Gifford Lectures 2, 2n, 4–5, 71n, 128n, 150n, 182n
global warming 25, 29, 121–2, 218
globalization 135, 135n, 278, 278n, 279
Globe 182
  confusion between science and 126–30
  curse of 245
  destruction of 144–5
  and feedback loops 136–44
  as impossible/deleterious 259
  Schmitt's concept 230–1
  skepticism concerning 238–9
  Sloterdijk's conception 122–6
  utopia of 136
Gnosticism 203–6, 203n, 205n, 207–10, 213n
God 11, 19, 46, 65, 76–7, 102n, 125, 128, 132, 133, 134, 134n, 145, 145n, 150n, 155–6, 157, 159, 169–73, 174, 176, 176n, 180, 188, 193, 201, 208, 208n, 222, 244, 254, 288
golden spike 114, 115, 116, 138n, 184n, 232
Golding, William 81
Grand Design 286
gravity 65–6, 69
Gravity (film, 2013) 80
great acceleration 3, 39, 192n, 233n
Great Emancipation Narrative 244
Great Enclosure 220–3, 233, 254

Haber–Bosch process 44n–45n
Hamilton, Clive 183n, 218
Harari, Yuval, *Homo Deus* 212n
Haraway, Donna 29, 66n, 72, 99n–100n, 121n
Hawking, Stephen 127
Heidegger, Martin 123
Henry IV 185, 186
Hergé
  The Adventures of Tintin: Explorers on the Moon 128n
  The Adventures of Tintin: Shooting Star 206
Hine, Dougald 73
Hitchcock, Dian 76
Hobbes, Thomas 178, 201, 227, 239, 245, 248, 282
  Leviathan 147–50, 274n
Holocene 112–13, 114
Homo oeconomicus 44, 107–8, 273n
Homo sapiens 44
human epoch 115–16
human history 74
humans 68
  at war 252
  as capable of responding 281, 282
  definitions of 290–1
  erosion caused by 117
  influence measured on same scale as natural phenomena 117
  irreversible actions by 39–40

and the land   250-1
provincial definition of   107-8
as sensitive, responsible, moral   141, 281
as untrustworthy   251
who, when, where, how   38-40
Hume, David, *Dialogues*   182
Hurricane Katrina (2005)   53
Hutton, James   95n

iconoclasm   177n, 202n
immanentization   176n, 200-6, 207, 242, 274-5
inanimate actors/actants   49-50, 52-4, 55-8
Intergovernmental Panel on Climate Change (IPCC)   26, 29, 43, 265n
International Commission on Stratigraphy (Subcommission on Quaternary Stratigraphy)   111
International Geological Congress (2012 & 2016)   111-16, 114n
International Geological Society   113, 119
Invisible Hand   102n, 134
IPCC *see* Intergovernmental Panel on Climate Change
Islam   204-5

James, William   2
Jameson, Fredric   108
Janda, Richard   235n
Jet Propulsion Laboratory, Pasadena   76
Joachim de Flore   197-9
Jonas, Hans   175n

Kant, Immanuel   126
Keeling, Charles David   41-2, 48, 59, 139, 215
Kennan, George   45
Kepler, Johannes   77
Kershaw, Ian, *The End*   72-3
Koerner, Joseph   220, 221, 222
Kopenawa, Davi   216

Koyré, Alexandre, *From the Closed World to the Infinite Universe*   77, 80

Lamartine, Alphonse de   192, 206n
land   243-4
language of the world   67-70
Laudato Sí encyclical   195n, 210n, 288n
laws of nature   23, 33-4, 37, 162, 260, 284
Le Corbusier   17
Le Monde   43
Lenton, Timothy   100-1
Leviathan   147-50, 190, 228, 248, 273, 282
Liebig, Justus von   88-9
Life of Pi (Yan Martel)   40, 40n
linguistics   47-8
Locke, John   85
Lorenzetti, Pietro   274-5
Louis XV, King   251
Lovelock, James   2, 59, 73, 270-1
  and adaptation to an environment   103
  chemistry explanation   91-4
  compatibility with Darwinian narratives   101-2
  concept of Gaia   86-7, 91-101
  difficulties expressing himself   95-8
  and Earth which is moved   79
  enigmatic name   81
  evolutionist criticisms of   102
  and intentionality   98-101
  Martian imaginings   77-8
  observational symmetry with Galileo   76
  and oxygen   105-6
  parallels with Pasteur   87-91, 100
  and "the point of view of nowhere"   78
  Tyrrell's argument against   131-6
  The Practical Science of Planetary Medicine   87-8
Luntz, Frank   25

Index

madness
  biblical assurance 11
  bipolar 12
  depressive 12
  discovering a course of treatment
    13–14
  escaping 13
  fanatical form 11
  and modernization 12
  quietist form 11
  scholarly term for 10–11
majesty 283, 283n
Mandeville, Bernard, *The Fable of the Bees: Private Vices, Publick Benefits* 131n
Mann, Michael 42, 59
  The Hockey Stick and the Climate Wars: Dispatches from the Front Lines 43
Margulis, Lynn 92, 99n–100n, 104–5, 105n, 288n
the Market 104, 104n, 131n, 134, 153, 188–9, 199
Mars 76
materiality 70–2, 109n, 200, 202, 207, 207n, 208, 210–12, 216–17, 251, 285
Mauna Loa 8, 139, 215
McPhee, John 52
  The Control of Nature 51–4
Melancholia (film, 2011) 144–5
Mercator, Gerardus 128
metamorphic zone 58, 58n, 59, 64, 66, 69–70, 74, 86, 119, 146, 163, 223
metaphysics 37, 86, 91, 143
Mississippi 52–4, 55, 56, 57, 58, 69
Modernity 116
  and the Apocalypse 194
  and Gnosticism 205–7
  and immanentization 207
  and the Promised Land 207
  as reflexive disinhibition 10n
  Toulmin's hidden agenda of 185, 188–90

Moderns
  and the apocalypse 199, 212
  certainty/uncertainty 193, 200, 200n
  and the Cosmocolossus 2
  and counter-religion 217, 242–3
  and deanimating the world 72
  and disinhibition 196, 196n
  and the exercise of terror 198
  and the future 242–3
  as heirs of the Mosaic division 156–7
  and immanentization 200, 204
  insensitivity of 211
  Latour's comment on 138n
  and Leviathan 147
  as looking backwards/up in the air 242
  neglect of materiality 211
  and the New Climate Regime 3
  and the question of "ends" 175–7
  and secularity 204
  and the State of Nature 149n
  temporality of 243, 243n
modes of existence (lectures) 2–3
Moltmann, Jürgen 287
Monod, Jacques 170n
monogenism 145n
Montaigne, Michel de 187

narcissistic wounds 79–80
nation-state 32, 225, 226, 232, 233, 237, 249, 262, 267, 271, 277–9
National Science Foundation 215
natural history 73–4, 117n, 291
natural law 20–1, 24, 35, 37, 63–4
natural religion 103, 150, 150n, 172–3, 178, 180, 182
  lectures (Edinburgh) 2, 2n
natural selection 103
natural theology 71n
natural world 160
  appeals to state of 27
  claims concerning 33
  and climate skeptics 24–8

Index

detachment of "natural" from
  "world" 35–8
gulf between what is/what must be
  22–4, 34, 63
and the human world 58
moral requirement 22
nature
  bifurcation of 85, 109
  as capitalized proper noun 38,
    39, 158, 178–80
  concept of 36–7
  contract with 63–4
  contradictions 127
  Creation as alternative to 286–7
  epistemological/anthropological
    168
  as established 90–1
  Hobbesian proposition 178–9
  and human nature 20
  instability of 21n, 34–5
  levels/strata 106
  living in harmony with 107
  and the meaning of "just" 22, 34
  moral connotation 21, 24
  nature/cultures prejudice 177
  no unity in 143
  normative dimension 20–2, 34
  polemical dimension 22, 24
  and politics 225–6, 241
  redefined 120–2
  second-degree normative dimension
    22–4
  two versions of 20–2, 37
  understanding of 216–17
  war and peace applied to 253–4
Nature journal 118–19, 146–7
nature/culture 68, 267
  and the Anthropocene 118
  art history comparison 16–17
  and collectives 143
  de facto/de jure paradox 21, 24–8
  definitions 15
  detachment of "natural" from
    "world" 35–8
  difficulties/objections 14
  disaggregation of 150–1

discrediting our relation to the
  world 85
as distribution of roles, functions,
  arguments 21
framing the notion of 37–8
and freedom of movement 108
marked/unmarked categories
  15–17
normative dimensions 20–4
panic concerning 15
role distribution 18–20
topic/resource shift 19
visually constructed subject/object
  16–19
networks 276, 276n
New Climate Regime 3, 3n, 4, 22,
  35, 43, 61, 74, 114, 118, 123,
  188, 216, 232, 238, 244, 247,
  248, 264, 266n, 279n, 284
Newton, Sir Isaac 65–6, 77
nomos 230, 230n, 232, 233–5,
  235n, 243, 250, 251, 255, 257

objectivity 33, 47, 47n
Old Climate Regime 120, 187n,
  190, 223, 226, 238, 279n, 284
Out-of-Which-We-Are-All-Born
  (OWWAAB) 159, 160–5
Outlook Tower, Edinburgh 128,
  128n
overanimation 68, 85, 87, 88, 97,
  169, 170, 172, 286, 286n
overlapping authorities 272–4
OWWAAB *see* Out-of-Which-We-
  Are-All-Born
ozone layer 139

painting, Western 16–19, 221–3
Pan 46, 46n
panic 46, 46n
Paradise 198–9
Paris Universal Exhibition (1900)
  128–9
Parliament of Things 262n
Pascal, Blaise 124
Pasteur, Louis 87–91, 100

peace 88, 129, 144, 148–9, 155, 166, 178–9, 227, 234, 235–6, 241
People of Creation 169, 170, 173
People of Gaia 213–17
People of Nature 159, 161–2, 166–7, 168, 169, 170, 173, 180–1, 214, 247
petromorphism 115–16
philosophy 125–6
phusimorphism 57, 65
planetary limits 3, 275
Plantatiocene 138n
Plus intra 291
political theology 199–200, 200n
politics
  Gaia's restoration of order 142
  gradual disappearance of 229
  and religion 198–9, 217, 278
  repoliticizing concept of ecology 223
  spatializing 249
  true/false distinction 229
  and war 240–1
Politics of Nature (Latour) 46–7
Pouchet, Félix-Archimède 90
Powers, Richard 55n
pre-Copernican system 60n
precautionary principle 10, 10n
prescription *see* description/prescription
principle of exception 229
Providence 100, 102, 103–4, 131, 131n, 133, 176

Quaternary Era 112, 114
Quesne, Philippe 255

Raum Labor 261n
reanimation 68, 170
Reclus, Élisée 128
religion 150, 284
  attributions, functions, origins of supreme authority 154–8
  battle with the secular 285
  definition 152
  difficulties talking about 194
  and ecological crisis 210–13
  features 178
  Mosaic division 155–6
  move to terrestrial via the secular 211–13
  naming of supreme authority 154, 156, 157–8, 159
  and origin of disinhibition 192–4
  and politics 198–9, 217, 278
  Science/Religion conflict 169–74
  supreme authority 151, 152–8, 159, 165, 176
  three types of truth 201
  translation tables 154–5, 157, 157–8, 202
  see also counter-religion
Rousseau, Jean-Jacques 64

Sagan, Carl 139
Sartre, Jean-Paul, *Nausea* 210, 210n
Schaffer, Simon 127n, 148n
Schmitt, Carl 199n, 228–32, 243, 253, 255, 272n, 276, 283–4, 285
  The Concept of the Political 235–41
  The Nomos of the Earth in the International Law of the Jus Publicum Europaeum 229–30, 230n, 232–5, 250–1, 252
science 126–30, 142–4, 199
  and agency 211
  and climate skeptics 215–16
  features 178
  and Gnostic truth 203n
  Science/Religion conflict 169–74
  studies 3–4, 167n
science community
  arguing about science 28, 28n
  attacks on 27–8
  as over-excited militants of a cause 28
  political/ideological dimensions 29–33
  quality, objectivity, solidity of disciplines 33
  and responsibility 26, 29

Index

Science of Nature   227, 260
scientific counter-revolution   189–90, 193
scientific narrative   68, 68n
scientific reports   54–8
scientific worldview   71–2
secular
  battle with religion   285
  and being in a secular world   156
  and collectives   153
  and counter-religion   176n, 179, 284
  and Darwin   103
  and definition of nature   174
  Gaia as   106–7, 133
  and iconoclasm   177n
  and immanentization   204
  internal argument within Christianity   201n
  materiality as   72
  meaning of term   87, 87n
  and the Moderns   204
  move from the religious to the terrestrial via the secular   211–13
  and the people of Nature   167
  and political theology   201n–202n
  in science   137
selfish gene   103, 103n, 104, 271
semiotics   99n
Serres, Michel   61–2, 79, 86, 152
  The Natural Contract   59, 63–5
Shapin, Steven   148n
Shelley, Mary, *Frankenstein*   109n
Shelley, Percy Bysshe, "Mont Blanc"   108–9, 109n
Sloterdijk, Peter   122–6, 139, 145, 181
social contract   64, 227, 245
sovereignty   225–6, 230, 248–9, 253, 270, 278, 279, 280–3
space
  localization in   271n
  political ecology connection   231, 231n, 232
space/time   106, 126, 271n
spheres   122–6, 136–7, 238

Spirit of the Laws of Nature (imagined title)   4
spiritualist/materialist distinction   171, 202–3
Spivak, Gayatri Chakravorty   46n
State   96, 148–9, 149n, 178, 187n, 190, 193, 199, 225, 233, 236, 248–9, 261
State of Nature   149, 149n, 226, 227–8, 239, 245, 260–1, 274
Stengers, Isabelle   141
Stern, Sir Nicolas   45
strategic essentialism   46, 46n
subject/object gaze   16–18
superorganism   94–5, 95n, 98, 102, 102n, 103, 125, 135
symmetry   54n

Tarde, Gabriel   98n, 135n
terrestrial
  control over   12
  and Gaia   133
  and the Gnostics   209
  indifference to   186
  and the Moderns   200, 204
  move from religion via the secular   211–13
  and natural religions   181
  and the people of Nature   167, 168
  vision of science   174
Territorial Agency project   268n
territory   38, 40, 143–4, 151, 157, 160, 166, 231, 241, 244, 245, 251–3, 262–3, 266–7, 270, 276
Theater of Negotiations (Théâtre des Amandiers)   255, 256n, 258
  art of diplomacy   270
  changes in balance of power   269–70
  conflicts between territories not nation-states   266–7
  delegations as "alone" not "unified"   261
  distribution of the sciences   264–5, 265n

Theater of Negotiations (Théâtre des Amandiers) (*cont.*)
 final scene involving nation-states 277, 279
 formation/language of delegations 262, 262n
 and Gaia 270, 271, 280
 geopolitical realism 270
 and globalization 267
 governing in period of ecological mutation 269
 higher common principles 259–60
 inclusions 267–8, 267n, 268n
 Laws of the Market 260
 lobbies/mafias 267
 no representation of Nature/voice of Gaia 265–6
 non-state delegations 262
 occupying territories 259
 quality of representation 264
 redefining of territory 262–3
 redistribution negotiations 263–4
 rule for composition 268–9
 scene of conflict 267
 simulations as realistic 261, 267–70
 and sovereignty 270, 279
Thomas, Yan 283n
Thomson, J. J. 126
three estates 281, 281n
thresholds, crossing of 41–4
tipping points 3, 39, 60, 74
Tolstoy, Leo, *War and Peace* 50–1, 52, 53, 54, 55, 56, 57, 58, 66, 69
Toulmin, Stephen 193, 199, 239
 Cosmopolis: The Hidden Agenda of Modernity 185, 186–90
tragedy of the commons 271–2
trait 64–5
Trier, Lars von 144–5
Tsing, Anna 246n
tsunami markers 277n
Twain, Mark 52

Tyrrell, Toby, *On Gaia: A Critical Investigation of the Relationship between Life and Earth* 131–6

Uexküll, Jakob von 122
utopia 124, 136, 198, 205, 243, 259, 262, 285

Veith, Johann Philipp 222
view from nowhere 77, 127, 181
Viveiro de Castro, Eduardo 73
Voegelin, Eric 175n, 176n, 200–6, 242, 288
 The New Science of Politics 196–9

war
 agreeing to 241
 and collectives 157–8, 159
 conspiring with our enemies 31
 explicitness of aims 246–7
 increased loss of life in Germany 72–3
 inevitability 236–7
 and inventions 9
 justification for 169–70
 and lack of disinterested/neutral third party 236–8
 as limitless 239–40
 link with climate 43, 43n, 136n, 139n
 and police operations 235–41
 political aspects 240–1
 precautionary principle 10, 45–6
 and resumption of hostilities 227
 return to total war 149, 233–4
 and role of Gaia 238
 for spatial order 253
 subjective/objective forces 50–1
 truth/neutrality 43
 waged in the name of reason 285
wars of religion 148, 150n, 175, 178, 186–7, 190, 193, 199, 211, 278
Watchmaker, blind 171, 172, 208
waves of action 101, 104–5

way of life
  challenges to industry 25
  as not negotiable 27, 27n
Westphalian system 187, 187n
Whitehead, Alfred North 2, 85, 164n
Whole Earth Catalog 266, 266n
Whyte, Lynn 210
Wilson, E. O. 102n

World Wildlife Fund 291n
world/worlding 35–8
  sublunary world 78
  as what is coming 91

Yack, Bernard, *The Longing for Total Revolution* 209n

Zalasiewicz, Dr Jan 111, 114, 265